T0327513

Optical Switching

Optical Switching

Device Technology and Applications in Networks

Edited by

Dalia Nandi
Department of Electronics and Communication Engineering,
Indian Institute of Information Technology Kalyani, West Bengal, India

Sandip Nandi
Department of Electronics and Communication Engineering,
Kalyani Government Engineering College, Kalyani, West Bengal, India

Angsuman Sarkar
Department of Electronics and Communication Engineering,
Kalyani Government Engineering College, Kalyani, West Bengal, India

Chandan Kumar Sarkar
Department of Electronics and Telecommunication Engineering,
Jadavpur University, Kolkata, West Bengal, India

Registered Office
John Wiley & Sons, Inc., 111 River Street, Hoboken, NJ 07030, USA

Editorial Office
John Wiley & Sons, Inc., 111 River Street, Hoboken, NJ 07030, USA

For details of our global editorial offices, customer services, and more information about Wiley products visit us at www.wiley.com.

Wiley also publishes its books in a variety of electronic formats and by print-on-demand. Some content that appears in standard print versions of this book may not be available in other formats.

Library of Congress Cataloging-in-Publication Data

Names: Nandi, Dalia, editor. | Nandi, Sandip, editor. | Sarkar, Angsuman,
   editor. | Sarkar, Chandan Kumar, editor. | John Wiley & Sons, publisher.
Title: Optical switching : device technology and applications in networks /
   edited by Dalia Nandi, Sandip Nandi, Angsuman Sarkar, Chandan Kumar Sarkar.
Description: Hoboken, NJ : Wiley, 2022. | Includes bibliographical
   references and index.
Identifiers: LCCN 2022010920 (print) | LCCN 2022010921 (ebook) | ISBN
   9781119819233 (cloth) | ISBN 9781119819240 (adobe pdf) | ISBN
   9781119819257 (epub)
Subjects: LCSH: Optical communications–Equipment and supplies. | Optical
   fiber communication. | Signal processing.
Classification: LCC TK5103.592.F52 O76 2022 (print) | LCC TK5103.592.F52
   (ebook) | DDC 621.38275–dc23/eng/20220412
LC record available at https://lccn.loc.gov/2022010920
LC ebook record available at https://lccn.loc.gov/2022010921

Cover image: © asharkyu/Shutterstock
Cover design by Wiley

Set in 9.5/12.5pt STIXTwoText by Straive, Pondicherry, India

Contents

Preface

The unprecedented proliferation in big data and IP traffic across interlinked data centers (DC) and high-performance computing systems (HPC) has made ultra-high link capacity and packet switching rate at network nodes necessary. To meet these very high demands, especially packet routing and forwarding, well-proven technologies, such as the optical switching and labelling technology, seem to provide appropriate solutions, not only by delivering high-bit rate data streams, but by achieving cost-efficiency and energy-efficient multi-Tb/s cross-connection throughputs. Experts predicted that optical switching would serve as the basis for the construction of seamless integrated networks to transmit all traffic via the same fiber line and offer other pathways for traffic to cross if a channel collapses. The net outcome is a quicker, more convenient internet, claim proponents. Therefore, up-to-date extensive information of optical switching and its application in modern optical networks is strongly needed for researchers to keep up with the recent technological advancement. Given this scenario, this book is planned to provide a comprehensive source of guidance on optical switching.

This book has several objectives. First of all, we lay the groundwork by offering a brief review of the optical switching and networking and the significant successes of recent decades, highlighting key enabling technologies and techniques that have paved the way for modern optical switching networks. We will next explore the broad overview of the future of optical switching and identify important milestones in the development of optical switches of the next generation. In the first section, a detailed introduction opens the way for the reader to learn the history and principles of the topic. The second section focuses on the switching devices, the concepts and theories of several optical switching technologies, characteristics, and application areas. The third section explores the optical switching applications in communication networks.

Today, optical switching covers a range of technologies and numerous scientific and engineering fields. It is not simple for one to master several areas in this subject, as with many modern domains of study. Thus, an edited book of its kind, with several writers, will give the reader better insight into the subject. We appreciate profoundly the effort, passion and quality of the work of the chapter authors. We are very thankful to many of them, who have not only given their chapters but were also gracious enough to provide important advice and recommendations on the preparation of this book.

Finally, this book is intended to offer readers with an in-depth overview and an in-depth grasp of state-of-the-art optical switching as well as recent and substantial advances in the field of optical switching and networks. It also will be a support for researchers, engineers, and experts in developing applications and services in this field. We hope that this book will appear as a one-volume reference for researchers, senior and postgraduate students, and professionals requiring quick expertise in the most recent technologies.

Dalia Nandi
Sandip Nandi
Angsuman Sarkar
Chandan Kumar Sarkar

About the Editors

Dalia Nandi is an Assistant Professor of Electronics and Communication Engineering at the Indian Institute of Information Technology Kalyani, West Bengal, India. She received her B.Tech., M.Tech., and PhD (Tech) degree from the University of Calcutta. She has over 30 research papers in international refereed journals and in international conferences. She is a senior member of the IEEE.

Sandip Nandi is an Assistant Professor of Electronics and Communication Engineering at Kalyani Government Engineering College, West Bengal, India. He received his B.Tech., M.Tech., and PhD (Tech) degree from the University of Calcutta. He has published more than 10 research papers in international refereed journals, and in international conferences. He is a senior member of IEEE and a life member of the Institute of Engineers India.

Angsuman Sarkar is presently serving as a Professor of ECE at KGEC, India. His current research interest spans the study of short channel effects of sub-100 nm MOSFETs and nano device modeling. He has credit for 200+ peer-reviewed publications.

Chandan Kumar Sarkar received a D.Phil. degree from the University of Oxford in 1983. He was an INSA–Royal Society Visiting Scholar at several UK universities. He served as the Chair of the IEEE Electron Devices Society, Calcutta Chapter, and was the past Chair of IEEE Section. He is serving as a Distinguished Lecturer of the IEEE Electron Devices Society and has been invited to several countries. He also serves as an Editor for the MOS Devices and Technology section of *IEEE Transactions on Electron Devices*.

List of Contributors

Arpita Adhikari
Department of Electronics and
Communication Engineering
Techno Main Salt Lake, Kolkata
West Bengal, India

Devlina Adhikari
Information and Communication Technology
School of Technology
Pandit Deendayal Energy University (PDEU)
Raisan, Gandhinagar
Gujarat, India

Rajan Agrahari
Department of Electronics and
Communication
Engineering
National Institute of Technology
Patna
Bihar, India

Nadir Ali
Department of Physics
Indian Institute of Technology Roorkee
Roorkee
India

Arighna Basak
Department of Electronics and
Communication Engineering
Brainware University, Kolkata
West Bengal
India

Aranya B. Bhattacherjee
Department of Physics
Birla Institute of Technology and
Science-Pilani
Hyderabad Campus
Telangana, India

Somak Bhattacharyya
Department of Electronics Engineering
Indian Institute of Technology (BHU)
Varanasi, Uttar Pradesh, India

Kalyan Biswas
Department of Electronics and
Communication Engineering
MCKV Institute of Engineering
Howrah, West Bengal, India

Nicola Calabretta
Eindhoven University of Technology
Eindhoven, Netherlands

Manash Chanda
Department of Electronics and
Communication Engineering
Meghnad Saha Institute of Technology
Kolkata, West Bengal, India

Arijit De
Department of Electronics and
Communication Engineering
Netaji Subhash Engineering College, Kolkata
West Bengal, India

Arpan Deyasi
Department of Electronics and
Communication Engineering
RCC Institute of Information Technology
Kolkata, West Bengal, India

Rajarshi Dhar
Department of Electronics and
Telecommunication Engineering
IIEST, Shibpur, Howrah
West Bengal, India

Mohammad Faraz Abdullah
Department of Physics
Indian Institute of Technology Roorkee
Roorkee, India

Sudipta Ghosh
Department of Electronics and
Telecommunication Engineering
Jadavpur University
Kolkata
West Bengal, India

Antony Gratus Varuvel
Indian Institute of Information Technology (IIIT)
Sri City
Chittoor
Andhra Pradesh, India

Bingli Guo
State Key Laboratory of Information Photonics
and Optical Communications (IPOC)
Beijing University of Posts and
Telecommunications
Beijing, China

Bheemappa Halavar
Indian Institute of Information Technology (IIIT)
Sri City
Chittoor
Andhra Pradesh, India

Shanguo Huang
State Key Laboratory of Information Photonics
and Optical Communications (IPOC)

Beijing University of Posts and
Telecommunications
Beijing, China

Rajesh Kumar
Department of Physics
Indian Institute of Technology Roorkee
Roorkee, India

Sambit Kumar Ghosh
Department of Electronics Engineering
Indian Institute of Technology (BHU)
Varanasi, Uttar Pradesh, India

Chandan Kumar Sarkar
Department of Electronics and
Telecommunication Engineering
Jadavpur University, Kolkata
West Bengal, India

Rashmi Kumari
Nano Bio Photonics Group, Department of
Electronics and Electrical Communication
Engineering
IIT Kharagpur
Kharagpur
West Bengal, India

Basudev Lahiri
Nano Bio Photonics Group, Department of
Electronics and Electrical Communication
Engineering
IIT Kharagpur
Kharagpur
West Bengal, India

Piyali Mukherjee
Department of Electronics and
Communication Engineering
University of Engineering & Management
Kolkata
West Bengal, India

Dalia Nandi
Department of Electronics and Communication
Engineering
Indian Institute of Information
Technology Kalyani
West Bengal, India

Sandip Nandi
Department of Electronics and Communication
Engineering
Kalyani Government Engineering College
Kalyani
West Bengal, India

Rajendra Prasath
Indian Institute of Information Technology
(IIIT) Sri City
Chittoor
Andhra Pradesh, India

Swarnil Roy
IEEE SSCS Kolkata chapter, Kolkata, West
Bengal, India

and

Department of Electronics and
Communication Engineering
Meghnad Saha Institute of Technology
Kolkata
West Bengal, India

Angsuman Sarkar
Department of Electronics and
Communication Engineering
Kalayni Government Engineering College
Kalyani,
West Bengal, India

Joydip Sengupta
Department of Electronic Science
Jogesh Chandra Chaudhuri College, Kolkata
West Bengal, India

K. Sujatha
Department of Electronics and
Telecommunication Engineering
Shree Ramchandra College of Engineering
Savitribai Phule Pune University
Pune, Maharashtra, India

Odelu Vanga
Indian Institute of Information Technology (IIIT)
Sri City
Chittoor
Andhra Pradesh, India

Chongjin Xie
Alibaba Cloud
Alibaba Group
Sunnyvale, CA, USA

Xuwei Xue
State Key Laboratory of Information Photonics
and Optical Communications (IPOC)
Beijing University of Posts and
Telecommunications
Beijing, China

Anjali Yadav
Nano Bio Photonics Group, Department of
Electronics and Electrical Communication
Engineering
IIT Kharagpur
Kharagpur
West Bengal, India

Surabhi Yadav
Department of Physics
Birla Institute of Technology and Science-Pilani
Hyderabad Campus
Telangana, India

Fulong Yan
Alibaba Cloud
Alibaba Group
Beijing, China

Part A

Introduction

Introduction

Sandip Nandi[1] and Dalia Nandi[2]

[1]Department of Electronics and Communication Engineering, Kalyani Government Engineering College, Kalyani, West Bengal, India
[2]Department of Electronics and Communication Engineering, Indian Institute of Information Technology Kalyani, West Bengal, India

Worldwide demand for enormous information transmission has created an immense need for expansion of capacity for higher generation communication networks. It is predicted that information traffic over the internet will double every year [1]. In this present scenario, Optical networks will be the favored solution to provide enhanced bandwidth for future communication system. Terabytes of information can be transferred between two nodes over the optical links. A switch is embedded in the optical communication system to route the information under the supervision of the control signals. The major limitation for high data communication over an optical link is the use of electronic switches, which will restrict the bandwidth of the link. With the passing of time, optical switches with their improved efficiency and low cost are replacing the traditional electronic switches to overcome the limitations. An optical switch is versatile in nature as it has a large number of usages in communication networks as well as in communication cores of modern-day high configuration computers with data rates of 1000 Gbps. However, the application of optical switches in networks has some of its own challenges in respect of signal impairments and network parameters. This book deals with the present status, advantages, and future developments of optical switches and networks.

This chapter provides a brief introduction to optical communication networks and optical switches. We start the journey with the historical perspective of each topic. Then gradually we provide the essential background, and clarify the key terminologies. Finally we present the organization of the book along with a brief overview of each chapter.

A. Optical Communication Networks

A.1 Historical Perspective

Optical fiber communication has advanced dramatically over the past 50 years due to the enormous advancement of scientific research in the fields of photonics and electronics starting from mid-1970s with the ground-breaking invention of Laser and low-loss optical fiber [2–4]. This advancement is widely reflected in the rate of data transmission increasing from megabits per

Optical Switching: Device Technology and Applications in Networks, First Edition. Edited by Dalia Nandi, Sandip Nandi, Angsuman Sarkar, and Chandan Kumar Sarkar.

second over several kilometers for the first ever point-to-point optical fiber connection to terabits per second wavelength division multiplexed optical communication networks. Also there is a continuous increase in productivity in terms of automation, flexibility, capacity, and cost reduction. Table 1 summarizes some significant developments in the progress of optical communication technology over the past few decades.

Today's optical networking shows the broadcast communication networks which can provide increased capacity, routing, and restoration based on recent optical communication technologies. The appreciable increase in capacity using optical networks offers cost reduction for bit transmission compared to other long-haul communication networks (e.g. wireless networks). This has a marvelous societal influence by revolutionizing the internet service. The optical network solution covers the entire transmission network range, from the main hub and metropolitan area network to the access network domain. Backbone networks and metropolitan area networks carry highly aggregated, high-bit-rate data traffic, ranging from the size of cities to entire continents. The access

Table 1 Historical developments in the field of optical communication technology.

Year	Significant developments
1880	Telephone invented by Alexander Graham Bell.
1948	Shannon's Limit described by Claude Shannon [5].
1957	Principle of Laser described first by Charles Townes and Arthur Schawlow [6].
1966	Concept of glass fiber with cladding that can carry light without much radiation described by Charles Kao and Hockham. This is the first proposal of optical fiber communication [7].
1970	First semiconductor lasers made separately by Zhores Alferov at Physical Institute in Leningrad and Mort Panish and Izuo Hayashi at Bell Labs.
1987	First report of erbium-doped optical fiber amplifier by David Payne at the University of Southampton [8].
1988	First demonstration of soliton transmission over 4,000 km of single-mode fiber by Linn Mollenauer of Bell Labs.
1991	First model based description of erbium-doped optical fiber amplifier by Emmanuel Desurvire and Randy Giles at Bell Labs [9].
1993	First transmission of data at 10 Gb/s over 280 km of dispersion-managed fiber Andrew Chraplyvy [10].
1996	Introduction of viable wavelength-division multiplexing (WDM) system.
2002	Nonlinearity compensation in fiber optic transmission was introduced for phase-modulated signals [11].
2003	ITU-T standardized gigabit-capable passive optical networks.
2009	Experimental demonstration of concept of Superchannel at 1.2 Tb/s [12].
2010	ITU-T standardized 10-gigabit-capable passive optical networks.
2011	Implementation of spatial multiplexing for optical transport capacity scaling by Peter Winzer [13].
2012	ITU-T standardized flexible-grid WDM [14].
2016	ITU-T specification for low-loss low-nonlinearity optical fibers [15].
2018	Development of low-loss MxN colorless-directionless-contentionless (CDC) wavelength-selective switch (WSS) [16].
2019	Demonstration of Super-C-band transmission with 6-THz optical bandwidth.

network carries different types of data streams in and out of private and commercial customers, and these data streams are multiplexed /de-multiplexed to the central transmission network in nodes with fixed backhaul connections. The use of the optical network undoubtedly occupies a dominant position in the core and metropolitan area network segments, and recently it has also begun to expand to the access network segment.

The era before 1990 is known for first-generation optical networking, which includes the use of synchronous optical network (SONET/ SDH). This is used as a structure for standardizing line speed, coding technique, bitrate, network elements, and O&M functions. This first-generation network runs on a single wavelength of each optical fiber and is opaque.

During the mid-1990s, a new technology called wavelength division multiplexing (WDM) was proposed and implemented in optical networks, which greatly enhances the capacity of the fiber by providing a technique to transmit dozens of optical signals over a single fiber. It also implements a wavelength routing network through the use of electronic or all-optical switching nodes. This multi-wavelength optical routing scheme establishes the second generation of optical networks. After the introduction of WDM technology, the central backbone and the metropolitan optical network greatly depended on the high-capacity wavelength division multiplexed link, while the traditional networks solely used the fixed capacity link between the nodes of networks with add/drop channels. Minimization of the number of repeaters correlating to minimizing the cost of the network is one of the common criteria for optimization. Hence network planning requires optimization of network design. The International Telecommunications Union (ITU) standardized a set of frequencies from which the wavelength should be chosen for the optimization.

However, due to a change in traffic or link/node failures, the network arrangement may change during the planning phase. Today the implementation of optical add/drop multiplexer and wavelength selective switching technique gives an extra level of opportunity to the execution of network design, permitting the restructuring of connections when there is any change in traffic. This approach has nominal interference with established traffic. In any case, a critical disadvantage actually stays unsettled and this is identified with the idle bandwidth issue because of the coarse and inflexible granularity of the framework.

The key element of the next generation of optical intelligent networks is the control plane, which is responsible for coordinating different network elements by introducing intelligent functions offering faster end-to-end link and traffic engineering with protection and restoration. Many regulatory bodies such as the Internet Engineering Task Force (IETF), the Optical Internetworking Forum (OIF), and ITU have been seriously chipping away at the applicable issues. Predicting standard of IP over WDM for the optical control plane can be well accomplished using generalized multiple protocol label switching (GMPLS). Although the central optical networks that are implemented over the past few years show good performance with respect to the present demands, there is a drawback regarding the very stringent specifications imposed on the equipment design for optical networks. This led to very complex design issues while improving the network performance. Hence future optical networks essentially require network dexterity and resource allocation flexibility that would empower a new generation of superior performance presented to the users with a lower cost. An upgraded version of GMPLS has been projected to launch in coming years which will allow more proficient use of network resources.

In addition to that, to improve the quality of service in the access domain of the network in terms of network availability, data loading speed etc., efforts are being made to shift the core infrastructure of connectivity to end users from the older copper or coaxial connection to a new optical fiber connection. The present day fiber-to-the-home (FTTH) system reflects this trend to expand the range and number of users associated to each access point. Application of wavelength division multiplexing in optical access networks is still being researched, yet it is normal that soon it will be good to go.

Optical fiber networks give a high-limit foundation to serving the developing traffic interest. In this regard, optical networking has an expanding crucial effect on our society and our personal satisfaction. Governments, research organizations, colleges, and the communications industry are putting vigorously efforts towards optical network-related innovations with the objective of determining developments that fundamentally will work on the performance of the upcoming networks while simultaneously expanding the expense and energy productivity of the services to be implemented.

A.2 Essential Background

Optical networking uses an encoded light signal for transmitting information. Local area networks (LAN), wide area networks (WAN), and metropolitan area networks (MAN) find major uses for optical networks and they also have an application in transoceanic communication. Optical networks finds their application in a well-organized and cost-effective way, taking advantage of the distinctive properties of fiber. Here we provide, in brief, some essential backgrounds of optical networks.

A.2.1 Optical Networks

Commonly, optical networks are of three types: point to point link, star network, and ring network. Another variant, the passive optical network, find its application in one to multi point connectivity with the use of optical splitters. A brief description of common optical network topologies is given below.

a) **Point-to-point Link:** At the very initial stage of optical networks, point-to-point transmission link between a pair of transmitter and receiver nodes was used. The transmitter node converts the information to be transmitted into an equivalent encoded optical signal by an electrical-optical (EO) conversion process and transmits it over the optical fiber. The purpose of the receiver node is to get back the original information from the optical domain by optical-electrical (OE) conversion process.

b) **Star Network:** In this topology, multiple nodes can be connected simultaneously. Star couplers plays a major part in establishing a combined connection of multiple point-to-point links. The optical signal received by the star coupler is forwarded to all the output ports of the coupler. Here also EO conversion is performed at the transmitter and OE conversion is performed at the receiver. This finally build an optical single-hop star network.

c) **Ring Network:** This type of network interconnects pairs of adjacent nodes with point-to-point fiber links. Each node in the ring network also makes use of an OE and EO conversion process for incoming and outgoing signals, respectively. Ring networks find wide application in Fiber Distributed Data Interface (FDDI).

A.2.2 SONET/SDH

One of the key benchmarks for optical networks is Synchronous Optical Network (SONET), associated with the Synchronous Digital Hierarchy (SDH) standard. It was first notified in 1985 and found its complete form in 1988. It aims to specify some criteria for optical point-to-point link interfaces. These criteria permit (i) optical transmitters to use different carriers for transmission,

(ii) directed optical interfacing, (iii) effortless access to branch signals, and (iv) the adding of new features in networks. To be specific, SONET explains the standard to be adopted for (i) the use of optical signals, (ii) the frame structure for synchronous time division multiplexing (TDM), and (iii) network strategy implementation and maintenance. SONET is essentially structured with digital TDM hierarchy with frame length of 125 μs. Ring network finds a wide application of SONET. The key elements to implement SONET in ring networks are add-drop multiplexers (ADM) and digital cross connect system (DCS). ADMs essentially connect multiple SONET end devices and also combine and split network traffic at different speeds. DCS is superior to ADM in the sense that DCS can connect larger number of links than ADM as well as being able to add or drop a single SONET channel at any location.

A.2.3 Multiplexing

It is well understood that bandwidth is a major constraint in every communication system. Efficient use of bandwidth can be realized by sharing the bandwidth between multiple traffics with the use of the multiplexing technique. Optical networks also implement the multiplexing technique for the efficient use of huge optical fiber bandwidth. The major approaches for optical multiplexing are briefly discussed below:

a) **Optical TDM Network:** Time Division Multiplexing (TDM) has been a well-established technique for network design for more than 50 years. Production of a short optical pulse enables TDM to apply in optical networks at 100 Gbps. This high-speed optical TDM (OTDM) needs significant attention to fiber transmission properties because dispersion can significantly limit the use of TDM to achieve a high bandwidth-distance product. Also the electro-optical bottleneck, which arises due to the modern faster processing technology at nodes in high speed networks, limits OTDM from full exploitation of bandwidth. OTDM networks are appropriate in short-range network application. However, a long-range network can deploy OTDM with the use of soliton transmission, where the nonlinear effect of fiber can eliminate the adverse dispersion effect. SONET/SDH is a significant example of OTDM networks.

b) **Optical SDM Network:** An easy solution to the electro-optical bottleneck is optical space division multiplexing (OSDM). Here a bunch of multiple parallel fibers replaces a single fiber for the network design. OSDM finds suitable application in short-range transmission networks. However, OSDM becomes more costly and thus less practical for long-range network applications.

c) **Optical WDM Network:** The conventional frequency division multiplexing can also be applied in the optical domain and is called wavelength division multiplexing (WDM). Here the information from each node is sent on separate wavelength. Multiplexer and demultiplexer operate in the wavelength domain. Since WDM can operate at any random line rate maybe below the overall TDM rate, it can avoid the limitation of TDM. Furthermore, WDM can take complete advantage of fiber bandwidth thus can avoid the need for multiple fibers as needed for SDM, and therefore is cost-effective. Hence, WDM emerges as the promising technology in optical networking. WDM fibers are installed between the nodes of optical network and can be implemented with or without electro-optic conversion. Optical WDM networks are classified as (i) Opaque WDM Network, (ii) Transparent WDM Networks, and (iii) Translucent WDM Networks.

A.2.4 All-Optical Networks

All-optical networks (AON) work on the principle of optical transparency, i.e., the transmitter and receiver nodes can be directly connected optically by fully avoiding the intermediary nodes. Like SONET/SDH, AON is also based on circuit-switched network. AON finds its application at the design level of network hierarchy. It implements all-optical node configurations which do not use optical-electrical conversion but instead work on the optical transparency principle. Switching and multiplexing in AON channels are accomplished by WDM. Like the ADM and DCS utilized in SONET/SDH, AONs also use a similar model known as optical add-drop multiplexer (OADM) and optical cross-connects (OXC). The AONs which make use of OADM and OXC are also stated as optical transport network (OTN). The use of optical bypassing in OADM and OXC makes AONs cost-effective and thus distinguishes it as a superior network application in the optical domain. The network agility of AONs can be enhanced by the use of reconfigurable OADM (ROADM) and reconfigurable OXC (ROXC).

A.2.5 Optical Transport Network

The International Telecommunication Union (ITU) designed the Optical Transport Network (OTN), a next-generation standard protocol, as a replacement to SONET/SDH. It is also called a "digital wrapper", which can give a productive and internationally acknowledged approach to multiplex various services in optical domain. Different traffic types, such as Ethernet, storage, and digital video, as well as SONET/SDH, can be transmitted over a single Optical Transport Unit frame thanks to OTN's enhanced multiplexing capacity. Traditional WDM transponder-based networks offer substantial advantages over OTN-based backbones and metro cores, including enhanced efficiency, dependability, and wavelength-based private services. The IP-over-OTN infrastructure also provides improved management and monitoring, fewer hops, better service protection, and lower equipment acquisition costs. OTN plays a significant role in making the network an open and programmable platform, allowing transport to become as important as computation and storage in intelligent data center networking, in addition to scaling the network to 100G and beyond.

B. Optical Switching in Networks

B.1 Historical Perspective

The nineteenth century was the revolutionary phase in communication technology because the first ever electrical communication was implemented with the evolution of the telegraphy system. At the same time the concept of switching was also introduced in communication. Although initially the switching was manual, gradually it was improved to the semi-automatic mode known as store and forward message switching [17]. The same principle of switching was adopted to further developed communication systems like telephony, facsimile, etc. till the computer network was evolved specifically for data communication. When computer networks were implemented, two new techniques of switching, circuit switching and packet switching, were developed gradually for better and hassle-free communication in and around 1960 [18–20]. Although earlier message switching was found to be efficient compared to circuit switching in terms of bandwidth utilization, it was not the better solution for modern data communication networks. In this scenario packet switching was found to be a better way because of its simplified data storage capability and the requirement for less of a retransmission process. Therefore it is understood that switching is an inherent process for communication technology to gain its application. The limitations of electrical communication systems have led people to move to optical communication systems. The optical communication systems have proven to be much more advantageous over the electrical systems as they overcome problems or limitations like bandwidth, speed, security, reduced system noise, and several other factors which are undesirable for a faithful and sustainable communication system [21, 22]. At the same time, the idea of optical switching gained attention and within three decades from 1970 to 2000, several attempts were made to design optical switches [23–25]. The invention of WDM technology to cope with the growing demand for internet creates a challenge in designing optical switching over a multi-wavelength channel. One of the problems lies in the optical processing. The optical processors that have been developed till now can only process low-speed signals or signals with low bit rates. The processors need electrical signals which are converted from optical signals are processed and then converted back to optical signals. This is done using optical-electrical-optical switches or OEO switches [26, 27]. Now this conversion and back-conversion takes up a lot of power and time, which are not desirable for high-speed systems. As a result, optical switching has been promoted as a remedy to these issues and a key to switching alleviation [28–30]. New optical switching technologies have been created as a consequence of extensive research and development, and existing ones have been improved as a result of advancements in material science and manufacturing processes. Many of these technologies became feasible options for implementation in networks. Obviously, the move to the optical domain would provide the network many significant benefits. Optical switching is the backbone of optical network and is vital for the network architecture of tomorrow. As such, it is the answer to many network challenges and an essential component of the long-term solution.

B.2 Essential Background

An optical switch is a device which can switch the optical signals between different circuits operating in optical domain. The capacity of the optical networks can precisely and completely be utilized with the help of optical switch technology. This section gives a brief overview of the basic optical switching technology.

B.2.1 Optical Switching in Networks

Electronics handle all of the additional functions in SDH/SONET networks, including multiplexing, cross-connection, add/drop, and control. Additionally, some equipment suppliers have created cross-connects based on electronic switching matrices with optical interfaces (OEO systems) and marketed them as optical cross-connects and/or optical switching systems. Switching systems based on optical switching fabrics, on the other hand, are known as OOO systems. While OOO systems are commonly referred to as transparent or all-optical, their OEO counterparts are frequently referred to as opaque. Transparency refers to a network's ability to convey any sort of data regardless of protocol and encoding schemes, data speeds, or modulation scheme. Transparency is inherent in optical networks, because data is transferred and switched end-to-end in the optical domain. Electrical networks, in contrast to transparent networks, are opaque since their performance is based on signal type and specifications. This is due to their capacity to read and analyze the signals they transmit, which is advantageous in many applications. WDM technology and optical switching on the one hand, and electronic control on another, are merged in a partial or complete transparent network. This integrates the greatest aspects of optics and electronics.

B.2.2 Optical Switching in Practice

Optical switching, like electrical switching, is divided into two types: optical circuit switching (OCS) and optical packet switching (OPS). Switching takes place at the granularity of an optical circuit in OCS. Because traffic load dominates the network and circuit switching is not optimal for traffic volume, the research community has placed a strong emphasis on OPS, which performs optical switching at the packet level. Although OPS is often envisaged as an electronically controlled system, it confronts problems in this optical form, such as the requirement for fast adaptable optical switching technologies, optical packet segmentation, and synchronization. Optical Burst Switching (OBS) was developed over the years as a balance between OCS and OPS, and has subsequently attracted a lot of research attention. Because OBS is packet-based, it may be more bandwidth-efficient than OCS. The necessity to read/write packet headers in the form of light as well as the extremely tiny switching granularity are both challenges for OPS. OBS minimizes the load on control by combining packets into bigger bursts. The requirement for optical header reading/writing is eliminated by sending data and control separately and handling control electronically. OBS, on the other hand, has its own technological and design issues.

B.2.3 Optical Switch Technology

Different classifications exist for optical space switching technologies. The optical electrical optical (OEO) switch and the all-optical switch (OOO) switch are the two most common types. Optical switching technologies can also be categorized depending on the fundamental physical effect that causes the switching process to occur. Numerous technology categories may be observed in this situation. Each of these classes contains a variety of technology types based on how the physical effect is exploited, the device design, the material used, and other factors. Some of these physical effects include: electro-optic, acousto-optic, thermo-optic, magneto-optic, liquid crystal-based, SOA-based, and many more. It should be emphasized that in many of the above category, regardless of the physical effect liable for the switching process, which is used as the foundation for this category, external control of the switching device is electrical.

On the other hand, optical signals are now used to transmit nearly all information due to their fast bit rate data transfer and huge bandwidth. Photonics has played a crucial part in this development since it allows for improved light-matter interaction to regulate optical signals for specific applications. Because there is no need to transform optical-electrical/electrical-optical signals at the interfaces, all-optical switches enable extremely efficient data transfer. This reduces power consumption and enhances operation speed. Photonic crystal-based switches play an important role in switching operations to satisfy the rising need for high data rate, bandwidth, low loss, and low power consumption.

C. Organization of This Book

In light of the realities discussed above, this book comprehensively and cohesively presents recent developments in optical switching and its application. Most of the recent challenges involved in the commercialization of optical switching are examined and discussed in this book. The book is organized in three parts. Part A provides the basic introduction of optical switching. Then we have presented the device technology for optical switching in Part B. Part B contains 13 chapters based on different device physics adopted for designing optical switches. A number of switching mechanisms associated with optical communication systems are summarized in a brief in Chapter 1. Chapter 2 gives a review of current electro-optic switches in terms of operating principle, fabrication material, and device structure along with the performance issues and subsequent challenges. The principles of thermo-optical switches including thermo-optic effect, trade-off between switching time and power consumption, cross-talk, as well as other merits are discussed in Chapter 3. Chapters 4 and 5 also describe the principle, design, and application area of magneto-optic and acousto-optic switches respectively. Further, some special switch fabrication, with underlying theory, principle of operation, and application area are described in this book. To be specific, Chapter 6, 7, and 8 focus on MEMS-based optical switches, SOA-based optical switches and liquid crystal optical switches, respectively. Chapters 9 and 13 specifically deal with photonic switches. Two special chapters (Chapter 10 and 12) are added in this book to illustrate the quantum and non-linear effect in designing optical switches. An additional Chapter 11 is added which focuses on the optical electrical optical (OEO) effect and its advancement. To summarize, Part B of this book will cover the wide range of possibilities of optical switch design and fabrication, and will certainly provide comprehensive guidance to students and researchers.

Part C of this book emphasizes the application of optical switches in optical networks. Five chapters are included in this section to discuss all the possibilities and challenges of the application of optical switches in networks. Starting with the switch fabric control in Chapter 14, this section gradually examines the reliability of switches in networks in Chapter 15 along with the detailed discussion of challenges and mitigation processes for protection and restoration of optical switches in networks covered in Chapter 16. Chapter 17 provides an illustrative idea for the application of optical switches in high-performance computing, which is currently a hot topic in computer networks. Finally, a special Chapter 18 is provided in this book with the discussion of software for optical network modelling, which will immensely help students and new researchers in this field to acclimatize with the available software.

Bibliography

1 K.G. Coffman and A.M. Odlyzko. Internet growth: Is there a "Moore's Law" for data traffic? In James Abello, Panos M. Pardalos, and Mauricio G. C. Resende, editors. *Handbook of Massive Data Sets*, 47–93. Boston, MA, Springer, 2002.

2 E. Agrell, M. Karlsson, A.R. Chraplyvy, D.J. Richardson, P.M. Krummrich, P. Winzer, K. Roberts, J.K. Fischer, S.J. Savory, B.J. Eggleton, and M. Secondini. Roadmap of optical communications. *Journal of Optics*, 18(6):063002, 2016.

3 I. Kaminow, T. Li, and A.E. Willner, editors. *Optical Fibre Telecommunications VI*. Elsevier, 2013.

4 P.J. Winzer, DT. Neilson, and A.R. Chraplyvy. Fiber-optic transmission and networking: the previous 20 and the next 20 years. *Optics Express*, 26(18):24190–24239, 2018

5 C.E. Shannon. A mathematical theory of communication. *The Bell System Technical Journal* 27(3): 379–423, 1948.

6 A.L. Schawlow and C.H. Townes. Infrared and optical masers. *Physical Review*, 112(6):1940, 1958.

7 K.C. Kao and G.A. Hockham. Dielectric-fibre surface waveguides for optical frequencies. In *Proceedings of the Institution of Electrical Engineers*, 113(7):1151–1158, 1966.

8 R.J. Mears, L. Reekie, I.M. Jauncey, and D.N. Payne. Low-noise erbium-doped fibre amplifier operating at 1.54 µm. *Electronics Letters*, 23(19):1026–1028, 1987.

9 C.R. Giles and E. Desurvire. Modeling erbium-doped fiber amplifiers. *Journal of Lightwave Technology*, 9(2):271–283, 1991.

10 A.R. Chraplyvy, A.H. Gnauck, R.W. Tkach, and R.M. Derosier. 8* 10 Gb/s transmission through 280 km of dispersion-managed fiber. *IEEE Photonics Technology Letters*, 5(10): 1233–1235, 1993.

11 X. Liu, X. Wei, R.E. Slusher, C.J. McKinstrie. Improving transmission performance in differential phase-shift-keyed systems by use of lumped nonlinear phase-shift compensation. *Optics Letters*, 27(18):1616–1618, 2002.

12 S. Chandrasekhar, X. Liu, B. Zhu, and D.W. Peckham. Transmission of a 1.2-Tb/s 24-carrier no-guard-interval coherent OFDM superchannel over 7200-km of ultra-large-area fiber. In *2009 35th European Conference on Optical Communication 2009*, September 20 2009, pp. 1–2. IEEE.

13 P.J. Winzer. Energy-efficient optical transport capacity scaling through spatial multiplexing. *IEEE Photonics Technology Letters*, 23(13):851–853, 2011.

14 ITU. Spectral grids for WDM applications: DWDM frequency grid. Recommendation G. 694.1. ITU, 2012.

15 L. Zong, H. Zhao, Z. Feng, and Y. Yan. Low-cost, degree-expandable and contention-free ROADM architecture based on M×N WSS. In *Optical Fiber Communication Conference 2016*, 20 March, pp. M3E–3. Optical Society of America.

16 P.D. Colbourne, S. McLaughlin, C. Murley, S. Gaudet, and D. Burke. Contentionless twin 8x24 WSS with low insertion loss. In *Optical Fiber Communication Conference 2018*, 11 Mar, pp. Th4A–1. Optical Society of America.

17 J. Bellamy. Subsections 1.2. 8 "Transmission Impairments" and 2.2. 1 "DSP Applications". In *Digital Telephony*, 3rd edition. New York, Wiley, 32–41 and 82–83, 2000.

18 P. Baran. On distributed communications networks. *IEEE Transactions on Communications Systems*, 12(1):1–9, 1964.

19 R.D. Rosner. Packet switching: tomorrow's communications today. Belmont, CA: Lifetime Learning Publications, 1982.

20 P. Baran. The beginnings of packet switching: some underlying concepts. *IEEE Communications Magazine*, 40(7):42–48, 2002.

21 D. Botez and G.J. Herskowitz. Components for optical communications systems: A review. *Proceedings of the IEEE*, 68(6):689–731, 1980.

22 R.C. Alferness. Guided-wave devices for optical communication. *IEEE Journal of Quantum Electronics*, 17(6):946–959, 1981.

23 R. Scarmozzino, A. Gopinath, R. Pregla, and S. Helfert. Numerical techniques for modeling guided-wave photonic devices. *IEEE Journal of Selected Topics in Quantum Electronics*, 6(1): 150–162, 2000.

24 Q. Chen, Y. Chiu, D.N. Lambeth, T.E. Schlesinger, and D.D. Stancil. Guided-wave electro-optic beam deflector using domain reversal in $LiTaO_3$. *Journal of Lightwave Technology*, 12(8):1401–1404, 1994.

25 M. Yamada, M. Saitoh, and H. Ooki. Electric-field induced cylindrical lens, switching and deflection devices composed of the inverted domains in $LiNbO_3$ crystals. *Applied Physics Letters*, 69(24):3659–3661, 1996.

26 N. Tsukada and T. Nakayama. Polarization-insensitive integrated-optical switches: A new approach. *IEEE Journal of Quantum Electronics*, 17(6):959–964, 1981.

27 A. Milton and W. Burns. Mode coupling in tapered optical waveguide structures and electro-optic switches (Invited Paper). *IEEE Transactions on Circuits and Systems*, 26(12):1020–1028, 1979.

28 J.M. Elmirghani and H.T. Mouftah, editors. *Photonic Switching Technology: Systems and Networks*. IEEE Press, 1999.

29 T.S. El-Bawab. Almost-all-optical core: motivations and candidate technologies. In *Technologies, Protocols, and Services for Next-Generation Internet*, 26 July 2001, 4527, pp. 172–176. International Society for Optics and Photonics.

30 T.S. El-Bawab. Potential of optical switching in future communication networks. In *Optical Transmission Systems and Equipment for WDM Networking II*, August 19 2003, 5247, pp. 111–114. International Society for Optics and Photonics.

Part B

Switch Characterization

1

Optical Switches

Rajan Agrahari[1], Sambit Kumar Ghosh[2], and Somak Bhattacharyya[2]

[1] *Department of Electronics and Communication Engineering, National Institute of Technology, Patna, Bihar, India*
[2] *Department of Electronics Engineering, Indian Institute of Technology (BHU), Varanasi, Uttar Pradesh, India*

1.1 Introduction

Any communication system needs three essential components, *viz.*, a source, a receiver, and a medium of transmission. The terminology changes depending on the mode of the communication systems. In modern-day optical communication, a semiconductor laser has been generally used as the source and optical fiber as the transmission medium. The demand for bandwidth is increasing with the introduction of new applications, *viz.*, grid computing, smart TV systems with live internet broadcasting services, etc. The emergence of three-dimensional (3D) movies in multiplexes, as well as in-home entertainment systems, demands more bandwidth of data transmission through optical communication. In order to enhance the transmission rate and bandwidth available on the optical fiber, the other ends of the optical network system need to be improved. The other ends of the optical network include a range of detectors, multiplexers, switches, and buffers [1, 2].

A switch is embedded in the optical communication system to route the message signal under the supervision of the control signals. The message signal could be large in size or a large block of multiplexed data traffic or a series of lower bit channels to be delivered to the users of an optical communication system. An optical switch is versatile in nature as it has a large number of usages in communication networks as well as in communication cores of modern-day high configuration computers with data rates of 1000 Gbps. With the advent of time, quantum computers have been developed for secure and ultrahigh-speed communications [3]. New computer architecture needs newly designed optical switches with the least interruption in the phase information of the quantum data packets. The main function of an optical switch is to selectively switch an optical signal transmitting through an optical fiber or one highly configured integrated optical circuit to another [4, 5]. A number of switching mechanisms are associated with optical communication systems. They are summarized in a brief way in different sections of this chapter.

Optical Switching: Device Technology and Applications in Networks, First Edition. Edited by Dalia Nandi,
Sandip Nandi, Angsuman Sarkar, and Chandan Kumar Sarkar.
© 2022 John Wiley & Sons, Inc. Published 2022 by John Wiley & Sons, Inc.

1.2 Electro-Optical Switching

Fiber-optic communication systems are experiencing a gradually increasing demand for digital smart TV, digital video and broadband internet services, and many other applications. These types of communication networks should be compatible enough to accommodate future demands of applications at the cost of robust, reliable, and cheap components for information transmission and switching. Optical switches are an integral part of optical networks and fiber-optic communication systems. Several compound semiconductors, *viz.*, gallium arsenide (GaAs) or indium phosphide (InP) with the integration of photodetectors and lasers have been excessively used as optical switches due to their flexible nature. But the practical realization of these types of devices is costly. Different types of integration technologies have been introduced thereafter with some of the demerits too, *viz.*, silica or/and glass on silicon configurations are used in integrated optical systems but monolithic integration with photodetectors and lasers is not possible. Optical switches made of polymer have also been used but their stability is not good at all [6]. This section will focus on the progress of the electro-optical switching technology as it is using an improved fabrication process. Recently, liquid crystals of graphene oxide have also been used as an electro-optical switch [7].

1.2.1 Working Principle of Electro-Optical Switches

Electro-optical switching works on various principles, as described below.

(a) Single-Mode Principle: A ridge waveguide is primarily taken for the formation of the optical switches. The design of a single-mode ridge waveguide has to fulfill certain design considerations. First, a ridge waveguide should have a numerical aperture equivalent in size to that of a single-mode fiber. Second, it must contain a large cross-sectional area equivalent to the core diameter of a single-mode fiber. Lastly, the waveguide supports the single-mode operation. A small pictorial description of a silicon-germanium-based ridge waveguide is shown in Figure 1.1.

The structure is formed using a Si-Ge layer of the refractive index of n_1 deposited on a Si substrate with refractive index n_2. The width (W) of the ridge is $2a\lambda$ following a height (h) = $2b\lambda$. The etching depth of the ridge is determined by the term, $h' = 2b(1 - r)\lambda$, where λ is a free-space optical wavelength, and r is the fractional height of the tapered portions of the rige waveguide. Here, n_0, n_1, and n_2 are the refractive index of free space, Si-Ge layer, and Si substrate, respectively. All these refractive indices are considered at the operational wavelength. The ratio of the lateral to transverse lengths must obey the condition stated in equation (1.1) and (1.2) [6] for the proper propagation of light within the input and output waveguides.

$$\frac{a}{b} \le \left(\frac{q+4\pi b}{4\pi b} \right) \frac{1+0.3\sqrt{\left(\frac{q+4\pi b}{q+4\pi rb} \right)^2 - 1}}{\sqrt{\left(\frac{q+4\pi b}{q+4\pi rb} \right)^2 - 1}}, \tag{1.1}$$

$$q = \frac{\gamma_0}{\sqrt{\left(n_1\right)^2 - \left(n_0\right)^2}} + \frac{\gamma_2}{\sqrt{\left(n_1\right)^2 - \left(n_2\right)^2}}, \text{ and} \tag{1.2}$$

Figure 1.1 Schematic of a Si-Ge/Si ridge waveguide.

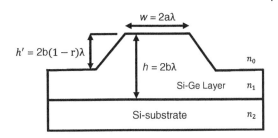

$$\gamma_{0,2} = \begin{cases} 1 & \text{for HE mode} \\ \left(\dfrac{n_{0,2}}{n_1}\right)^2 & \text{for EH mode}. \end{cases} \tag{1.3}$$

(b) Multimode Principle: The evolution of the multimode interference principle has come into the picture owing to its benefits in terms of easy fabrication, compact size, low loss, and fabrication tolerances. The multimode interference (MMI) switch works on the self-imaging principle. In this process, an input field is regenerated in terms of one or many images at periodic instances during the propagation of the light. The effective width of a multimode waveguide is represented in equation (1.4) [6], where W_M stands for the width of the multimode waveguide, n_c is the effective refractive index of the cladding region, and n_r is denoted as the effective refractive index of the waveguide. The values of σ are zero and one for TE mode and TM mode, respectively [6].

$$W_e = W_M + \left(\frac{\lambda_0}{\pi}\right)\left(\frac{n_c}{n_r}\right)^{2\sigma}\left(\left(n_r\right)^2 - \left(n_c\right)^2\right)^{-\left(\frac{1}{2}\right)}, \tag{1.4}$$

(c) Plasma Dispersion Effect: In this method, the carrier concentration of the materials determines the refractive index. Equation (1.5) can validate this fact clearly [6].

$$\Delta n = -\left(q^2\lambda^2 \middle/ 8\pi^2 c^2 n\varepsilon_0\right) \cdot \left[\left(\Delta N_e / m_{ce}\right)^* + \left(\Delta N_h / m_{ch}\right)^*\right]. \tag{1.5}$$

Here q is a charge of an electron, n is the refractive index of the less concentrated SiGe/Si material, c stands for the speed of light in free space, ε_0 is free-space permittivity, ΔN_h and ΔN_e are the change in concentrations of holes and electrons and m_{ch} and m_{ce} are conductivity effective masses of the holes and electrons of the SiGe/Si material. For example, taking SiGe/Si material when the concentration of Ge is less than 20%, the plasma dispersion effect comes into play to alter the values of refractive indices [6].

1.2.2 Realization of Electro-Optical Switches

Electro-optical waveguide switches have been developed using Si-based components, polymers and lithium niobate (LiNbO₃), etc. Silicon-germanium (SiGe) has been used in the production of multifunctional photonics-based switching technology because of its less propagation loss within the wavelength region 1.3–1.55 μm. Epitaxial growth of SiGe material is in demand as this process of device realization is compatible with large-scale silicon integration. Germanium (Ge) is preferable in the electro-optical switching technology as it has a shorter bandgap and the higher mobility of the electrons. Silicon-based electro-optical switches were usually developed using chemical vapor deposition (CVD) techniques and molecular beam epitaxial (MBE) growth. The details of these fabrication processes can be found elsewhere [8].

1.3 Acoustic-Optical Switching

Acoustic-optical switching deals with the acoustic-optic effect, where mainly the refractive index of the medium involved with the optical communication system is altered by the acoustic waves [9, 10]. The periodical strain is generated within the optical communication medium in terms of simultaneous compressions and rarefactions. The acoustic waves are generated with the help of the piezoelectric materials either in the bulk of the material or on the surface of the said material. An RF signal has been used as the driver excitation for producing the acoustic-optic effect. The electrodes between which the RF electric field has been applied are called the acoustic transducers.

1.3.1 Types of Acoustic-Optical Switching

Several types of acoustic-optical switching have been evolved through the years, such as modulators, tunable filters, switches, deflectors, frequency shifters, etc. [11].

(a) Acoustic-Optical Modulator: An acoustic-optical modulator is used for the ultimate control of the power of a laser beam while the input excitation signal is an electrical signal. In this case, the refractive index of the optical medium, such as glass or some other crystal is altered by the mechanical strain generated by the oscillation of a sound wave, also called a photo-elastic effect.

$$\Delta\eta_{ij} = \Delta\left(\frac{1}{\left(\eta_{ij}\right)^2}\right) = p_{ijkl}S_{kl}, \tag{1.6}$$

where $\Delta\eta_{ij}$ is the change in refractive index due to mechanical strain, S_{kl} is the strain (a tensor generating from the propagating acoustic wave within the medium), and p_{ijkl} is the fourth-rank optical tensor.

The main part of an acoustic-optical modulator is a transparent crystal object or glass. Light is launched through this glass. Transducers made of piezoelectric material are attached to the glass. Thereafter, the RF excitation signal is applied. The piezoelectric transducers produce an oscillating electrical signal, which further excites a sound wave at frequency 100 MHz or a multiple of 100 MHz with an acoustic wavelength of 10 µm to 100 µm (Figure 1.2). This sound wave produces a strain wave within the material which is known as photo-elastic effect. This photo-elastic effect produces a grating in the refractive index profile; Bragg diffraction occurs at these types of the grating. Therefore, the acoustic-optical modulators are also called Bragg cells [12, 13].

(b) Acoustic-Optical Frequency Shifter: The working principle of this device is related to the acoustic-optical modulator. In the acoustic-optical modulator, when the light beam is diffracted within the traveling refractive index grating the diffracted light encounters a shift in the optical frequency which is more or less than the excitation signal frequency. This is normally called a Doppler shift, which is utilized in acoustic-optical frequency shifters. Acoustic-optical frequency shifters are different from acoustic-optical modulators in terms of input excitation RF signal. These frequency shifters are usually operated with constant drive power where the frequency is fixed sometimes.

Figure 1.2 Setup of a non-resonant acoustic-optical modulator; a sound wave is generated using the transducer, the light beam is partially diffracted.

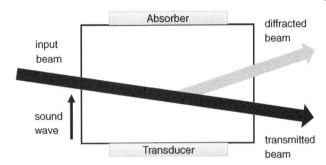

(c) Acoustic-Optical Deflectors: These types of acoustic-optical devices switch the input laser beam in one particular direction using a variable angle. The electrical drive signal usually controls this variable angle. This device comes from the acoustic-optical modulators family but it is operated with a constant power electrical drive signal with variable frequencies. In this case, the direction of the diffracted beam is determined by fulfilling the Bragg condition as shown in equation (1.7).

$$\theta = \frac{\lambda f}{v},$$ (1.7)

where λ is the wavelength of the vacuum, v stands for the velocity of the sound wave passing through the acoustic-optical material and f is the frequency of the excitation or drive signal.

1.3.2 Acoustic-Optical Device Materials and Applications

General materials for developing acoustic-optical switches are crystalline quartz, fused silica, and tellurium dioxide, etc. Germanium (Ge) and indium phosphide (InP) are used in the making of these types of switches for infrared applications. Lithium niobate ($LiNbO_3$) and gallium phosphide (GaP) are used for high-frequency optical communication [14]. Acoustic-optical modulators are used for active mode-locking purposes to modulate the loss of a resonator in terms of round-trip frequency or their multiple forms. They can also be used to modulate the power of the laser beams. Acoustic-optical switches are also used as noise-eater devices where the diffraction loss is controlled with feedback circuitry.

1.4 Thermo-Optical Switching

Thermo-optical switching is the main building block of optical communication and switching networks [6]. They are in high demand because they are smaller in size, high in scalability, and they have the potential to integrate with the waveguide multiplexers and demultiplexers. They also have a great impact on optical communication systems as they are an essential part of the optical add-drop multiplexing systems (OADM). Thermo-optical switches are made of silica and polymers. These two materials are primarily chosen as a flexible range of refractive index values can be selected to have fewer coupling losses with the optical fiber. Within the channel waveguide, a bottom cladding surface is grown. Thereafter, a core layer is grown on the cladding surface. A portion of the core ridge is etched following a coating layer deposited on the cladding region. Commercially, thermo-optical switches are produced with silica and polymer technology. Silicon substrates are chosen as they are compatible with modern IC technology, good heat conduction characteristics, and great surface quality. The second property is essential for thermo-optical switching as the substrate itself should have a good heat sink property. A specific type of stripe electrodes is deposited on the cladding surface for inducing switching by generating an exception to the effective refractive index of the medium [6].

1.4.1 Working Principle of Thermo-Optical Switches

The optical property of any medium can be characterized by the complex form of the refractive index function of that medium. Alternatively, that can also be presented by the complex-valued dielectric constant $\epsilon = \epsilon_1 - \epsilon_2 j$. This dielectric constant (ϵ) is a function of the refractive index (n), they are related as $\epsilon = n^2$; therefore ϵ_1 and ϵ_2 can be derived when the N and k are known: $\epsilon_1 = N^2 - k^2$ and $\epsilon_2 = 2Nk$, where N, k, ϵ_1 and ϵ_2 are the optical constants. They are dependent on the energy of a photon, i.e., $E = \dfrac{\hbar\omega}{2\pi}$, where ω = frequency of a photon. These functions are known as the relations of the optical dispersion [15, 16]. The Clausius–Mossotti derivation for the isotropic materials is given by the expression in (1.8) [6]:

$$\frac{(\epsilon - 1)}{(\epsilon + 2)} = \frac{4\pi\alpha_m}{3V} \tag{1.8}$$

where V is the volume of the small sphere, α_m is the polarizability of the macroscopic sphere. Three factors are responsible for the temperature dependence of a dielectric constant. Now, one can have differentiation of equation no. (1.8) with respect to the temperature under the constant pressure [6],

$$\frac{1}{(\varepsilon - 1)(\varepsilon + 2)\left(\dfrac{\partial \varepsilon}{\partial T}\right)_P} = -\frac{1}{3V}\left(\frac{\partial V}{\partial T}\right)_P + \frac{1}{3\alpha_m}\left(\frac{\partial \alpha_m}{\partial V}\right)\left(\frac{\partial V}{\partial T}\right)_P + \frac{1}{3\alpha_m}\left(\frac{\partial \alpha_m}{\partial T}\right)_V = a + b + c, \tag{1.9}$$

where:
a: when the temperature increases, the specific volume will increase, the inter-atomic spacing will also increase and the dielectric constant will decrease with constant pressure;
b: when the volume of the lattice will increase, the polarizability of the macroscopic sphere increases;
c: it decides how polarizability changes with the temperature when the volume is constant [6].

1.4.2 Realization of Thermo-Optical Switches

The first realized thermo-optical switch is a digital optical switch (DOS). It is useful for space switching systems where multiple wavelength operation is needed. The operating principle of DOS is based on the waveguide heating process. A Y-shaped junction is placed on the substrate integrated core-cladding stack. Two electrodes are deposited on the two arms of the Y-shaped junction. The refractive index has been varied by heating one single arm at a time or heating both arms at the same time. Improved functionalized DOS has been reported in several literatures [17].

1.4.3 Thermo-Optical Switch Materials and Applications

The thermo-optical switch can be implemented from specific materials which have low optical losses, low polarization-dependent losses, low wavelength dispersion, resistance to humidity, mechanical properties with thermal stability, etc. Polymeric materials, benzocyclobutene, fluoroacrylate, polyimide, polyurethane, amorphous silica, silicon, silicon nanocrystals, III-V semiconductors, lithium niobate ($LiNbO_3$), tantalum pentoxide (Ta_2O_5), and aluminum oxide (Al_2O_3) have been used for the production of the thermo-optical switches [6]. Recently, the photonic crystal-based thermo-optical switch has also been reported [18].

1.5 Liquid Crystal-Optical Switching

Liquid crystal bears two opposite meanings words in its nomenclature. This is because they exhibit intermediate phases in which the state of the matter shows characteristics like liquids as well as crystals [8]. The phase of the liquid crystal is represented in terms of 'nematic' in electro-optical applications. The tensor of the refractive index of nematic is presented similarly to the dielectric tensor of a material.

$$n = \begin{bmatrix} n_0 & 0 & 0 \\ 0 & n_0 & 0 \\ 0 & 0 & n_e \end{bmatrix}, \tag{1.10}$$

where n_e stands for the extraordinary index of refraction, and n_0 is the ordinary index of refraction. The birefringence of a nematic liquid crystal can be defined as $\Delta n = n_e - n_0$. The values possible for Δn reside between 0.01 and 0.3. This large range of values of Δn coupled with the rotation of molecules of the liquid crystal helps in the polarization of light in optical communication systems. The refractive index of the extraordinary ray is given by (1.11) [19, 20].

$$n_{eff} = \sqrt{\frac{\left(n_e\right)^2 \left(n_0\right)^2}{\left(n_e\right)^2 \left(\cos\theta\right)^2 + \left(n_0\right)^2 \left(\sin\theta\right)^2}}, \tag{1.11}$$

where θ is the angle between the propagation vector of the extraordinary ray and the director axis [21].

1.5.1 Types of Liquid Crystal-Optical Switches

In liquid crystals, the refractive index and birefringence are calculated from the arrangement of the molecules, which further helps in the formation of the switches for the optical communication systems. The arrangement of the molecules in liquid crystals can be controlled by applying an electric field across the liquid crystal.

(a) Liquid Crystal Optical Switch Based on Birefringence: These types of switches are made by controlling the direction of the light propagation along the liquid crystals to modify the birefringence property of the medium without changing the polarization state, but controlling the reflection and refraction of light coming out of the surface. A setup of this type of liquid crystal switch has been presented in Figure 1.3. The refractive index of the liquid crystal (LC) molecules can be monitored by placing them accordingly the light propagation vector through the LC material.

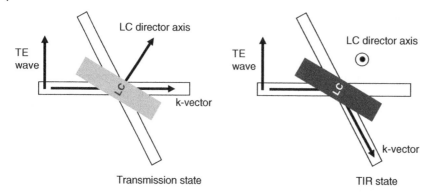

Figure 1.3 States of liquid crystal for transmission and total internal reflection (TIR).

The change in refractive index in different cases causes transmission of the optical beam or total internal reflection (TIR) of the optical beam [22].

Two other important types of liquid crystal optical switches are (b) liquid crystal optical switch based on polarization and (c) wavelength-selective liquid crystal optical switch.

1.5.2 Liquid Crystal-Optical Switch Applications

Liquid crystal-based switches offer vast use in telecommunication applications because they need shorter response time and larger efficient optical networks. Modern-day displays called LCD are in use everywhere. LCD technology eradicates the demerits of temperature dependence, power consumption, etc. Liquid crystal optical switches find applications in beam steering applications. In the photonics domain, liquid crystal optical switches have been used as photonic crystal fiber (PCF) [23]. Liquid crystal optical switches are used to achieve polarization diversity features in smart antenna applications to set different focal points from the same system [24]. They have been used in logic devices also [25]. These types of switches have also been used in polarization switching applications in communication systems.

1.6 Photonic Crystal Optical Switching

Most optical switches operate on the principle of refractive index change or phase shift between the two ports [8]. An optical switching occurs when the phase difference of the light between the two ports becomes π, corresponding to the path difference of $\lambda/2$. The change in refractive index is very small in most of the all-optical switches. Therefore, a very long path length device is needed for making the required phase shift. Due to large light-matter interaction and bandgap property, photonic crystal-based optical switches have the advantage of smaller size devices and higher integration [26].

Photonic crystals are the periodic arrangement of the dielectric rod or air hole perforated in a dielectric slab to manipulate the light within a crystal platform. Depending on the periodicity direction, photonic crystals are classified as one-dimensional (1D PCs), two-dimensional (2D PCs), and three-dimensional (3D PCs) [27]. Photonic crystals can be arranged in a square or triangular lattice. It prohibits/guides the light in a certain defined range of frequency called bandgap which is obtained from the dispersion diagram of the unit cell of the periodic structure. The bandgap of the periodic structure can be controlled by varying the dimensions of the periodic unit cells and

(a)

(b)

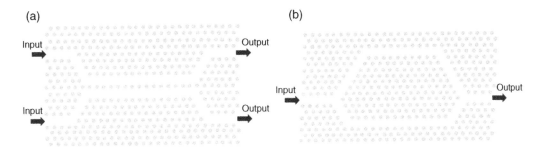

Figure 1.4 Realization of (a) directional coupler, and (b) Mach–Zehnder interferometers from photonic crystal. Source: (a) Modified from [34] and (b) Modified from [37].

the refractive index difference of the substrate and surrounding material. Due to ease in fabrication, mostly two-dimensional photonic crystals are used for the optical switches [28–31]. All-optical switch in 2D PCs can be realized by introducing one of the following in PCs:

- directional coupler structures [32–34],
- Mach–Zehnder interferometer [35–37], and
- resonators [38–41].

Directional coupler photonic crystal can be realized by introducing the line defect waveguide, i.e., removing a row of the dielectric rod or air hole in the dielectric slab from photonic crystals which create a waveguide to guide the electromagnetic wave. Figure 1.4(a) shows two-line defect waveguides separated by a row of rods [34]. An electromagnetic wave, launched in one waveguide, is coupled or switched to other waveguides under suitable conditions.

Mach–Zehnder interferometers (MZIs) as an optical switch in photonic crystals can be implemented by making the difference in the effective path lengths between two optical signals, which results in a difference of phase shift. In MZIs, when an optical signal enters the photonic crystal, it splits into two different optical signals and propagates through each arm of the MZI as shown in Figure 1.4(b) [37]. A phase shift between the two optical signals occurred due to the difference in the path length. When these two signals are further added, depending on the phase shift, the signal intensity at the output is either high or low and therefore acts as a switch.

Due to the long device length in the directional coupler and MZI structure, they are difficult to integrate with the other optical components [42]. By using the resonator-based photonic crystal, the size of the optical switch can be reduced efficiently.

Resonator-based photonic crystals have nanocavities that restrict the light in a very small region that results in higher interaction of light with matter and reduction in size. To design the multiport optical switch, either square-shaped [38] or circular-shaped resonators [28] with waveguide structure are mostly used. The schematic of the photonic crystal with square- and circular-shaped resonator are shown in Figure 1.5. When the optical signal enters in photonic crystal waveguide, the refractive index of the material in the cavity changed which shifts the resonance frequency, and switching occurs at the output.

1.7 Semiconductor Optical Amplifier (SOA) Optical Switching

Semiconductor optical amplifier (SOA) based all-optical switches have multi-fold advantages: compact size, mass production, low switching power, larger nonlinear coefficients, and low latency due to monolithic integration [43]. A schematic of the SOA is shown in Figure 1.6, which is used

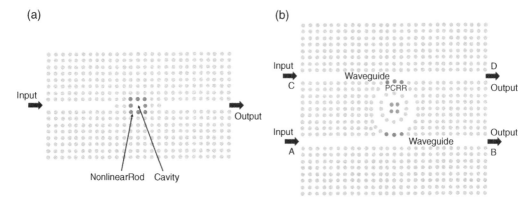

Figure 1.5 Photonic crystal switching by: (a) square-shaped and (b) circular-shaped ring resonator.

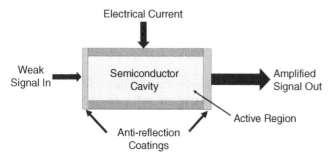

Figure 1.6 Schematic of SOA.

for amplification of optical signal based on the gain of the semiconductor medium. The principle of SOA is similar to that of a laser diode where light is amplified by stimulated emission [44]. But in the structure, the mirrors at both the facets of the laser are replaced by antireflection coatings to eliminate the resonator structure. In addition, to reduce the reflectance at the end, an inclined waveguide can also be used.

In SOA, weak signal input is passed through the single-mode waveguide, which has significant overlap with the active region. An electric current is used to drive the active region of the SOA, which generates the large carrier density in the conduction band. These generated carriers take part in the optical transition from the conduction band to the valence band.

SOAs can be categories into two types [45]:

(a) Fabry–Pérot Amplifiers (FPA)

In FP amplifiers, due to mirrors at the facets, light is reflected back to the active region and acts as a resonant cavity. Therefore, the light gets amplified due to reflection between the two facets. FPA is sensitive to the frequency of the input signal and temperature.

(b) Traveling-Wave Amplifiers (TWA)

In contrast to FPA, TWA has antireflection coatings at the facets. Therefore, multiple reflections in the active region do not take place for TWA, and input light gets amplified in a single pass through the amplifier. TWA is widely used due to its large bandwidth and low polarization sensitivity.

SOA-based optical switches can be made by integrating SOA with gating elements. The switching function can be achieved by making the gating element sensitive to the property changed by the SOA, like wavelength, phase, and amplitude.

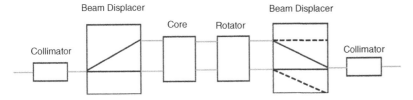

Figure 1.7 Working of MO switch.

1.8 Magneto-Optical (MO) Optical Switching

All-optical switches, e.g., MEMS switches, thermo-optic switches, and electro-optic switches, have high insertion loss or low switching speed. To overcome these limitations a magneto-optical (MO) switch has been developed which has low insertion loss and fast switching time. The magneto-optical switch worked on the principle of Faraday rotation of polarized light. By changing the external magnetic field, the polarization of the incident polarized light changed, thus switching the optical path [46–48].

Generally, the MO switch is composed of beam displacer, core, and Faraday rotator, as shown in Figure 1.7. The working principle of MO switch is as follows:

When the light has been passed through the collimator and entered into the beam displacer, light is divided into mutual orthogonal polarization. Then, by changing the external magnetic field, the rotator changes the polarization plane of incidence polarized light, and thus the light route is switched.

Optical switches with MO materials have several advantages, which include high integration, low insertion loss, large optical cross-section, low operating voltage, polarization-independence, and non-reciprocity of the induced rotation. Further, due to no physical movement, MO switch has better reliability.

1.9 Micro Electro-Mechanical Systems (MEMS) Optical Switching

A system that consists of micrometer-sized electrical and mechanical components called MEMS and its applications include actuators, accelerometers, sensors, switches, and gyroscopes [49–51]. The MEMS system is fabricated by photolithography, ion beam etching, chemical etching, wafer bonding, etc.

The structure of the MEMS-based optical switch is shown in Figure 1.8. It consists of an array of a small mirror on the silicon crystal. The array of the mirror is driven by electromagnetic or electrostatic force to switch the direction of the input light. When the light is passed through the input collimator array towards the mirror array, the angle of the mirror is changed by the driven force. Thus, the input light changes its path to the different output terminals of the optical switch to realize the ON and OFF function of the light path. In general, MEMS-based optical switches are categories into two types based on spatial structure: 2D switches and 3D switches as shown in Figure 1.8 [49].

By using micromechanical manufacturing technology, the mirror array of the 2D MEMS optical switch is integrated on the silicon substrate. Mirrors in the array have CROSS and BAR states as shown in Figure 1.8(a). Input light is reflected to the output collimator by using CROSS state mirrors, while input light passes through uninterrupted from the mirror array by using BAR state mirror.

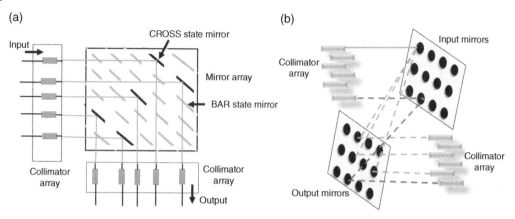

Figure 1.8 MEMS-based optical switch (a) 2D, and (b) 3D. Source: [49]/IEEE.

In the 3D MEMS optical switch, the micro-mirror can rotate along two axes which results in the multiple positions of each mirror and direct light to multiple angles (changing the optical path) as shown in Figure 1.8(b). 3D MEMS switch contains pair of mirror arrays: input mirror and output mirror. The input light first reaches the array of input mirror, then reflected towards output mirror array, and finally reflected to the output collimator array.

The MEMS optical switches are widely used due to their advantages of high integration, low power consumption, and low cost.

1.10 Metasurfaces Switches

Metasurfaces are normally used to manipulate the electromagnetic wave or light. Metasurfaces have a diverse field of applications in light absorbers, polarization converters, phase shifters, cloaking, and filters, etc. [52–56]. Recently, lumped elements have been incorporated with the metasurfaces to achieve switching characteristics. Most of the developments have been made for the microwave and milimeter wave domain [57, 58]. Graphene has also been used as a switch in a recent literature for terahertz applications [59]. Metasurfaces-based switching applications for the optical domain have not been explored explicitly.

1.11 Conclusion

Optical switches are one of the most important devices for enhancing the transmission rate and bandwidth and reducing the network cost in optical communication systems. An overview of various types of optical switching has been discussed in this chapter. The electro-optic switches are advantageous for their tiny size due to single mode of operation and are independent on polarization of the input signal. In acoustic switches, change in refractive index occurred by acoustic wave but has the disadvantage of low sensitivity, poor response to optical signal, and high crosstalk. The optical switches based on thermo-optic effect has been mostly fabricated by silicon-on-insulator (SOI) technology due to its larger thermo-optic coefficient. These switches consume less power and have fast response for the optical signal. Photonic crystal switches based on nonlinear material have ultra-high response due to the fast response of nonlinear material. Due to the small size of the photonic crystals, these switches are easy to integrate with future high-speed optical networks. However, higher coupling of these switches with optical fiber is still challenging. Magneto-optic switches utilize the Faraday rotation for the switching application. These switches have the advantages of higher integration, low insertion loss, large optical cross-section, low operating voltage, and polarization-independence. But the area for the suitable materials for magneto-optic switches needs more research. Finally, MEMS switches have been discussed which offer several advantages like high integration, low power consumption, and low cost. These switches are most widely used optical switches. MEMS architectures with 2D and 3D structure have been described. Metasurface-based optical switches are in the embryonic stage and need to be further explored.

Bibliography

1 G.I. Papadimitriou, C. Papazoglou, and A.S. Pomportsis. *Optical Switching*. Wiley-Interscience, 2006.

2 V. Sasikala and K. Chitra. All optical switching and associated technologies: a review. *Journal of Optics*, 47(3):307–317, 2018.

3 F. Lecocq, F. Quinlan, K. Cicak, J. Aumentado, S.A. Diddams, and J.D. Teufel. Control and readout of a superconducting qubit using a photonic link. *Nature*, 591(7851):575–579, 2021.

4 C.Y. Jin and O. Wada. Photonic switching devices based on semiconductor nano-structures. *Journal of Physics D: Applied Physics*, 47(13):133001, 2014.

5 P. Andreakou, S.V. Poltavtsev, J.R. Leonard, E.V. Calman, M. Remeika, Y.Y. Kuznetsova, L.V. Butov, J. Wilkes, M. Hanson, and A.C. Gossard. Optically controlled excitonic transistor. *Applied Physics Letters*, 104(9):091101, 2014.

6 S.J. Chua and B. Li, eds. *Optical Switches: Materials and Design*. Elsevier, 2010.

7 J.K. Song. Electro-optical switching of liquid crystals of graphene oxide. In *Liquid Crystals with Nano and Microparticles* (pp. 817–846), 2017.

8 T.S. El-Bawab. *Optical Switching*. Springer Science & Business Media, 2006.

9 A. Yariv and P. Yeh. *Optical Waves in Crystals* (Vol. 5). New York, Wiley, 1984.

10 C.S. Tsai. *Guided-Wave Acousto-Optics*. Springer-Verlag, 1990.

11 H. Nishihara, M. Haruna, and T. Suhara. *Optical Integrated Circuits*. McGraw-Hill, 1989.

12 J.Y Son, S.A. Shestak, V.M. Epikhin, J.H. Chun, and S.K. Kim. Multichannel acousto-optic Bragg cell for real-time electroholography. *Applied Optics*, 38(14):3101–3104, 1999.

13 J. Blanche Pierre-Alexandre. *Field Guide to Holography*. SPIE, 2014.

14 R.S. Weis and T.K. Gaylord. Lithium niobate: summary of physical properties and crystal structure. *Applied Physics A*, 37(4):191–203, 1985.

15 N.W. Ashcroft and N.D. Mermin. *Solid State Physics*. New York, London, Holt, Rinehart and Winston, 1976.

16 P.Y. Yu and M. Cardona, *Fundamentals of Semiconductors; Physics and Materials Properties*. Berlin and Heidelberg, Springer-Verlag, 1996.

17 Y.T. Han, J.U. Shin, S.H. Park, S.P. Han, C.H. Lee, Y.O. Noh, H.J. Lee, and Y. Baek. Crosstalk-Enhanced DOS Integrated with Modified Radiation-Type Attenuators. *ETRI Journal*, 30(5):744–746, 2008.

18 K. Cui, Q. Zhao, X. Feng, Y. Huang, Y. Li, D. Wang, and W. Zhang. Thermo-optic switch based on transmission-dip shifting in a double-slot photonic crystal waveguide. *Applied Physics Letters*, 100(20):201102, 2012.

19 P.G. de Gennes. *The Physics of Liquid Crystals*. Oxford, Clarendon Press, 1974.

20 S. Chandrasekhar. *Liquid Crystals* (2nd ed.). Cambridge, Cambridge University Press, 1992.

21 J.C. Chiao, K.Y. Wu, and J.Y. Liu, September. Liquid-crystal WDM optical signal processors. In 2001 IEEE Emerging Technologies Symposium on BroadBand Communications for the Internet Era. Symposium Digest (Cat. No. 01EX508), 53–57, 2001.

22 K. Noguchi, T. Sakano, and T. Matsumoto. A rearrangeable multichannel free-space optical switch based on multistage network configuration. *Journal of Lightwave Technology*, 9(12):1726–1732, 1991.

23 K.R. Khan, S. Bidnyk, and T.J. Hall. Tunable all optical switch implemented in a liquid crystal filled dual-core photonic crystal fiber. *Progress in Electromagnetics Research*, 22, 179–189, 2012.

24 Y. Sakamaki, K. Shikama, Y. Ikuma, and K. Suzuki. Wavelength selective switch array employing silica-based waveguide frontend with integrated polarization diversity optics. *Optics Express*, 25(17):19946–19954, 2017.

25 K. Asakawa, Y. Sugimoto, Y. Watanabe, N. Ozaki, A. Mizutani, Y. Takata, Y. Kitagawa, H. Ishikawa, N. Ikeda, K. Awazu, and X. Wang. Photonic crystal and quantum dot technologies for all-optical switch and logic device. *New Journal of Physics*, 8(9):208, 2006.

26 D.M. Beggs, T.P. White, L. O'Faolain, and T.F. Krauss. Ultracompact and low-power optical switch based on silicon photonic crystals. *Optics Letters*, 33(2):147–149, 2008.

27 J.D. Joannopoulos, S.G. Johnson, J.N. Winn, and R.D. Meade. *Photonic Crystals*. Princeton University Press, 2011.

28 R. Rajasekar, K. Parameshwari, and S. Robinson. Nano-optical switch based on photonic crystal ring resonator. *Plasmonics*, 14(6):1687–1697, 2019.

29 Y. Zhang, P. Li, Y. Chen, and Y. Han. Four-channel THz wave routing switch based on magneto photonic crystals. *Optik*, 181, 134–139, 2019.

30 L. O'Faolain, D.M. Beggs, T.P. White, T. Kampfrath, K. Kuipers, and T.F. Krauss. Compact optical switches and modulators based on dispersion engineered photonic crystals. *IEEE Photonics Journal*, 2(3):404–414, 2010.

31 M. Djavid, M.H.T. Dastjerdi, M.R. Philip, D.D. Choudhary, T.T. Pham, A. Khreishah, and H.P.T. Nguyen. Photonic crystal-based permutation switch for optical networks. *Photonic Network Communications*, 35(1):90–96, 2018.

32 C.C. Chen, C.Y. Chen, W.K. Wang, F.H. Huang, C.K. Lin, W.Y. Chiu, and Y.J. Chan. Photonic crystal directional couplers formed by InAlGaAs nano-rods. *Optics Express*, 13(1):38–43, 2005.

33 F. Cuesta-Soto, A. Martínez, J. Garcia, F. Ramos, P. Sanchis, J. Blasco, and J. Martí. All-optical switching structure based on a photonic crystal directional coupler. *Optics Express*, 12(1):161–167, 2004.

34 A. Granpayeh, H. Habibiyan, and P. Parvin. Photonic crystal directional coupler for all-optical switching, tunable multi/demultiplexing and beam splitting applications. *Journal of Modern Optics*, 66(4):359–366, 2019.

35 A. Martínez, P. Sanchis, and J. Martí. Mach–Zehnder interferometers in photonic crystals. *Optical and Quantum Electronics*, 37(1):77–93, 2005.

36 H.C. Nguyen, S. Hashimoto, M. Shinkawa, and T. Baba. Compact and fast photonic crystal silicon optical modulators. *Optics Express*, 20(20):22465–22474, 2012.

37 C.Y. Liu and L.W. Chen. Tunable photonic-crystal waveguide Mach–Zehnder interferometer achieved by nematic liquid-crystal phase modulation. *Optics Express*, 12(12):2616–2624, 2004.

38 M. Shirdel and M.A. Mansouri-Birjandi. Photonic crystal all-optical switch based on a nonlinear cavity. *Optik*, 127(8):3955–3958, 2016.

39 J. Bravo-Abad, A. Rodriguez, P. Bermel, S.G. Johnson, J.D. Joannopoulos, and M. Soljačić. Enhanced nonlinear optics in photonic-crystal microcavities. *Optics Express*, 15(24): 16161–16176, 2007.

40 M. Ghadrdan and M.A. Mansouri-Birjandi. Concurrent implementation of all-optical half-adder and AND & XOR logic gates based on nonlinear photonic crystal. *Optical and Quantum Electronics*, 45(10):1027–1036, 2013.

41 V.D. Kumar, T. Srinivas, and A. Selvarajan. Investigation of ring resonators in photonic crystal circuits. *Photonics and Nanostructures-Fundamentals and Applications*, 2(3):199–206, 2004.

42 D.A. Miller. Device requirements for optical interconnects to silicon chips. *Proceedings of the IEEE*, 97(7):1166–1185, 2009.

43 R. Konoike, K. Suzuki, T. Inoue, T. Matsumoto, T. Kurahashi, A. Uetake, K. Takabayashi, S. Akiyama, S. Sekiguchi, S. Namiki, and H. Kawashima. SOA-integrated silicon photonics switch and its lossless multistage transmission of high-capacity WDM signals. *Journal of Lightwave Technology*, 37(1):123–130, 2018.

44 H. Ghafouri-Shiraz. *Principles of Semiconductor Laser Diodes and Amplifiers: Analysis and Transmission Line Laser Modeling*. World Scientific, 2003.

45 G. Keiser. Optical fiber communications. *Wiley Encyclopedia of Telecommunications*, 2003.

46 R. Bahuguna, M. Mina, J.W. Tioh, and R.J. Weber. Magneto-optic-based fiber switch for optical communications. *IEEE Transactions on Magnetics*, 42(10):3099–3101, 2006.

47 S. Kemmet, M. Mina, and R.J. Weber. Current-controlled, high-speed magneto-optic switching. *IEEE Transactions on Magnetics*, 46(6):1829–1831, 2010.

48 T. Murai, Y. Shoji, N. Nishiyama, and T. Mizumoto. Nonvolatile magneto-optical switches integrated with a magnet stripe array. *Optics Express*, 28(21):31675–31685, 2020.

49 M. Džanko, B. Mikac, and V. Miletić. Availability of all-optical switching fabrics used in optical cross-connects. In *2012 Proceedings of the 35th International Convention MIPRO*, 568–572, 2012.

50 G.I. Papadimitriou, C. Papazoglou, and A.S. Pomportsis. Optical switching: switch fabrics, techniques, and architectures. *Journal of Lightwave Technology*, 21(2):384, 2003.

51 P. De Dobbelaere, K. Falta, and S. Gloeckner. Advances in integrated 2D MEMS-based solutions for optical network applications. *IEEE Communications Magazine*, 41(5):pp.S16–S23, 2003.

52 S.K. Ghosh, V.S. Yadav, S. Das, and S. Bhattacharyya. Tunable graphene-based metasurface for polarization-independent broadband absorption in lower mid-infrared (MIR) range. *IEEE Transactions on Electromagnetic Compatibility*, 62(2):346–354, 2019.

53 S.K. Ghosh, S. Das, and S. Bhattacharyya. Graphene based metasurface with near unity broadband absorption in the terahertz gap. *International Journal of RF and Microwave Computer-Aided Engineering*, 30(12):e22436, 2020.

54 V.S. Yadav, S.K. Ghosh, S. Das, and S. Bhattacharyya. Wideband tunable mid-infrared cross-polarisation converter using monolayered graphene-based metasurface over a wide angle of incidence. *IET Microwaves, Antennas & Propagation*, 13(1):82–87, 2019.

55 V.S. Yadav, S.K. Ghosh, S. Bhattacharyya, and S. Das. Graphene-based metasurface for a tunable broadband terahertz cross-polarization converter over a wide angle of incidence. *Applied Optics*, 57(29):8720–8726, 2018.

56 S.K. Ghosh, S. Das, and S. Bhattacharyya. Transmittive-type triple-band linear to circular polarization conversion in THz region using graphene-based metasurface. *Optics Communications*, 480, 126480, 2021.

57 R. Phon, S. Ghosh, and S. Lim. Novel multifunctional reconfigurable active frequency selective surface. *IEEE Transactions on Antennas and Propagation*, 67(3):1709–1718, 2018.

58 S. Ghosh and K.V. Srivastava. A polarization-independent broadband multilayer switchable absorber using active frequency selective surface. *IEEE Antennas and Wireless Propagation Letters*, 16, 3147–3150, 2017.

59 P. Kumar, S. Rai, S. Bhattacharyya, and A. Lakhtakia. Graphene-sandwich metasurface as a frequency shifter, switch, and isolator at terahertz frequencies. *Optical Engineering*, 59(11):110501, 2020.

2

Electro-Optic Switches

Arpita Adhikari[1], Joydip Sengupta[2], and Arijit De[3]

[1] Department of Electronics and Communication Engineering, Techno Main Salt Lake, Kolkata, West Bengal, India
[2] Department of Electronic Science, Jogesh Chandra Chaudhuri College, Kolkata, West Bengal, India
[3] Department of Electronics and Communication Engineering, Netaji Subhash Engineering College, Kolkata, West Bengal, India
(Presently associated with CNES/ ONERA project)

2.1   Introduction

Superfast transmission of large data sets is now needed to cope with the increasing demand for high-quality audio and video signals in mobile phones, smart TV etc. [1]. Currently, the advanced fiber-optic communication networks serve as the building blocks of the information superhighway. Reliable routing, fast switching, and errorless detection are the three essential goals that need to be fulfilled to accommodate the huge demands within the purview of fiber-optic communication technology. One of the indispensable components of fiber-optic communication is optical switches and, for the current scenario, they need to be reliable, robust, yet inexpensive. A good-quality optical switch can increase the capacity of optical fiber and reliably distribute optical signals and subsequently reduces the overall cost. Moreover, optical switches can convert or redirect light without any electronic to optical conversions and vice versa. In the electro-optic effect, the refractive index (RI) of non-centrosymmetric crystals can be changed upon application of an electric field due to the rearrangement of the position and/or density of the charge carriers and by inducing slight deformations in the crystal lattice. There are two categories of electro-optic effect, namely the linear (Pockels) effect (RI varies in proportion to the field) and the quadratic (Kerr) effect (RI varies in proportion to the square of the field). Electro-optic modulation has distinct advantages over the other mechanisms that can be used for 1.31–1.55 µm fiber-optic communications and optical networks. Moreover, light can be restrained within a tiny area by using guided-wave-type electro-optic modulators, which are much superior to their conventional bulk counterparts in terms of switching speed, power consumption, and compatibility with optical fiber [2]. Thus, an electro-optic modulator-based efficient optical communication system is evidently required for telecommunication, remote detection systems [3], military applications, etc. In this article, a review of current optical switches in terms of operating principle, fabrication material, and device structure is presented. The performance issues and subsequent challenges are also briefly discussed.

2.2 Operating Principles

2.2.1 Operating Principles of the Single-Mode Switch

The fundamental building block of the optical switch is known as the ridge waveguide [4]. The parting between the cut-off numbers of the dominant and first higher-order mode of a ridge waveguide is much larger in comparison to that for a conventional rectangular waveguide. Structurally, if one or two ridges are inserted in the top or bottom or both of the walls of the rectangular waveguide, then it becomes a ridge waveguide (Figure 2.1). In a ridge waveguide, the cut-off number can be altered by amending the dimensions of the ridge without modifying the outside dimensions. Moreover, the characteristic impedance value of the ridge waveguide can be varied in such a way that it lies in between that of a regular rectangular waveguide and coaxial cable. From the design perspective, the numerical aperture (NA) of the ridge waveguide must be matched with that of single-mode fiber. Secondly, the cross-section of the ridge waveguide must be enough wide to accommodate the core of the single-mode fiber. Considering the points discussed above, a ridge waveguide can be principally operated as a single-mode switch.

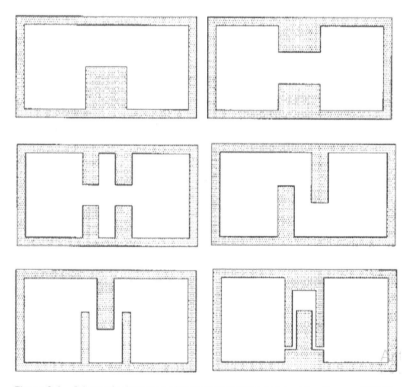

Figure 2.1 Schematic diagrams of rectangular ridge waveguides. Source: [5]/IET.

2.2.2 Operating Principles of the Multimode Switch

Recently, optical switches based on multimode interference (MMI) [6] have gained global attention because of their novel characteristics, such as low loss, compactness, polarization insensitivity, simple design, ease of fabrication, and large optical bandwidth. The self-imaging principle is the primary theory on which the optical MMI switch is operated. The self-imaging principle was first reported nearly 200 years ago [7]. The principle was beautifully described in Soldano's article [8]: "Self-imaging is a property of multimode waveguides by which an input field profile is reproduced in single or multiple images at periodic intervals along the propagation direction of the guide". The pictorial representation of the self-imaging principle is shown in Figure 2.2.

Structurally, an MMI device is a waveguide that is designed in such a way that it is capable of facilitating lots of modes (typically ≥ 3). More than one waveguide is positioned at the beginning and the end of the MMI device for launching light into it and recovering light from it. These types of devices are designated as N × M MMI couplers, where N and M are the numbers of input and output waveguides, respectively (Figure 2.3).

Figure 2.2 Field distribution of MMI.
Source: [9]/SPIE.

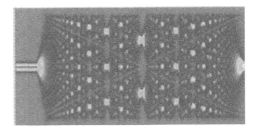

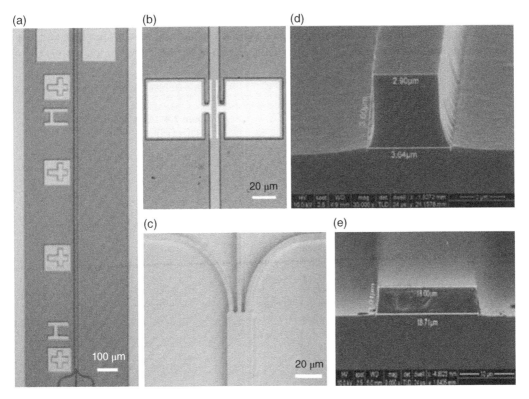

Figure 2.3 Optical micrographs of (a) a 3 × 3 MMI coupler electro-optic switch device, (b) electrode contact pads A and B deposited on the two index-modulated regions, including two extended square metal pads for easy contact, and (c) a top view of the MMI and coupling waveguide parts. Scanning electron microscope images of the cross-section of (d) a ridge input (output) waveguide, and (e) an MMI waveguide. Source: [10]/Optica Publishing Group.

2.3   Materials for the Fabrication of Electro-Optic Switch

2.3.1   Ferroelectric Materials

Several ferroelectric materials such as lead zirconate titanate – Pb(Zr, Ti)O$_3$ or PZT [11] – lead lanthanum zirconate titanate – (Pb, La)(Zr, Ti)O$_3$ or PLZT [12] – and lithium niobate (LiNbO$_3$) [13] are used for the fabrication of electro-optic switches. However, due to the large electro-optic coefficient [14], LiNbO$_3$ is widely used for the fabrication of electro-optic switches, demonstrating the Pockels effect with typical RI value 2.2 [15]. LiNbO$_3$ crystal possesses electric dipoles even in the absence of an electric field. Another advantage of LiNbO$_3$ is that it is capable of modifying the intensity of light without any perturbation on the phase, which is an important requirement for high-bitrate optical signals. Czochralski's (CZ) process is generally used to grow LiNbO$_3$ and as the CZ process is a matured technology high-quality LiNbO$_3$ is readily available in the market. In addition, the physical properties of LiNbO$_3$ are in line with the cleanroom process. The high Curie temperature [16] of LiNbO$_3$ maintains the electro-optic properties during high-temperature annealing. Since the LiNbO$_3$ crystal is anisotropic, the RI for light polarized along the c-axis (Figure 2.4) is thus smaller than that in the perpendicular direction to the c-axis. Moreover, the RI in the directions normal to the c-axis has strongly influenced the composition of the LiNbO$_3$, the operating temperature, and the wavelength used, rather than those along the c-axis. LiNbO$_3$ also possesses a wide transparency band [17] with low dispersion loss [18], which offers a wide range of applications from visible to mid-infrared region. Thus LiNbO$_3$ is employed widely for its ever-lower propagation losses and minor coupling losses with fibers, low power consumption, and very fast switching speeds while maintaining ever higher electro-optic efficiency.

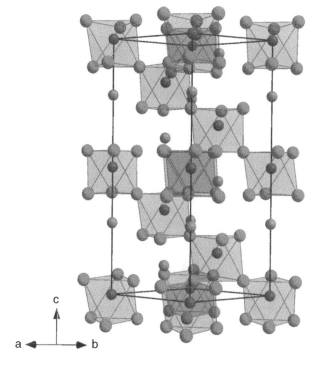

Figure 2.4   Crystal structure of LiNbO$_3$ [19].

c

a ◄─────► b

2.3.2 Compound Semiconductors

Compound semiconductors possess a linear electro-optic coefficient, though it is much smaller in comparison to that of $LiNbO_3$. However, a high RI value (typically 3.5 [20]) compared to $LiNbO_3$ compensates for a lower electro-optic coefficient. The RI can be controlled either by the introduction of an electric field or the injection of carriers. Moreover, in compound semiconductors, because of the dispersion of the dielectric constant, a much larger electric field can be employed for the same applied voltage, which in turn significantly increases the modulation efficiency. The dependence of the dielectric constant on the frequency of the applied signal is also much less in the case of compound semiconductors. Moreover, quadratic electro-optic effects can be used for enhancing the electro-optic efficiency of compound semiconductors. High-quality compound semiconductors can be grown employing metal-organic chemical vapour deposition (MOCVD) and molecular beam epitaxy (MBE), and the resulting modulator can be potentially integrated with optical sources, detectors which are also fabricated similar technologies. Thus, it is feasible to fabricate compound semiconductor electro-optic switches with improved drive voltage and bandwidth characteristics compared to a $LiNbO_3$-based electro-optic switch, but compound semiconductor switches have higher propagation loss in comparison to $LiNbO_3$.To date, GaAs [21], AlGaAs [22], InP [23], InGaAs [24], InAlAs [25], InGaAsP [26], or their mixed system [27] have been widely used for the fabrication of electro-optical switches (Figure 2.5).

2.3.3 Polymers

Electro-optic polymers contain nonlinear molecules arranged in a non-centrosymmetric structure. The state of polarization, phase, and frequency of light beams can be controlled by varying the nonlinear optical properties of materials during fabrication. High speed, low drive in voltage, and wide bandwidth are the major factors for which electro-optic polymers are gaining global attention [29]. Another significant aspect of electro-optic polymers is the tunability of their RI, which facilitates high-density compact structures. The negative thermo-optic coefficient of electro-optic polymers are the main factor behind low power consumption [30]. The electro-optic coefficient of

Figure 2.5 Bandgap energy (300 K) vs. lattice constant for III-V compound semiconductors commonly used for optoelectronic devices. Source: [28]/Elsevier.

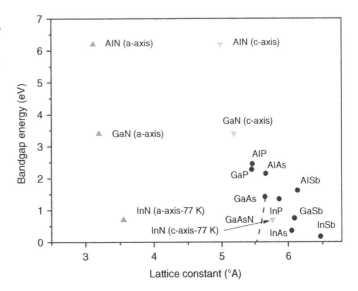

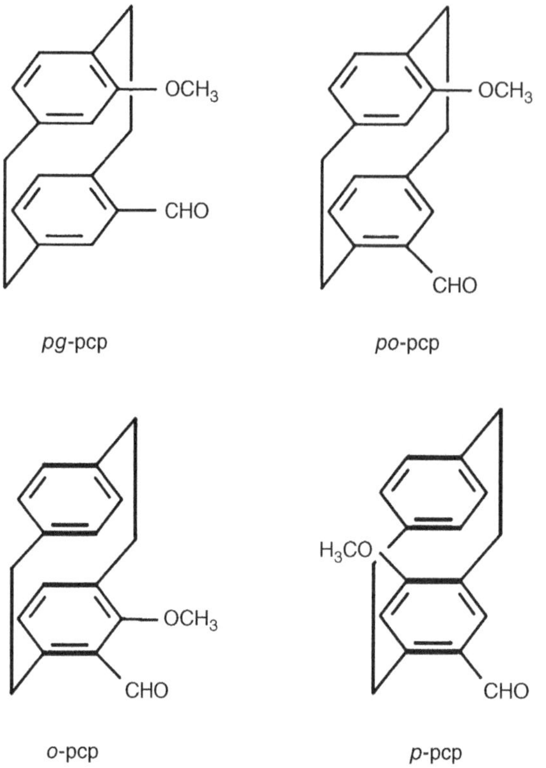

Figure 2.6 Chemical structures of selected [2] paracyclophanes. Source: [32]/Taylor & Francis.

these polymers can match that of LiNbO$_3$. Moreover, as the RI value is almost constantly low within DC to optical frequency range, the speed of the electrical and optical waves can thus be synchronized at ease [31]. The electro-optic polymers are based on either a side-chain system or a guest-host system. Currently, [2] paracyclophanes are mostly used as the electro-optic polymer for the fabrication of switches (Figure 2.6).

2.4 Device Structures of Electro-Optical Switches

2.4.1 1 × 1 Switch

1 × 1 optical switch may be regarded as a 1 × 1 optical modulator. The fabrication can be done in III-V materials as well as in LiNbO$_3$, Si(Ge), or polymers. A single-mode n-type silicon waveguide of width W is formed on a SiO$_2$ layer. The carriers are injected into the waveguide using an abrupt p$^+$-n junction at the forward bias condition. As a consequence, the RI decrease due to the plasma dispersion effect and the guided mode will be converted into the radiation mode. The specifications are shown in Table 2.1.

Table 2.1 Specifications for 1 × 1 optical switch.

Switching current	Modulation depth (injection current = 45 mA)	Insertion loss (1.3 μm)	Switching time
45 mA	96%	3.65 dB	160 ns

2.4.2 1 × 2 Switch

1×2 optical switches can be arranged as Y-shaped waveguides. The input and output of the switch consist of one and two waveguides, respectively. A branching point having a small angle is used to connect the input to the output waveguides. This branching switch is also called digital photonic splitting (DPS). The output power of the switch is independent of the voltage variations. At a certain voltage light will be propagated through the output optical waveguide. The DPS switch has the advantages of low insertion loss, small switching voltage, minute crosstalk, high switching speed, and good reliability. A Y-branch DPS with varied branch angles is shown in Figure 2.7. LiNbO$_3$ substrate is used to fabricate by diffusing Ti to a depth of 60 nm and width of 7 μm at 1050°C temperature for 8 hours. The SiO$_2$ layer has a thickness of 250 nm. The length of the electrode is 10 mm. A schematic diagram of DPS is shown in Figure 2.8. Figure 2.9 shows the configuration of

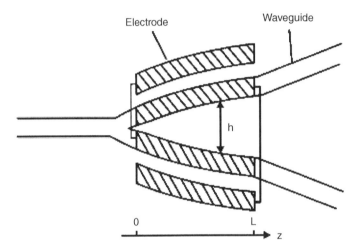

Figure 2.7 Diagram of Y-branch digital optical switch with varied local branch angle. Source: [33]/IET.

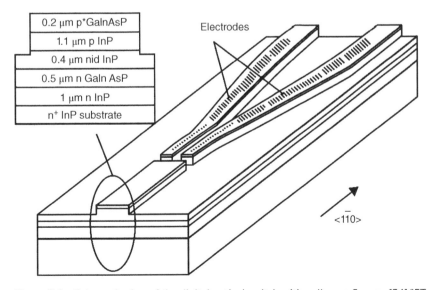

Figure 2.8 Schematic view of the digital optical switch with epilayers. Source: [34]/IET.

the silicon 1 × 2 optical switch. Silicon single-mode ridge waveguide is used to form a Y-junction. An impurity induced at the n/n$^+$ interface can be used for the vertical confinement of the optical wave whereas a Si/Si$_3$N$_4$ ridge wall is used for lateral confinement. The double-etch design can be used for efficient switching and low radiation loss. A double-etch Y-branch switch/modulator may be designed using InGaAsP multiple-quantum wells (MQWs). Figure 2.10 shows a configuration of a 1 × 2 digital optical switch (DOS) waveguide. Chemical beam epitaxy in InGaAsP-InP is mainly used for the DOS p-i-n waveguide heterostructure. Si$_3$N$_4$ can be used as the insulating layer. Power consumption can be reduced by choosing a wide carrier injection region. The etching depth

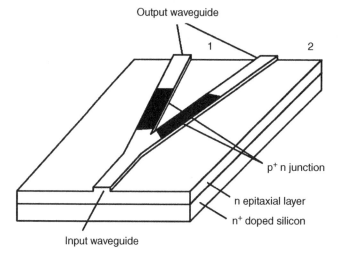

Figure 2.9   1 × 2 optical waveguide switch. Source: [35]/IET.

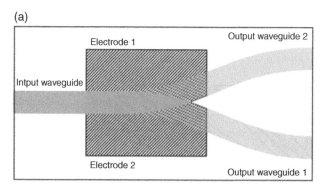

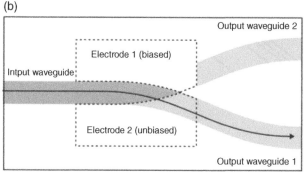

Figure 2.10   (a) Configuration of the digital optical switch Y-junction switch, and (b) effective waveguide configuration when the index under the biased electrode matches the RI of the surrounding slab region. The shaded areas indicate a high effective index, while the clear areas indicate a low effective index. Source: [37]/IET.

of the ridge waveguide is 1.2 μm with 4% of Ge content. To analyze the mode propagation behaviour, the effective index method can be used. This will help to determine the branch angle. The beam propagation method (BPM) is used to understand the device performance. In the ON condition of the switch, the 47.6% output optical power has been seen in the two arms of the Y switch. The insertion loss in this condition is 0.2 dB. The output optical powers are 58.7% and 3% at the two arms for the ON and OFF state respectively. The insertion loss in both cases is 2.1 dB. The threshold voltage and current are 1.0 V and 85 mA, respectively. The insertion loss in this condition is 3.2 dB. At 2 V forward bias voltage, the active branch is cut off. If zero and forward bias voltage are applied at branch 1 and branch 2 respectively, the output optical power at branch 1 and 2 increases and decreases respectively. The cut-off voltage is 2 V. The injection current density and the power consumption is 6.3 kA/ cm^2 and 190 mW, respectively. The measured crosstalk and the insertion loss are 29.6 dB and 5.2 dB, respectively.

2.4.3   2 × 2 Switch

The 2 × 2 Si-based optical switches are the basic device in integrated optics. These switches are operated on the total internal reflection (TIR) principle. RI and carrier injection may control the reflection interface of the optical switch [36]. The input of the 2 × 2 SiGe waveguide is connected to the output structure of the Y branch through an intersection region.

At the top surface p-n$^+$ junction at forward biased condition is used to inject the carriers. This is called the carrier injection region. Because of the plasma dispersion effect in SiGe, the RI decreases in the carrier injection region. This is the reason for the formation of the region of reflection at the intersection. By adjusting the drift distance, the carrier utilization ratio can be improved. This is also called an asymmetric switch. The input light at port 1 is in the forward bias condition of the p-n$^+$ junction. As a consequence, the laser beam is reflected to port 3 from port 1, and the switching operation is established. Most 2 × 2 optical switches are operated based on the TIR principle. 2 × 2 switches are constructed using two single-mode waveguides. The two waveguides intersect at an angle θ with an electrode. An optical intersecting switch with a wide deflection angle can be used to control the reflection angle [38, 39]. Two ridge waveguides can be used to form an intersectional ridge optical waveguide switch. At zero bias the input beam will propagate to port 4. At the forward bias condition, TIR occurs and the incident light beam will reflect at port 3, as a result, the switching operation is achieved. The specifications are shown in Table 2.2.

The 2 × 2 optical switches can also be worked on the principle of the free-carrier plasma dispersion effect and the multimode interference (MMI) principle. The MMI switch is operated on the self-imaging principle [40]. A two-mode interference (TMI) photonic switch having a double carrier injection can be devised for application in fiber-optic communication. In the forward bias condition, two p-n$^+$ junctions are used to inject. As a consequence, the injection current at both junctions increases [41]. This structure is called double carrier injection. Due to the change of RI the TMI region can be achieved. The input beam A is coupled to the single-mode ridge

Table 2.2   Specifications for 2 × 2 optical switch.

Branch angle (θ)	Bow-tie angle ($θ_w$)	Thickness of the ridge waveguide	Width of the ridge waveguide	Etching depth	Bow-tie electrode length is (L)	Length of the switch
2°	1.5°	2.6 μm	9 μm	1.0 μm	600 μm	5 mm

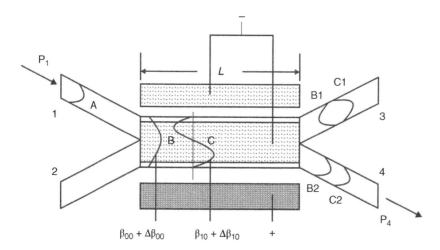

Figure 2.11 Schematic plan view showing the two-mode interference photonic switch with double carrier injections output from a waveguide at forward bias. Source: [42]/Elsevier.

waveguide 1, as designated by arrow P_1 (Figure 2.11). As a consequence, mode B (fundamental mode) and C (first-order mode) are excited. β_{00} and β_{10} are the propagation phase constants of modes B and C respectively. The above two modes may interfere along the direction of propagation. The light power at a propagation length L is given by

$$P_3/P_1 = \sin^2\left(\Delta\varphi/2\right) \tag{2.1}$$

and

$$P_4/P_1 = \cos^2\left(\Delta\varphi/2\right) \tag{2.2}$$

where P3 and P4 denote the output optical power at waveguides 3 and 4, respectively. $\Delta\varphi$ denotes the phase difference between mode B and mode C and follows the condition

$$\Delta\varphi = \Delta\beta L, \tag{2.3}$$

where $\Delta\beta = (\beta_{00} - \beta_{10})$.

Without forwarding bias the light will be propagated through the waveguide 3. The switch is in the OFF state. At the Forward bias condition, carriers in large numbers $(I_1 + I_2)$ will be introduced into the optically modulated section. The RI in the region will decrease, which will cause the propagation constants β_{00} and β_{10} to be changed. If the following condition satisfies,

$$\left(\beta_{00} - \beta_{10} + \Delta\beta_{00} + \Delta\beta_{10}\right)L = 0, \tag{2.4}$$

the light beam at waveguides 3 and 4 will be cut off (B1+C1 = 0). Now the switch is in the ON state. β_{00}, β_{10} are the propagation constants at zero forward bias of modes B and C respectively. $\Delta\beta_{00}$ and $\Delta\beta_{10}$ are the respective changes of β_{00} and β_{10} under forward bias respectively.

2.4.4   2 × 3 Switch

2 × 2 optical switches are mainly used for multichannel switches. 2 × 3 switches are designed for multi-ports optical switches. A 2×3 switch structure is made of three sections: input, central, and output sections. The input section is made of two waveguides (A and B) whereas the output section

has three output waveguides 1, 2, and 3. The central section has an MMI waveguide with two regions I and II. These waveguides are single-mode waveguides. Both the index-modulation regions are designed as a p-n junction. Under the application of forward bias to the p-n junction RI in the regions I and II will decrease. This happens because of the plasma dispersion effect. As a consequence, the propagation constant of the optical waveguide will change. As a result, the input optical signal will be switched to the three output ports. The thickness of the ridge waveguides is 2.6 µm p-SiGe with 4% Ge content. The MMI section will facilitate a large number of modes. If a 1.55 µm light is coupled to the input waveguide A, and if the p-n junctions of the region I and II are at unbiased condition, then the RI will remain unchanged. The input optical signal will be propagated through the output port 3. At a particular forward bias applied to the region I, the RI decreases by 0.3. The output light will be propagated to port 1. Applying forward bias to I and II will cause the output light to switch to port 2. When a 1.55-µm optical signal is applied to waveguide B the refractive indices will remain unchanged. The input optical signal will be propagated through port 1. Applied forward bias to the region I, will cause the output light to switch to output port 3. Application of forward bias to both the regions will cause the output light to switch to output port 2. If two similar optical signals of wavelength 1.55-µm having the same amplitude, phase, and polarization are applied to the input A and B, they will switch to port 3 and port 1, respectively. However, when these two 1.55-µm optical signals are applied into ports A and B with a forward bias to the region I, the output will switch to port 1 and port 3, respectively. In this condition, the switch will be at optical cross-connection. The switch will act as an optical power combiner. The power at the output ports 1, 2, and 3 is 0.9%, 70%, and 1.0%, respectively.

2.4.5 3 × 2 Switch

Both the 3 × 2 and 2 × 3 are applied for optical combiner or cross-connector and for multi-wavelength [43]. It has three sections: input, central, and output. The input section consists of two S-shaped waveguides (2 and 3) and one straight waveguide (1). The output section is made of two S-shaped waveguides (4 and 5). The central section is the main component of the photonic switch, which is an intersecting part. The central section has two carrier injection regions that are electronically controlled. The two regions can control the flow of the optical signal between the input ports and the output ports. Mirror I and II are the two reflecting interfaces that can be triggered by forward bias voltage. Both the mirrors have an ON and OFF state. Under the application of forward bias to the carrier injection region, the carrier concentration will increase which in turn causes to reduce the RI. This causes a reflecting interface to form. This is called the mirror-ON state. If no forward bias is applied, the RI of the material will remain unchanged. This is called a mirror-OFF state.

The switches are designed using SiGe material. For the single-mode operation, it has raised the ridge waveguide structure. The thickness and width are 2.5 µm and 10 µm respectively. The etching depth is 1.1±0.1 µm for a Ge content of 4%. The behaviour of the mode propagation is analyzed employing the effective index method. In the optimum design, the input and output branching angles have been chosen as 1.5° and 3° respectively. An impurity-induced p-SiGe/p-Si can be used for the vertical confinement of the input light. The operation is based on the principle of TIR due to the plasma dispersion effect. The RI of SiGe is related to the carrier concentration. At forward bias condition, huge numbers of electrons are injected into the p region of the waveguide, which enhances the carrier concentration. As a consequence, a reduction in the RI of the SiGe layer can be observed. This will help to form a reflecting mirror at branch 4 or 5. The samples of the switch can be prepared using a UHV-chemical vapor deposition (CVD) system. It has a speed of 100-200 ns.

The advantages of the device are wavelength insensitivity, polarization-independence, single-mode operation, ultracompact, low-cost, and highly reliable. The applications are in photonic integrated circuits, fiber-optic communications systems, and wavelength division multiplexed networks. The device combines multi-wavelength coming from different input channels.

2.4.6 3 × 3 Switch

High-capacity optical networks are required shortly for optical digital information processing. 3×3 switch or all-optical switch combines the advantages of the MMI principle and Si-based optical waveguide device. It is made of input, central, and output sections. The input and the output sections are made of three waveguides. The central section consists of an MMI coupler.

2.4.7 1 × 4 Switch

The 1 × 4 optical switches are mainly fabricated using InGaAsP which has been shown in Figure 2.12. It consists of a beam-steering section and output waveguide. The beam steering section is made of a steering region and an input waveguide. The steering region is composed of two parallel layer waveguide contacts which are mainly formed by Ti/Au/Zn/Au. The two layers are separated by 22 μm. The length and width of the contact stripes are 800 μm and 10 μm respectively. The output has two waveguides. The width and length of the two waveguides are 4 μm and 500 μm respectively. At the initial condition, two waveguides are separated by 2 μm. However, the value may increase up to 5 μm. The n^+- InP substrate is used as a wafer structure. A 1-μm thick n-type InP layer can be grown on 14 pairs of 10-nm thick and undoped InGaAsP (E_g= 0.816 eV) quantum wells having interspacing of 10-nm InGaAsP (E_g = 1.08 eV) barriers. The slab waveguide is grown using the metal-organic chemical vapor deposition (MOCVD) technique. The principle of carrier-induced RI change is used in the device. A laser beam of 1.51-μm wavelength may pass through the

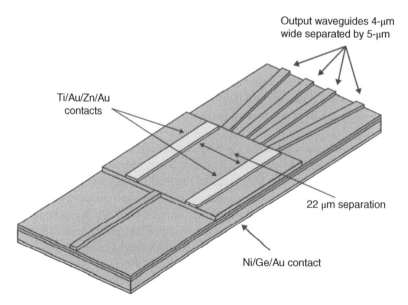

Figure 2.12 Schematic of the 1 × 4 InP-based optical switches. Source: [44]/Elsevier.

steering region if there is no current in the contact stripes. Zinc can be diffused below the contact stripes to control the carrier spreading within. This diffused region is used to channel the electrons which enhance their efficiency. The highest carrier concentration can be seen below the stripes and the carrier concentration will be decreased with lateral distance. This will cause a graded-index channel to form between the two stripes. The waveguide can be shifted and steer the signal beam by adjusting the ratio between the currents of the two parallel stripes. At this condition, the input beam will be directed to the output. Silicon nitride diffusion mask is used for device fabrication. The plasma-enhanced chemical vapor deposition (PECVD) technique can be used for 200-nm-thick Si_3N_4 film deposition on the substrate surface.

2.4.8   2 × 4 Switch

2 × 4 optical switches are designed based on the free-carrier plasma dispersion effect and MMI principle. It is made of two single-mode waveguides in the input, an MMI section, and four single-mode waveguides in the output. Two index-modulation regions are formed in the MMI section. When forward bias applied to the index-modulation region, the input optical signal can be propagated to any one of the four waveguides.

The switch is made of three parts: input, central, and output section. The input and the output section consist of two and four waveguides respectively. The central section is made of a rectangular air groove and two index-modulation regions. The p-n junctions are formed into the index modulation. The control signals VC1 and VC 2 are the forward bias voltages. The applied control signal will decrease the RI due to the plasma dispersion effect. This phenomenon will cause an alteration in the transmission of the input signals. As a consequence, the input signals will be propagated to one of the four output waveguides. The device can be performed as a 2 × 4 decoder switch.

2.5   Conclusions

Speed, capacity, and integrability with low power consumption and errorless transmission are the key requirements of next-generation data communication. Silicon-based photonics extends an efficient platform in optical communication. However, the technology needs to surmount certain hurdles like controlling the optical transmission path as well as data extraction at desirable locations. The solution can be availed through the proper usage of optical switches. An optical switch in silicon-based phonic chips extends several advantages offering state-of-the-art of modern optical communication. Moreover, selective switching without intermediate conversion to an electronic counterpart is the unique feature of optical switches. Optical switches usually utilize the carrier-induced RI modulation technique. Such switching can be either based on linear or nonlinear effects depending upon the interaction between the applied electric field and induced changes in the refractive index. Optical switching principles based on nonlinear effects are associated with high energy consumption. The carrier injection within the switching devices can either be vertical or lateral. Both modes of operation of optical switches are associated with some drawbacks. Vertical injection of carrier results in high insertion loss at the interface between the waveguide layer and n^+/p^+ substrate. Whereas such loss can be avoided in the case of lateral carrier injection as in that case the carriers are not necessarily transmitted through the n^+/p^+ substrate that is utilized in the case of the vertical one for realizing ohmic contacts. However, the control of injection current or injection current density is beyond the capability of either mode of operation.

Bibliography

1 R. Maher, A. Alvarado, D. Lavery, and P. Bayvel. Increasing the information rates of optical communications via coded modulation: a study of transceiver performance. *Scientific* 6(1):21278, 2016.

2 J.H. Wülbern, S. Prorok, J. Hampe, A. Petrov, M. Eich, J. Luo, A. K.-Y. Jen, M. Jenett, and A. Jacob. 40 GHz electro-optic modulation in hybrid silicon-organic slotted photonic crystal waveguides. *Optics Letters*, 35(16):2753–2755, 2010.

3 H. Liu, Y. Dong, R. Gao, Z. Luo, and G. Jin. Principle demonstration of the phase locking based on the electro-optic modulator for Taiji space gravitational wave detection pathfinder mission. *Optical Engineering (Redondo Beach, Calif.)*, 57(05):1, 2018.

4 S. Fattah poor, T.B. Hoang, L. Midolo, C.P. Dietrich, L.H. Li, E.H. Linfield, J.F.P. Schouwenberg, T. Xia, F.M. Pagliano, F.W.M. van Otten, and A. Fiore. Efficient coupling of single photons to ridge-waveguide photonic integrated circuits. *Applied Physics Letters*, 102(13):131105, 2013.

5 J. Helszajn. *Ridge Waveguides and Passive Microwave Components*. IET, Stevenage, UK, 2000.

6 H. Xiao, L. Deng, G. Zhao, Z. Liu, Y. Meng, X. Guo, G. Liu, S. Liu, J. Ding, and Y. Tian. Optical mode switch based on multimode interference couplers. *Journal of Optics*, 19(2):025802, 2017.

7 H.F. Talbot. LXXVI. Facts relating to optical science. *No. IV. The London and Edinburgh Philosophical Magazine and Journal of Science*, 9(56):401–407, 1836.

8 L.B. Soldano and Pennings, E.C.M. Optical multi-mode interference devices based on self-imaging: principles and applications. *Journal of Lightwave Technology*, 13(4):615–627, 1995.

9 A.M. Al-hetar. Multimode interference photonic switches. *Optical Engineering (Redondo Beach, Calif.)*, 47(11):112001, 2008.

10 S. Mu, K. Liu, S. Wang, C. Zhang, B. Guan, and D. Zou. Compact InGaAsP/InP 3×3 multimode-interference coupler-based electro-optic switch. *Applied Optics*, 55(7):1795–1802, 2016.

11 M.M. Zhu, Z.H. Du, and J. Ma. Defect enhanced optic and electro-optic properties of lead zirconate titanate thin films. *AIP Advances*, 1(4):042144, 2011.

12 A. Mukherjee, S.R. Brueck, and A.Y. Wu. Electro-optic effects in thin-film lanthanum-doped lead zirconate titanate. *Optics Letters*, 15(3):151–153, 1990.

13 H. Okayama. Lithium niobate electro-optic switching. In *Optical Switching*. Springer US, Boston, MA, 39–81, 2008.

14 M. Luennemann, M. Luennemann, and K. Buse. Electrooptic properties of lithium niobate crystals for extremely high external electric fields. *Applied Physics. B, Lasers and Optics*, 76(4): 403–406, 2003.

15 D.F. Nelson and R.M. Mikulyak. Refractive indices of congruently melting lithium niobate. *Journal of Applied Physics*, 45(8):3688–3689, 1974.

16 H. de Castilla, P. Bélanger, and R.J. Zednik. High temperature characterization of piezoelectric lithium niobate using electrochemical impedance spectroscopy resonance method. *Journal of Applied Physics*, 122(24):244103, 2017.

17 M. Leidinger, S. Fieberg, N. Waasem, F. Kühnemann, K. Buse, and I. Breunig. Comparative study on three highly sensitive absorption measurement techniques characterizing lithium niobate over its entire transparent spectral range. *Optics Express*, 23(17):21690–21705, 2015.

18 A. Chirakadze, S. Machavariani, A. Natsvlishvili, and B. Hvitia. Dispersion of the linear electro-optic effect in lithium niobate. *Journal of Physics D: Applied Physics*, 23(9):1216–1218, 1990.

19 File:LiNbO3.png (no date) Wikimedia.org. Available at: https://commons.wikimedia.org/wiki/File:LiNbO3.png (Accessed: July 7, 2021).

20 D.T.F. Marple. Refractive Index of GaAs. *Journal of Applied Physics*, 35(4):1241–1242, 1964.

21 N. Dagli. Compound semiconductor electro-optic modulators for microwave photonics applications. In *2013 IEEE Photonics Conference*. IEEE, 2013.

22 P. Bhasker, J. Norman, J. Bowers, and N. Dagli. Low voltage, high optical power handling capable, bulk compound semiconductor electro-optic modulators at 1550 nm. *Journal of Lightwave Technology*, 38(8):2308–2314, 2020.

23 M. Stepanenko, I. Yunusov, V. Arykov, P. Troyan, and Y. Zhidik. Multi-parameter optimization of an InP electro-optic modulator. *Symmetry*, 12(11):1920, 2020.

24 J. Shin, H. Kim, P. Petroff, and N. Dagli. Enhanced electro-optic phase modulation in InGaAs quantum posts at 1500 nm. *IEEE Journal of Quantum Electronics*, 46(7):1127–1131, 2010.

25 Z. Xu, C. Wang, W. Qi, and Z. Yuan. Electro-optical effects in strain-compensated InGaAs/InAlAs coupled quantum wells with modified potential. *Optics Letters*, 35(5):736–738, 2010.

26 B. Knüpfer, P. Kiesel, A. Höfler, P. Riel, and G. H. Döhler. Electroabsorption in InGaAsP: electro-optical modulators and bistable optical switches. *Applied Physics Letters*, 62(17):2072–2074, 1993.

27 N. Agrawal, D.Franke, N.Grote, F.W.Reier, and H.Schroeter-Janssen. MOVPE growth and characterization of InGaAs/In(GaAs)P and InGaAsP/InP/InAlAs multi-quantum-well structures for electro-optic switching devices. *Journal of Crystal Growth*, 124(1–4):610–615, 1992.

28 S. Mokkapati and C. Jagadish. III-V compound SC for optoelectronic devices. *Materials Today (Kidlington, England)*, 12(4):22–32, 2009.

29 R.S. Lytel. Applications of electro-optic polymers to integrated optics. In Peyghambarian, N. (ed.) *Nonlinear Optical Materials and Devices for Photonic Switching*. SPIE, 1990.

30 I. Rau, L. Puntus, and F. Kajzar. Recent advances with electro-optic polymers. *Molecular Crystals and Liquid Crystals*, 694(1):73–116, 2019.

31 A. Tsarev, R. Taziev, E. Heller, and M. Chalony. Polymer electro-optic modulator efficiency enhancement by the high permittivity dielectric strips. *Photonics and Nanostructures: Fundamentals and Applications*, 25:31–37, 2017.

32 I. Rau, L. Puntus, and F. Kajzar. Recent advances with electro-optic polymers. *Molecular Crystals and Liquid Crystals*, 694(1):73–116, 2019.

33 H. Okayama, T. Ushikubo, and M. Kawahara. Low drive voltage Y-branch digital optical switch. *Electronics Letters*, 27(1):24–26, 1991.

34 J.F. Vinchant, M. Renaud, A. Goutelle, M. Erman, P. Svensson, and L. Thylén. Low driving voltage or current digital optical switch on InP for multiwavelength system applications. *Electronics Letters*, 28(12):1135, 1992.

35 Y.L. Liu, E.K. Liu, S.L. Zhang, G.Z. Li, and J.S. Luo. Silicon 1×2 digital optical switch using plasma dispersion. *Electronics Letters*, 30(2):130–131, 1994.

36 B. Li, G. Li, E. Liu, Z. Jiang, C. Pei, and X. Wang. 1.55 μm reflection-type optical waveguide switch based on SiGe/Si plasma dispersion effect. *Applied Physics Letters*, 75(1):1–3, 1999.

37 S. Abdalla, S. Ng, P. Barrios, D. Celo, A. Delage, S. El-Mougy, I. Golub, J. He, S. Janz, R. Mckinnon, P. Poole, S. Raymond, T. Smy, and B. Syrett. Carrier injection-based digital optical switch with reconfigurable output waveguide arms. *IEEE Photonics Technology Letters*, 16(4):1038–1040, 2004.

38 J. Nayyer and H. Hatami-Hanza. Optical intersecting-waveguide switches with widened angle of deflection. *IEEE Photonics Technology Letters*, 4(12):1375–1377, 1992.

39 B. Li and S.-J. Chua. Reflection-type optical waveguide switch with bow-tie electrode. *Journal of Lightwave Technology*, 20(1):65–70, 2002.

40 C.F. Janz, B. Keyworth, W. Allegretto, R. MacDonald, M. Fallahi, G. Hillier, and C. Rolland. Mach-Zehnder switch using an ultra-compact directional coupler in a strongly-confining rib structure. *IEEE Photonics Technology Letters*, 6(8):981–983, 1994.

41 B. Li, Y. Zhang, L. Teng, Y. Zhao, S. Chua, and X. Wang. Symmetrical 1×2 digital photonic splitting switch with low electrical power consumption in SiGe waveguides. *Optics Express*, 13(2):654–659, 2005.

42 B.J. Li. Electro-optical switches. In *Optical Switches*. Elsevier, E5–E8, 2010.

43 B. Li, J. Li, Y. Zhao, X. Lin, S. Chua, L. Miao, E. Fitzgerald, M. Lee, and B. Chaudhari. Ultracompact, multifunctional, and highly integrated 3×2 photonic switches. *Applied Physics Letters*, 84(13):2241–2243, 2004.

44 D.A. May-Arrioja and P. Likamwa. Reconfigurable 1×4InP-based optical switch. *Microelectronics*, 39(3–4):644–647, 2008.

3

Thermo-Optical Switches

Fulong Yan[1], Xuwei Xue[2], and Chongjin Xie[3]

[1] *Alibaba Cloud, Alibaba Group, Beijing, China*
[2] *State Key Laboratory of Information Photonics and Optical Communications (IPOC), Beijing University of Posts and Telecommunications, Beijing, China*
[3] *Alibaba Cloud, Alibaba Group, Sunnyvale, CA, USA*

3.1 History of Thermal Optical Switching

The electro-optical effect has been used to design and implement fast optical switching from 1970s [1]. Although people noticed the thermo-optical effect almost at the same time, the thermal effect was regarded as undesirable and is generally considered to cause a deterioration in operation performance. In addition, due to the slow switching speed of the thermo-optic effect, it was not until the 1980s that people began to study the area of thermo-optic switching [2].

In 1981, lithium niobate (LiNb03) [2] and ion-exchanged glass [3] were used to fabricate thermo-optical devices. However, the thermo-optical effects of these two materials were relatively weak, resulting in inefficient devices. Benefiting from its strong thermo-optical effect and low thermal conductivity, polymer was regarded a suitable material for fabricating thermo-optical devices [4, 5], and many researchers focused on the manufacturing of polymer-based and silica-based thermo-optical switches [6, 7].

Recently, a compact (15 μm × 15 μm) and a highly-optimized 2 × 2 silicon thermo-optic switch device was proposed. It achieved power consumption of <3 mW and a switching time of 1 μs due to the efficient design [8]. On a small 11×25 mm^2 die, a 32 × 32 non-blocking Si-wire-based thermo-optic switch integrating 1024 thermo-optic Mach–Zehnder switches was proposed. The switching time (<30 μs) was fast enough for dynamic optical path allocation [9].

3.2 Principles of Thermo-Optic Switch

3.2.1 Thermo-Optic Effect

Thermo-optic switching is implemented based on the thermo-optic effect, namely the refractive index of waveguide varies as the temperature changes. The optical properties of any medium can be described by the complex refractive index n ($n = n_1 - n_2 j$). The refractive index depends on the

Optical Switching: Device Technology and Applications in Networks, First Edition. Edited by Dalia Nandi, Sandip Nandi, Angsuman Sarkar, and Chandan Kumar Sarkar.
© 2022 John Wiley & Sons, Inc. Published 2022 by John Wiley & Sons, Inc.

frequency of the light. The thermo-optical coefficient (TOC) could be deduced from the Clausius-Mossotti relation:

$$\frac{\varepsilon - 1}{\varepsilon + 2} = \frac{4\pi\alpha_m}{3V} \tag{3.1}$$

where ε is the complex dielectric constant ($\varepsilon = \epsilon_1 - \epsilon_2 j$), and ε can be calculated by $\varepsilon = n^2$. V is the volume of a tiny sphere in a macroscopic view, and its polarizability is denoted by α_m. Under the constant intensity of pressure, we have the following equation under the differential of equation (1.1).

$$\frac{\partial n}{\partial T} = \left(n^2 - 1\right)\left(n^2 + 2\right)\left(-\frac{1}{3V}\frac{\partial V}{\partial T} + \frac{1}{3\alpha_m}\frac{\partial \alpha_m}{\partial V}\frac{\partial V}{\partial T} + \frac{1}{3\alpha_m}\frac{\partial \alpha_m}{\partial T}\right) \tag{3.2}$$

In the above equation, the first two terms are related with the volume expansion. While the last term is the relation between polarizability and temperature under fixed volume. For polymers, TOC (thermo-optic coefficient) is negative since the value of the first term contributes mainly due to the thermal expansion, while the TOC of silica is positive since it is largely determined by the thermal change of polarizability in the second term [10]. However, the absolute value of TOC for silica is more than ten times smaller than polymer.

3.2.2 Trade-Off Between Switching Time and Power Consumption

Given Fouriers law of heat conduction [11],

$$J_T = -\kappa\Delta T = -\kappa\left(\vec{i}\frac{\partial T}{\partial x} + \vec{j}\frac{\partial T}{\partial y} + \vec{k}\frac{\partial T}{\partial z}\right) \tag{3.3}$$

where κ is the thermal conductivity. We know that the rate of heat transfer through the given cross-section of a material is proportional to the negative gradient of the temperature that is perpendicular to this cross-section and the area of this cross-section. Moreover, the direction of heat flows is opposite to the direction of temperature increasing. It is clear that the materials heat conductivity increases as the value of thermal conductivity κ becomes large. From equation 3.3, we can derive the heat conduction equation:

$$\rho c_p \frac{\partial T}{\partial t} = \kappa\nabla^2 T + Q\left(x, y, z, t\right) \tag{3.4}$$

Where Q is the rate of heat generation in an unit volume, and the density of the material is denoted by ρ. κ, and c_p are the thermal conductivity and specific heat capacity, respectively. While κ is generally taken as unchanged.

Without loss of generality, we could assume that the starting temperature distributes as follows:

$$T\left(x, y, z, t\right)_{t=0} = \bar{T} \tag{3.5}$$

The boundary conditions on the top, lateral, and bottom surfaces are shown in the following Eqs. 3.6, 3.7, and 3.8 respectively.

$$\frac{\partial T}{\partial s} = -h\left(T_s - T_A\right) \tag{3.6}$$

$$\frac{\partial T}{\partial s} = 0 \tag{3.7}$$

$$\bar{T} = T_B \tag{3.8}$$

When the first item in equation 3.4 is 0, the solution will be independent of both time t and boundary conditions. And we will obtain the steady-state temperature distribution. Meanwhile, we could also obtain the transient part of the solution.

The switching time τ could be calculated by the following equation.

$$\tau = \frac{0.47 p c_p H^2}{\kappa} \tag{3.9}$$

τ shows the time period needed for reaching a ratio of $1 - 1/e$ on the temperature value in a steady state, where H is the thickness of waveguide layer stack. Considering the fact that the power consumption per unit length means a certain temperature difference, the power consumption must be proportional to the thermal conductivity. Therefore, given the initial and stationary temperal gap ΔT, the power consumption can be defined as $P = \kappa \Delta T$.

Based on the above two equations, to implement a thermo-optical switch, we should take the trade-off between the switching time and the power consumption into consideration. Namely, the higher the thermal conductivity is, the shorter the switching time will be. Meanwhile, the power consumption of switching increases as material's thermal conductivity increases.

3.2.3 Merits of Thermo-Optic Switch

The crosstalk is defined as the ratio denoted by dB of the optical power which passes through the output port during the "on" state with respect to the "off" state. For a practical thermo-optic switch to be deployed in networks, the crosstalk should be less than −20 dB. To optimize the crosstalk, optical attenuator could be utilized and the resistance of attenuator region could be tuned so as to achieve a crosstalk as low as −70 dB. Similar to optical switches implemented by other materials and principles, thermo-optical switches also have figure of merits including propagation loss, insertion loss, polarization-dependent loss (PDL), and wavelength-dependent loss (WDL). PDL and WDL are calculated by the measurement of the peak-to-peak difference in transmission of the switch with respect to the possible states of polarization and wavelengths, respectively. Those parameters can be determined based on the requirements of various network applications.

3.3 Category

The thermo-optical switches can be classified into various categories based on the adopted materials, the implementation principle, and device architecture. To choose the proper material to implement a thermo-optical switch, low optical loss (≤ 0.1 dB/cm) and high TOC ($\geq 10^{-4} K^{-1}$) must be taken into consideration. In addition, the cost, wavelength dispersion, and low polarization-dependent loss may also be considered.

3.3.1 Material

In general, three kinds of materials including polymeride, semiconductor and crystalline materials are adopted to implement a thermo-optical switch.

Table 3.1 Proprieties of polymers employed for the fabrication of TOSes (BCB: benzocyclobutene, BPA: bisphenol, FA: fluoroacrylate, FPAE: fluorinated poly(arylene ether sulfide), PMMA: poly(methyl methacrylate), PI: polyimide, PUR: polyurethane).

Material	TOC ($10^{-4} \times K^{-1}$)	$\partial n/\partial T$, TEC, and Δn	Description
BCB [12, 13]	$-0.25 \sim -1.15$	$0.65 \times 10^{-4} K^{-1}$	Benzene ring (C_8H_8)
BPA [14]	$-0.96 \sim -1.33$	–	Organic compound
FA [15]	-2.8	–	Temperature feature (20 °C)
FPAE [16]	-1.0	$0.0040 \sim 0.0045$	Interlayer dielectric materials
PMMA [17]	-1.3	$-1.17 \sim -1.26 \times 10^{-4} K^{-1}$	Thermoplastic
PI [18]	$-0.46 \sim -1.04$	–	Good chemical resistance
PUR [19]	-5.3	$-3.9 \sim -4.1 \times 10^{-4} K^{-1}$	Versatile polymer

The fabrication of polymer waveguides is simple and flexible, which includes etching, molding, and laser delineation. Therefore, polymer is widely used to make a complex PLC (planar lightwave circuit). Optical polymers are low in cost and have good optical and mechanical properties. Morevoer, the loss of the polymer waveguide is less than 0.1 dB/cm in the 3 communication windows of fiber-optics, which are around 850 nm for multi-mode fiber as well as 1310 nm and 1550 nm for single-mode fiber.

There are several companies, including DuPont, General Electric, NTT, and ChemOptics, developing polymers for integrated optics. The TOC of integrated optical polymer at room temperature is about -10^{-4} K^{-1}, and the power consumption of the drive device can be <100 mW. In order to limit the overall thermal drift, it is necessary to consider the operation of the device at different temperatures. A substrate such as silicon or aluminum can be used as a heat sink to relax the low speed of macroscopic thermal process. On such a relatively small substrate, a heat-optical exchange device with a switching time of far less than 1 ms can be realized. The Table 3.1 shows the thermo-optical properties of various optical polymers.

As an amorphous material, it is well known that silica (SiO_2) is used to manufacture optical fibers due to the stability and fine steerability of refractive index. As the result of its good compatibility with fibers, the low-loss waveguide based on silica is widely used in the fiber transmission systems and fiber sensor systems. In addition, silica-based single-mode waveguides can also be used for active optical devices such as optical switches.

The TOC of silica glass is in the range of 0.62×10^{-5} K^{-1} and 1.28×10^{-5} K^{-1} generally. A large number of thermo-optical switches have been fabricated under silica despite the relatively low value of TOC. Most of the research has focused on the improvement of operational efficiency under the original design.

Hydrogenated amorphous silicon (a-Si:H) maintains compatibility with various microelectronic technologies. It is an ideal material to implement planar waveguide for routing and modulator. The refractive index of a-Si:H fluctuates in the range from 2.6 to 3.6. At room temperature, the a-Si:H-based waveguide presents a TOC of 2.3×10^{-4} K^{-1} at 1300 nm.

Crystalline silicon (c-Si) is silicon in crystalline forms, including polycrystalline silicon (Poly-Si), which comprises small crystals, and monocrystalline silicon (Mono-Si), which is a continuous crystal. At room temperature, the TOC of Mono-Si is 1.9×10^{-4} K^{-1}, and this is independent of sample doping or crystal orientation. Meanwhile the TOC of Poly-Si is 2.3×10^{-4} K^{-1}.

LiNbO$_3$ (lithium niobate) has usually been used for the fabrication of optical waveguide. The $\partial n/\partial T$ of LiNbO$_3$ is 3.3×10^{-5} K^{-1} at a wavelength of 1523 nm. III-V semiconductors are broadly used to fabricate passive and active devices such as waveguides and modulators, as well as thermo-optical switches. The TOC of GaAs and InP are 2.4×10^{-4} K^{-1} and 2.0×10^{-4} K^{-1}, respectively.

3.3.2 Implementation Principle

Based on the implementation principles, thermo-optic switches can be classified into two categories: interferometric and digital [10]. For the interferometric thermo-optical switch, the input optical signal is split into two parallel waveguides. The phase difference between the two waveguides could be generated by equipping a heating electrode in one of the waveguides. Either constructive or destructive interference will happen under matched phase conditions. Therefore, the optical signal is switched to one of the two outputs. To achieve satisfied switching performance, accurate control of phase as well as the thermal tolerance are important to implement interferometric thermo-switch.

Unlike the interferometric thermo-optic switch, digital thermo-optic switches utilize mode coupling of waveguides. The branches of digital thermo-optic switch are mounted with a heater to vary the refractive index. As we heat one waveguide branch, the refractive index decreases and it will be smaller than its counterpart. That is to say, the coupling of fundamental mode from the branch with lower refractive index to the one of higher index.

3.3.3 Device Architecture

Vairous materials can be used to realize a Mach–Zehnder interferometer, and a Mach–Zehnder interferometer is probably the most widely chosen architecture to implement a thermo-optic switch due to the simplicity [20–22]. As shown in Figure 3.1, the traditional thermo-optical switch based on Mach–Zehnder interferometer consists of a 3 dB splitter and a 3 dB combiner, which are connected by two channels. And a heater is deposited on the waveguide of the interferometer to generate phase shifting. Due to its compactness and wide bandwidth, MMI couplers are usually used to replace the 3 dB splitters and 3 dB combiners.

In the year of 2004, Harjanne used a multi-level voltage circuit to control the switching of thermo-optic switch and achieved a switching time of <1 μs [23]. To reduce the size of the MZI as well as the power consumption, silicon photonic wire waveguides with a silicon core with nanometer cross-sectional dimensions was considered. Kiyat developed a thermo-optical switch based on a racetrack resonator, which achieved a switching time of around 4.8 μs and a power consumption of 17 mW [24]. In 2007, Gu demonstrated a PhC waveguide-based MZI featured with switching time of about 20 μs and power consumption of 78 mW [25]. Pruessner used the Fabry–Perot cavity to achieve a trade-off between thermal mass and device performance, and the device obtained a switching time of approximately 640 ns and a power consumption of around 10 mW [26].

Figure 3.1 1 × 2 Mach–Zehnder interferometer based thermo-optic switch.

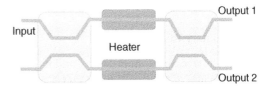

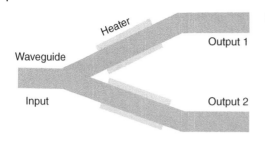

Figure 3.2 1 × 2 Y-junction thermo-optic switch.

As shown in Figure 3.2, in the linear Y-junction 1 × 2 branch, The heater is used to cause the temperature of the waveguide to change, therefore the refractive index is changed. When the temperature of the two ports is the same, the optical power is equally distributed, which is equivalent to a splitter. Y-junction use the mode coupling of waveguide. When the Y-branch thermo-optical switch works above the switching threshold, changes in polarization and wavelength will not significantly affect the switching capacity of digital optical switch (DOS). In addition, to ensure the coupling of adiabatic mode, the design of Y-junction needs to be elaborated to achieve a small angle. The design for a small angle is difficult to fabricate, and it requires a large device length.

For network applications, crosstalk needs to be below −20 dB. To increase the level of crosstalk, an optical attenuator was attached to the end of the Y branch arm [27]. The resistance of the attenuator area can be adjusted to achieve a crosstalk of −70 dB. For the device which utilized this architecture [28], the power consumption was about 200 mW. While the switching time and the PDL were around 10 ms and 0.1 dB, respectively. Chen proposed a thermo-optic switch consisting of two vertically coupled polymer waveguides. When the current passed through the electrodes of a waveguide, the refractive index changed and the switch started. The device achieved −23 dB of crosstalk, approximately 50 mW of electrical power consumption, and approximately 2 ms of switching time [29].

3.4   Scalability

Thermo-optic switch can be scaled out by interconnecting multiple 1 × 2 TO (thermo-optic) switching modules. By integrating high radix thermo-optic switch on a single chip, the insertion loss and cost can be minimized. To maintain the switching performance, only non-blocking architectures are considered and used. Non-blocking switching could be divided into three categories: strict non-blocking, rearrangeable non-blocking, and generalized non-blocking.

Strict non-blocking switching means that as long as the input port and the output port are idle, a connection for the port pair can be established in the switching network at any time without having an influence on other established connections. A rearrangeable non-blocking network relaxes the requirement for strict non-blocking. The existing connections can be reconfigured directly or indirectly in the switching network at any time to establish a connection for the idle port pair. While a generalized non-blocking network exists there is the possibility of blocking. However, all the possible blocking can be avoided by selecting routes with a smart method.

3.4.1   Binary Tree

A binary tree features short depths, and it can be utilized to scale out the radix of a thermo-optic switch. Figure 3.3 shows a 4 × 4 and $N \times N$ thermo-optic switch built by 1 × 2 TO switching

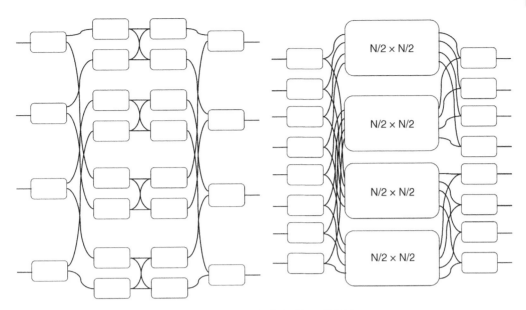

Figure 3.3 4 × 4 binary tree and *N* × *N* binary tree built by *N/2* × *N/2* binary trees.

modules. 4×4 [30, 31], 8×8 [32] and 16×16 [33] thermo-optic switches have been demonstrated under binary tree architecture. For the 4×4 thermo-optic switch, the power consumption and the switching time are 2.1 W and <3 ms.

3.4.2 Modified Crossbar

Regarding the 1×2 TO switching modules, crossbar could be utilized to scale out the radix of thermo-optic switch due to the short depths. Figure 3.4 shows a $N \times N$ thermo-optic switch built

Figure 3.4 4 × 4 Cross-bar architecture.

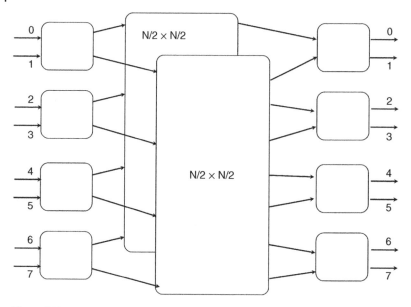

Figure 3.5 *N × N* Benes switch architecture with *N* of 8.

by 1 × 2 TO switching modules. The 16 × 16 thermo-optic switch has been demonstrated with silica waveguide under crossbar fabric [34]. The switching power and the crosstalk are 17 W and −33 dB, respectively.

3.4.3 Benes

Benes belongs to rearrangeable non-blocking multi-stage switch architecture. It can be built with two Banyan networks. To be specific, one Banyan network and one reversed Banyan network are connected back to back where the two adjacent stages in the middle are merged into one level. Since each Banyan network has log2N stages, the Benes network has a total of 2log2N-1 levels. Similar to the switching modules in Banyan network, each switching module has two states: direct connection and crossover. The topology of Benes architecture is shown in Figure 3.5.

To scale out the thermo-optic switch, the loss, crosstalk, and power consumption must be considered. For example, the loss and crosstalk in polymer are large. While binary tree has the advantage of short depths and provides the optimized switch matrix design so as to minimize the crosstalk. Therefore, binary would be an ideal architecture for polymer to scale out. Regarding silica waveguides, moderate power consumption could be obtained despite the inefficiency of silica TO switches.

3.5 Application Scenarios

By comparing the characteristics of polymer and silica based 16 × 16 thermo-optic switch matrices, it is clear that the power consumption of polymer is quite low regardless the waveguide loss of polymer based thermo-optic switch is higher. While the crosstalk of polymer and silica based matrix switches is comparable (in the order of −30 dB).

The main advantages of thermo-optic switches are the polarization-insensitive operation and switching speed on the order of millisecond and sub-millisecond. The volume of thermo-optic switch is relatively small, and it can even realize the switching speed of sub-microsecond level. Its disadvantages are higher insertion loss, more serious crosstalk, lower extinction rate, higher power consumption, and a good heat sink is required. Currently, the crosstalk for TOS with radix of larger than 16 is high with respect to other optical switching technologies, resulting in the inferior position for the high radix TOS. However, it is proper to implement TOSes with radix of small and medium sizes. The low radix optical cross-connects, small port count optical protection switches, and low size of optical add and drop multiplexers could all be the application scenarios for TOSes. And DOS has already acted as a quite promising device for space switching optical communication system applications supporting multi-wavelength.

Bibliography

1 M. Kawabe, S. Hirata, and S. Namba. Ridge waveguides and electro-optical switches in linbo 3 fabricated by ion-bombardment-enhanced etching. *IEEE Transactions on Circuits and Systems*, 26(12):1109–1113, 1979.

2 M. Haruna and J. Koyama. Thermo-optic effect in LiNbO3, for light deflection and switching. *Electronics Letters*, 17(22):842–844, 1981.

3 M. Haruna and J. Koyama. Thermooptic deflection and switching in glass. *Applied Optics*, 21(19):3461–3465, 1982.

4 M.B.J. Diemeer, J.J. Brons, and E.S. Trommel. Polymeric optical waveguide switch using the thermooptic effect. *Journal of Lightwave Technology*, 7(3):449–453, 1989.

5 J.-M. Cariou, J. Dugas, L. Martin, and P. Michel. Refractive-index variations with temperature of pmma and polycarbonate. *Applied Optics*, 25(3):334–336, 1986.

6 N. Keil, H.H. Yao, C. Zawadzki, and B. Strebel. 4×4 polymer thermo-optic directional coupler switch at 1.55 μm. *Electronics Letters*, 30(8):639–640, 1994.

7 M.B.J. Diemeer. Polymeric thermo-optic space switches for optical communications. *Optical Materials*, 9(1-4):192–200, 1998.

8 Z. Han, G. Moille, X. Checoury, J. Bourderionnet, P. Boucaud, A. De Rossi, and S. Combrié. High-performance and power-efficient 2×2 optical switch on silicon-on-insulator. *Optics Express*, 23(19):24163–24170, 2015.

9 K. Tanizawa, K. Suzuki, M. Toyama, M. Ohtsuka, N. Yokoyama, K. Matsumaro, M. Seki, K. Koshino, T. Sugaya, S. Suda, et al. Ultra-compact 32×32 strictly-non-blocking si-wire optical switch with fan-out lga interposer. *Optics Express*, 23(13):17599–17606, 2015.

10 G. Coppola, L. Sirleto, I. Rendina, and M. Iodice. Advance in thermo-optical switches: principles, materials, design, and device structure. *Optical Engineering*, 50(7):071112, 2011.

11 H. Fröhlich. *Theory of Dielectrics.* 1949.

12 T. Nikolajsen, K. Leosson, and S.I. Bozhevolnyi. Surface plasmon polariton based modulators and switches operating at telecom wavelengths. *Applied Physics Letters*, 85(24):5833–5835, 2004.

13 D. Sun, Z. Liu, Y. Zha, W. Deng, Y. Zhang, and X. Li. Thermo-optic waveguide digital optical switch using symmetrically coupled gratings. *Optics Express*, 13(14):5463–5471, 2005.

14 Y. Song, J. Wang, G. Li, Q. Sun, X. Jian, J. Teng, and H. Zhang. Synthesis, characterization and optical properties of fluorinated poly(aryl ether)s containing phthalazinone moieties. *Polymer*, 49:4995–5001, 2008.

15 H. Zou, K.W. Beeson, and L.W. Shacklette. Tunable planar polymer Bragg gratings having exceptionally low polarization sensitivity. *IEEE Journal of Lightwave Technology*, 21(4):1083–1088, 2003.

16 M. Oh, H. Lee, M. Lee, J. Ahn, and S.G. Han. Asymmetric X-junction thermo-optic switches based on fluorinated polymer waveguides. *IEEE Photonics Technology Letters*, 10(6): 813–815, 1998.

17 X. Li, Z. Cao, Q. Shen, and Y. Yang. Influence of dopant concentration on thermo-optic properties of PMMA composite. *Materials Letters*, 60:1238–1241, 2006.

18 Y. Terui and S. Ando. Anisotropy in thermo-optic coefficients of polyimide films formed on Si substrates. *Applied Physics Letters*, 83(23):4755–4757, 2003.

19 F. Qiu, D. Yang, G. Cao, R. Zhang, and P. Li. Synthesis, characterization, thermal stability and thermo-optical properties of poly(urethane-imide). *Sensors and Actuators B*, 135:449–454, 2009.

20 B. Lin, X. Wang, J. Lv, Y. Cao, Y. Yang, Y. Zhang, A. Zhang, Y. Yi, F. Wang, and D. Zhang. Low-power-consumption polymer mach–zehnder interferometer thermo-optic switch at 532 nm based on a triangular waveguide. *Optics Letters*, 45(16):4448–4451, 2020.

21 Y. Gao, Y. Xu, L. Ji, L. Sun, Y. Yi, F. Wang, Y. Wu, and D. Zhang. Thermo-optic mode switch based on an asymmetric Mach–Zehnder interferometer. *IEEE Photonics Technology Letters*, 31(11):861–864, 2019.

22 S. Wang and D. Dai. Polarization-insensitive 2×2 thermo-optic mach–zehnder switch on silicon. *Optics Letters*, 43(11): 2531–2534, 2018.

23 M. Harjanne, M. Kapulainen, T. Aalto, and P. Heimala. Sub-μs switching time in silicon-on-insulator mach-zehnder thermooptic switch. *IEEE Photonics Technology Letters*, 16(9):2039–2041, 2004.

24 I. Kiyat, A. Aydinli, and N. Dagli. Low-power thermooptical tuning of soi resonator switch. *IEEE Photonics Technology Letters*, 18(2):364–366, 2006.

25 L. Gu, W. Jiang, X. Chen, and R.T. Chen. Thermooptically tuned photonic crystal waveguide silicon-on-insulator Mach–Zehnder interferometers. *IEEE Photonics Technology Letters*, 19(5):342–344, 2007.

26 M.W. Pruessner, T.H. Stievater, M.S. Ferraro, and W.S. Rabinovich. Thermo-optic tuning and switching in SOI waveguide Fabry–Perot microcavities. *Optics Express*, 15(12):7557–7563, 2007.

27 U. Siebel, R. Hauffe, J. Bruns, and K. Petermann. Polymer digital optical switch with an integrated attenuator. *IEEE Photonics Technology Letters*, 13(9):957959, 2001.

28 H.J. Lee, Y.H. Won, Y.O. Noh, and M.C. Oh. Polymer waveguide thermo-optic switches with −70 db optical crosstalk. *Optics Communications*, 258 (1):18–22, 2006.

29 P.L. Chu , K. Chen, and H.P. Chan. A vertically coupled polymer optical waveguide switch. *Optics Communications*, 244 (1–6):153–158, 2005.

30 R. Moosburger and K. Petermann. 4×4 digital optical matrix switch using polymeric oversized rib waveguides. *IEEE Photonics Technology Letters*, 10(5):684–686, 1998.

31 L. Guiziou, P. Ferm, J-M Jouanno, and L. Shacklette. Low-loss and high extinction ratio 4×4 polymer thermo-optical switch. In *Proceedings 27th European Conference on Optical Communication (Cat. No. 01TH8551)*, volume 2, pages 138–139. IEEE, 2001.

32 A. Borreman, T. Hoekstra, M. Diemeer, H. Hoekstra, and P. Lambeck. Polymeric 8×8 digital optical switch matrix. In *Proceedings of European Conference on Optical Communication*, volume 5, pages 59–62. IEEE, 1996.

33 F.L.W. Rabbering, J.F.P. Van Nunen, and L. Eldada. Polymeric 16×16 digital optical switch matrix. In *Proceedings 27th European Conference on Optical Communication (Cat. No. 01TH8551)*, volume 6, pages 78–79. IEEE, 2001.

34 T. Goh, M. Yasu, K. Hattori, A. Himeno, M. Okuno, and Y. Ohmori. Low loss and high extinction ratio strictly nonblocking 16 16 thermooptic matrix switch on 6-in wafer using silica-based planar lightwave circuit technology. *Journal of Lightwave Technology*, 19(3):371, 2001.

4

Magneto-Optical Switches

K. Sujatha

Department of Electronics and Telecommunication Engineering, Shree Ramchandra College of Engineering,
Savitribai Phule Pune University, Pune, Maharashtra, India

4.1 Introduction

The most comprehensive optical switch device is used to open or close an optical circuit. It can be of a mechanical, opto-mechanical, or electronic type. An optical switch has one or more available input ports and two or more output ports that we commonly call a $1 \times N$ or $N \times N$ optical switch.

An optical switch (Figure 4.1) is a multiport network bridge which connects multiple optical cables to each other and controls data packages route between input and outputs. Some electric switches convert lights into electrical data before moving forward and switch them to make the light switch again. Some optical switches, called all-optical, can route and move forward in the lights themselves without an electrical switch [1, 2].

Fiber-optic switches connect the wiring to the output fibers on a one-to-one basis. They also establish and release connections between fiber pathways. The fiber-optic switches available today are all low-voltage exposed. Again, they have less impact on the operator signals and fiber-optic connectors.

4.1.1 Types of Optical Switch

Depending on the variety of fabrication process and technologies, optical switches can be divided into mechanical switches, MEMS (microelectromechanical system) switches, electro-optical switches, thermo-optical switches, magneto-optical switches, acousto-optic switches, photonic crystal all-optical switches, optical-electrical-optical (OEO) switches and others. In addition, thermo-optical switches, electro-optic switches, and acousto-optic optical switches are also used for some specific applications.

The opto-mechanical switch can be considered to be the oldest type of optical switch, and it was the most widely deployed. Mainly due to the way it functions, it is slow, with switching times of 10–100 ms. However, they can achieve excellent reliability, insertion loss, and crosstalk. In general,

Optical Switching: Device Technology and Applications in Networks, First Edition. Edited by Dalia Nandi,
Sandip Nandi, Angsuman Sarkar, and Chandan Kumar Sarkar.
© 2022 John Wiley & Sons, Inc. Published 2022 by John Wiley & Sons, Inc.

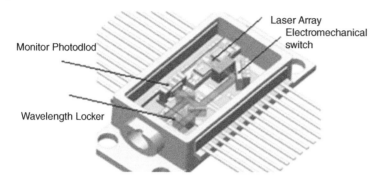

Monitor Photodlod

Laser Array
Electromechanical
switch

Wavelength Locker

Figure 4.1 Optical switch.

opto-mechanical optical switches separate the optical beam from each input and the output fiber and transfer them inside the device. This allows a distance between the input and the output fiber without any detrimental effect to achieve lower optical loss.

Microelectromechanical switch (MEMS) have attracted a lot of attention because of their diversity. MEMS switches can be considered as subdivisions of opto-mechanical switch [2]. But they differ in the process of their fabrication, their special tiny nature, their characteristics, their performance, and their reliability. The obvious point is that a opto-mechanical switch is heavier, but replacing MEMS overcomes this drawback.

Consider the switch technology available today. Studies show that opto-mechanical switches influence the market today. In operation, they use a moving mechanism to revert fiber or optical material. The technique is usually a solenoid, piezoelectric device, or stepper motor. The activity is either manual or electric. The resulting instructions for these types of opto-mechanical technologies are similar and show low insertion loss and switching speeds in the milliseconds range.

The performance of switching devices has significantly increased in recent years. Consider the details of this standard data for single mode and multimode switches: insertion loss of 0.5 dB; −60 dB back reflection; 0.005 dB repetition; 80 dB isolation; 10 million rounds; and 20 ms conversion speed. These specifications are consistent even with fiber-optic components. Opto-mechanical switching technology has matured. Major technological revolutions are not expected. However, step-by-step improvements continue to change, and several improvements have been made in recent years. For example, advances in manufacturing technology have allowed for more fiber balance. As a result, switches have a loss of less than 0.4 dB – almost the lowest of fiber-optic connectors.

Many industry analysts have predicted that opto-mechanical fiber-optic switches will be replaced in the future using faster integrated configurations or photonic switches. To date, this change has not occurred. As for electro-mechanical switch equipment, although there is increasing difficulty in devices, opto-mechanical switches (relays) are expected to work with optical components integrated in the long term.

On the other hand, no matter how reliable they are, an optical transceiver will always provide a conversion speed of milliseconds. This is not suitable for high-speed networks that require nanosecond or picosecond switching. Thus, according to technical literature, most research efforts focus on the combination of optical and photonic switching. They employ an electro-optic device that transforms the dynamic range of the electric field presence. Several switches have been developed from silicone and lithium niobate waveguide technology.

Fiber pigtailing of these devices have also been improved. The remaining improvement hurdles for electro-optic switches are their relatively high insertion loss (5 dB), high crosstalk (20 dB), and high cost.

Acousto-optic devices use sound waves to produce sinusoidal refractive index wave that conducts light, a principle known as Bragg diffraction. This effect is used to control the intensity and position of the laser beam. Several factors indicate acousto-optic effects, including mixed quartz, iron oxide, germanium, indium phosphide, lithium niobate, and tellurium dioxide.

"Acousto-Opics" are often used as laser beam modulators, deflectors, and frequency shifters. For switching applications, both splitting and multi-folding are available, with faster conversion speeds possible. However, to date, it has not been successful in its use of fiber-optic substitutes due to poor light coupling resulting from splitting and fiber-optic latency.

Other devices considered fiber-optic substitutes are magneto-optic devices, liquid crystal modulators, holographic objects, and photo-refractive materials. However, out of low cost, low insertion loss, high frequency, low back reflection, and accurate return, none showed as much promise or success as technology.

In applications, fiber-optic switches are widely used for communication (usually single mode fiber), data communication (usually with multimode fiber), and fiber sensing (usually with large core fibers). Most switches are used to replace network security, fiber and component testing, and fiber monitoring.

Switching network security includes bypass switching and computer security switching. The bypass switching consists of two 2 switches of optical bypass terminals, one for the primary FDDI ring and one for the second FDDI ring. The two-way exchange protects the network by disconnecting the FDDI center dual attachment, and preserving the integrity of the ring, in the event of a malfunction, loss of power, or removal of the permanent base [3].

On/Off switches are widely used to replace fiber security in classified networks. Some computer networks used by banks and the military will need special permission and easily accessible fiber security switches. For example, turning on the fiber-optic cable in a secure key switch can only be used by authorized personnel to activate computer access in certain areas.

4.1.2 How Does an Optical Switch Work?

Replacing optical fiber is a technology that works in a fiber-optic circuit to operate in the same way as traditional electrical network switches. The optical switches are powered by a mechanical mechanism that physically moves fiber or other large objects (Figure 4.2). For example, an opto-mechanic switch reverses the optical signal by moving the fiber using a mechanical device usually driven on a stepper motor. It moves a mirror (prisms, or directional couplers) that illuminates directly from the input of the desired product.

4.1.3 Applications of Optical Switches

The optical switch is an important component of optical networks these days. It is generally used for a variety of applications such as providing light paths, switching protection, and objects that allow high-speed packets for circular networks. It is also very useful for fiber-optic components or system testing and measurement, as well as multipoint fiber sensor system applications [4].

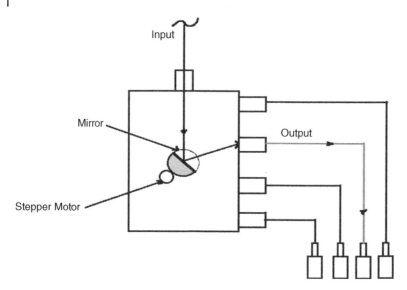

Figure 4.2 Optical switch.

4.2 All-Optical Switch

Today's service providers are looking to set up new smart networks based on a new generation of optical segments to reduce operating costs and increase revenue. These components are characterized by new optical function as well as high-performance interactions. One such component is the all-optical switch (Figure 4.3). The switch has evolved from a simple, state-of-the-art mechanical device to an integrated component that operates at high speed to add new functionality [5]. The addition of new capabilities, such as the dynamic variable optical attenuation and optical multicast, in the optical transformation component, along with the increase in operating speed, offers great benefits and enables a variety of new applications

- Free-space switches suffer from a significant insertion loss compared to all-fiber switches, because they require a straight beam collimation and balance free-space components [6]. This also affects their compatibility with fiber networks. The all-optical switch controls the path between multiple optic cables without any electrical data being converted. Switching All-Optical switches goes through all the light signals from the optical input and transmits them all to the optical output without changing or altering the IP data packets. Due to the lack of switching power, all-optical switches do not release latency data or jitter time.
- No packet routing is essential in an all-optical switch as it is circuit switch.

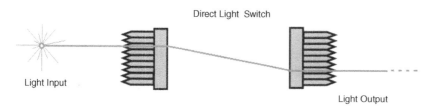

Figure 4.3 All-optical switch.

- IP packets will not route selectively.
- Optical input, optical switch core, and optical output are used in all-optical switches.
- OEO (Optical-Electrical-Optical) conversion is not used.
- Whole light signal switches from one fiber to another.
- No conversion needed in switching data.
- No data corruption or latency, no timing jitter.

Single-mode signals with 1260–1675 nm wavelength and multimode signals with 850 and 1300 nm wavelength propagating through all-optical switches can be transferred and routed. All-optical switches allow data to transfer at any rate and in any format, no data conversion routing method is required. Video, audio, data, and optical sensor signals can route through all-optic switches with 400 Gbps+ bandwidth [7]. Conventional copper switches are surpassed by fiber switches.

The static fiber layer is used in optical switches. Patching or manual intervention is not needed in optic switch as cabling needs to be done only once and can be reconfigured remotely. Optical switches create transparent paths with near zero latency, which reduces cost, power usage, and electrical conversion delays. Fiber-optic switching uses management software in addition to maintain an inventory of cross-connects routing.

There are three types of all-optical switches

In the first example (Figure 4.4), the switch is shown on the right. The object here, a single light beam (white), is now released from indirect material in addition to the exit routes. The item here is currently in one of the two switching regions; the "off" state. The current switch turns "on" side, when a second light (black), is imposed onto a nonlinear optical switch and the output beam now changes its direction.

The second example (Figure 4.5) is where one beam of light (white) passes through an indirect device in addition to the exit in a certain direction. The item is in "on" mode. The "off" state of the beam is now reached when the current weak (black) light is injected into the indirect glass device, which will change the direction of the white light.

In the third example (Figure 4.6), a single beam of light (white) passes through an indirect device in addition to the exit in a certain direction. The item here is in one of the two switching regions: the current switch turns to the "off" state when the second light beam (black), is now injected into the indirect material [8]. For this reason, the light of both the first light (white) in addition to the circuit light (black) is present and absorbs the material; in addition, there is currently no output light for either.

Figure 4.4 All-optical switch that changes the direction of output light.

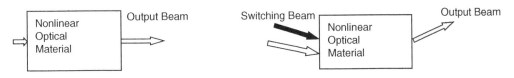

Figure 4.5 All-optical switch that redirects light.

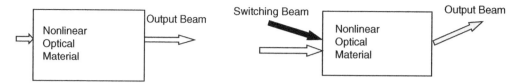

Figure 4.6 Absorptive all-optical switch

4.2.1 Why is an All-Optical Switch Useful?

Electronic models were used in the same way as existing electronic devices. The most important power supply at the moment is two opposite wires so the current can be disconnected or added to the communication – "Off" or "On" respectively. In an electric switch, electricity is used for the on/off operation. An all-optical switch performs the same operation using light.

Undoubtedly, all-electrical switches are very useful at the moment. At present there is a large presence capable of switching off the "on" power supply in addition to "turning off" the power supply in addition to the use of a switch to control the signal stream in the network. Interestingly, even more important at the moment is the use of all-electric power that is enjoyed as a transistor usage for digital smart circuits. The concept is now available in both variables ("on" plus "off") and all electronics because the current switching condition is controlled by other electronic signals.

All-optical switches are capable of carrying out the same functions as those of all-electronic switches, for example, directing the streams signal around optical networks or serving as the building blocks of optical computers.

An area connected to all-optical switches that are now essential for communication exists, taking into account the fact that these days most long-distance phones are added to the existing internet communications carried by optical cables. These thin sections of glass allow a lot of information to travel long distances near the speed of light. Although we all currently use a lot of optical cables, the computer data available is not an optical signal. Wherever the data signal is switched by wires to reach their destination, the signal must be switched on and converted to electricity so that the current address is able to be read in addition to the data now capable of existing pointing in the right direction. The task of replacing the signal from light to electricity as well as back again uses more power (and causes more heat) and products currently in place will be able to judge whether there is a transfer to occur immediately or several times in a row [6]. The effectiveness of optical communication is increased if the devices are now designed to guide signals while they are currently in optical mode.

All-optical computing is still the technology of the future. It has some important advantages compared to electronic computers, such as the small size/high density, high speed, and low heating of the joints and the substrate. However, we are very interested in the possibility of using an improved model as our all-optical switch is an alternative for computing and communication-based. An all-optical switch alternative for effective computing and communication should have the ability to identify or deal with single photons.

In the air or in vacuum, the light beams move easily and without interruption. Therefore, in the case of light, it is not possible to change the path of the same light. On the other hand, in a nonlinear material, a strong enough light beam changes the optical properties of material, which can affect any light beam also propagating through the material. Therefore, one beam on the material can control the interaction of the material and another beam. Therefore, one beam can cause a second beam to change direction.

The problem is, this kind of light-to-light control happens; the lights and material should interact continuously. Materials usually react as needed only when there are strong light sources. This means that a high-power beam requires monitoring for even small light by light interactions. These high forces reduce the production design for all-optical devices as more advanced measures are needed, which increases the cost. Therefore, the difficulty is in finding the right light-matter interaction necessary to make all-optical switches.

In order to create a useful all-optical switch, one has to adjust the operating conditions of the light. Several mechanisms are in place to increase the intensity of light interactions. Research includes the use of light whose frequency is proportional to the amount of light emitted by an atom when the electron is resting from a excited state to a ground state. This technique is called the "resonant enhancement" of light interactions, which simply means that by illuminating the same light source on the atom it emits, the light and the atom interact more easily than if another lights were used.

A nonlinear material formed by a different type of atom can have a strong interaction of light-matter due to the resonant enhancement [1]; for example, the use of a material that includes only rubidium atoms allows it to have the strong light-to-light interactions required for all-optical switching. On the other hand, most modern television devices are made from semiconductor materials, where resonant enhancement does not work. Therefore, further research is needed to find ways to apply the principles that allow for all-optical switching in our current devices and available technology.

Increased performance lights up the wavelength in small electrical currents. The effect of each individual increase on the previous channels is irrelevant. Therefore, balance can be maintained. The downside is the length of time required for the connection to be made and for increasing the complexity to the full design.

The concept of multicast packet-oriented networks has been widely studied in the past due to the increasing number of additional bandwidth-powered applications [7]. By adding that focus to optical domains, packet-based applications such as broadband video, high-definition TV, storage space and network, and multimedia can be provided with additional functionality. In addition, other benefits such as improved performance can occur. This has led to interest in the latest developments in optical multicast networks.

Change operation speed is also important. New wavelength lighting should be done quickly or slowly and with increasing speed performance that allows the optical path to stabilize before the protection mechanism is triggered. They need to stabilize the time in the sub-millisecond range, which may not be possible with all-optical components. The effect of each individual increase on the previously associated steps is irrelevant. Therefore, calming down with time can be a relief. The disadvantage is the length of time required for the connection to be made and for increasing the complexity to the full structure. Various tests use optical switches. Time-consuming tests such as polarization-dependent-loss or insertion-loss measurements are performed on a multiport optical system, and we find significant time savings when manufacturing done by fast, reliable optical switches to turn between the systems port and test equipment. The evolution of the switch in a multifunctional all-optical device has enabled many novel styles that may not have been available before. These applications originate from user networks such as dynamic all-optical networks and optical burst switching to test equipment and delay lines.

4.3   Magneto-Optical Switches

Magnetic-optical switches use Faraday magneto-optical switching. Capable of changing the external magnet, this type of magnetic-optical switch uses Faraday rotation to change the polarization plane of the event that controls the lights to change the optical modes. Compared to the traditional optical switch, the magneto-optical switch has the advantages of a high-speed switch and strong stability. Compared to other non-electronic optical switches, it also handles low voltage better and has low crosstalk. Naturally, it is now slowly gaining widespread use in fiber networks.

4.3.1   Magneto-Optical Switch Features

- Solid-state high speed
- High stability, high reliability
- Low driving voltage
- Low cross talk
- Epoxy-free on optical path
- Fail-safe latching
- Built-in circulator and isolator functions

Magnetic-optical switches are widely used in the running of any optical network. They generally consist of a PBS (polarization beam splitter), FR (Faraday rotator), separator lamp removing polarization, and polarization beam connector [5]. During the usage of a magnetic-optical switch, the lamp is first divided into orthogonal polarization PBS, and then the magneto-optical switch will change the polarization plane of this phenomenon, polarize the lamp, and change the external magnet to finally determine the function of the switch

Fiber-optic beam splitters are used to separate light from one fiber to other fibers. The light from the insertion fiber is first collected, then sent through a fiber-optic splitter to split it in two. The result of the output beams will be traced back into the output fiber. Both 1XN and 2XN splitters can be built in this fashion, with as many as eight or more outputs with both low return loss and minimal insertion loss. This design is highly flexible, allowing one to use different types of fibers at different channels, and different ports and different beam splitter optics inside. The custom design includes circulators, polarizing splitters, and non-polarizing plates in one commonly manufactured package. Splitters are made up either of fiber that is completely attached to each port (pigtail system) or of equipment at each station where one can inject the fiber inside (reception method).

The common types of splitters offered are polarizing beam splitters and polarization to protect beam splitters. The operating modes are described below.

In polarization conservation splitters, plates use a reflective mirror to transfer part of the light from the insertion fiber to the main output fiber, and to show the remaining light to the second output fiber. Separated polarization uses a well-designed multi-parting light with the exact ratio regardless of the incoming polarization. Due to the nature of this coating, the performance will vary in length, and therefore it is recommended for a working wavelength range of about ±10 nm. Broadband beam splitters are supplied, but with significant variation of the split component they have to do with setting polarization. Splitters that only divide a small part of the input light are commonly known as taps. These splitters are commonly used in power tests. There is a very low-cost alternative solution, including tap operation and photodiode testing in one unit.

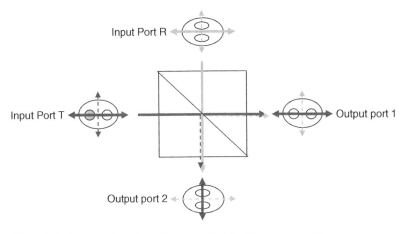

Figure 4.7 Polarization orientation on polarizing fiber beam splitters.

In split beam splitters, polarizing beam splitters divide incoming lights into two orthogonal states. They can also be used to combine light from two fibers to form a single output fiber. When used as a connecting beam, each input signal transmits on a different output polarization axis. It is important when using these splitters with polarization to preserve the fiber that one understands how polarization axes fit on each channel. Figure 4.7 one shows the standard configuration. With this configuration, the following behavior will be observed.

1) The slow axis of T input port induces light to slow axis of output port 1.
2) The fast axis of T input port induces light to slow axis of output port 2.
3) The slow axis of R input port induces light to fast axis of output port 1.
4) The fast axis of R input port induces light to fast axis of output port 2.

This configuration may change as per customer need.

In the polarizing splitters there are two factors to consider: the ability of the splitter to prevent the light arranged at port 1 to reach port 2 and to maintain the polarization extension of light emerging fibers at each port. The level of crosstalk is equal to or greater than that of the split extinction level. For example, splitter can be made with the high crosstalk ratio, but if the output fibers have has a medium shape, then the output polarization extension ratio intend be lower.

4.3.2 Principles of Magneto-Optical Switches

Compared to the traditional mechanical-optical switch, the magneto-optical switch [9] has the advantages of being a high-speed switch and having strong stability. Compared to other non-electronic optical switches, the magneto-optical switch has the advantages of lower driving voltage and smaller crosstalk. In the near future, the magneto-optical switch aims to be a competitive product.

4.3.2.1 The Design Core of the Magneto-Optical Switch

The Faraday rotation effect was applied to the magneto-optical switch. Adjust the action of the magnetic-optic crystals in the event of lighting and a polarized plane by adding the external magnet, to gain the ability to change the optical path. In other words, the Faraday rotation effect is the event that the separation plane rotates when linearly polarized light passes through the center in an external magnetic field. There is a polarized lamp with a line next to M in the center, which can be divided into two circular lights on the other side. And then the two circular flashing lights scatter forward in two different velocities without any interaction. After illumination, there is only a step difference between them. Therefore, the light output is still refined, but the separation plane rotates in a certain direction compared to the light-paced event. This means that when the outer magnet acts on the magneto-optical crystal, its separation state can be changed after the illumination of the light.

The magneto-optical switch can detect the e-optical switching function that is required for all-optical communication.

4.3.3 Magneto-Optic Effect

Magnetic-optic effect refers to the physical contact between light and magnetism. Once the light source enters the source material, the light emissions can be adjusted by the interaction of the light-object, such as the polarization, spectrum, phase, and intensity of light. The magneto-optic system [9] has played important roles in the evolutionary history of electrical engineering, as it provides a solid basis for experimenting to understand electromagnetic theory, classical theory of matter, and quantum theory. In 1845, Michael Faraday observed that when a light passed through a glass toward the outer magnet, its polarization angle was rotated by angle. This is the first time that the magnetic-optic performance (Faraday effect) was tested and provides some solid evidence that light is an electromagnetic wave in nature [10]. In 1876, John Kerr observed a plane of light reflected from a magnetic field. Demonstrating such a magneto-optic effect will be called the magneto-optic Kerr effect (MOKE). The sensitivity of MOKE is very high as it can be affected by sub-atomic layer magnetism, so this is an ideal tool for reflecting magnetic ground-side systems, especially for atomic-thickness two-dimensional (2D) materials and other nanostructures. Subsequently, the Zeeman endings and the magneto-birefringence endings (Voigt effect and Cotton–Mouton effect respectively) were obtained. Currently, giant magnetic-optic effects have been found in many magnetic objects, such as rare-ground element-doped garnets, rare-ground element-doped magneto-optical glasses, rare-ground element-doped iron garnet films and magnetic alloy films. Due to its high sensitivity, low power output or even no power supply (if magnets are constantly used) and high-efficiency, magneto-optic results are attracting great interest from scientists and engineers in different fields. In particular, as the future of electronics grows to focus on the atomic level, conventional devices, or technologies for magnetic resonance imaging, such as vibrating magnetometer sampling and superconducting quantum interference technology (SQUID), may lose performance. Therefore, it is essential to use the MOKE or Zeeman results as high-throughput systems for analyzing the physical properties of micro, nano, or even systemic systems, as they show a meaningful retaliation to the magnetism of the subatomic layer. Meantime, improvements to compact and sensitive magnetic-optic systems are also promising the realization of future lighting modulation with the advantages of non-interference, low power consumption, and non-damage.

The Cotton–Mouton effect refers to the behavior of 2D materials diffused in liquid. The other three effects are characterized in the form of solids. The 2D material shown is in xy-plane. In contrast, 2D products are a class of sheet-type non-standard, with a thickness of hundreds of nanometer or sub-nanometer. Depending on the specified measurement $r = l/t$, where l and t are the back size and thickness, 2D materials have a very large form anisotropy r. Since the discovery of graphene in 2004, it has been found that the 2D substance family has more than 1000 elements, such as iron dichalcogenides (TMDCs), black phosphorus (BP), hexagonal boron nitride (h-BN), and iron oxides. In general, most of the objects are designed with a strong intra-buffer connection while weak van der Waals forces between the components, enabling the surface preparation of 2D object from multiple suspensions and extensions. Meanwhile, chemical carbon deposition (CVD) is widely used to create high-quality 2D crystals. Currently, bands of 2D products cover a wide range, and that family of 2D products contains protective, semiconducting, and metallic materials. Again, by creating separate 2D elements on top of each other, the new and magnetic properties of 2D heterostructures can be enhanced. Thus, the magneto-optic output of 2D objects has become one of the areas of research in recent years, as it reflects a wide range of physical phenomena. Figure 4.8 shows the fourth case-based analysis of magneto-optic effects of 2D objects, related to the interaction between light event and 2D objects in front of an external magnetic field. In 2011, the giant Faraday effect of graphene was reported. More recently, due to its increased Dirac electron dispersion and chiral characteristics of the electromagnetic field, graphene was found to contain half-dimensional Hall effect, which results in a higher Faraday effect and MOKE. Even with 2D magnetic field devices, such as TMDCs, the large Zeeman effects determined by the magnet field caused a clear valley crack of 3030 μeV/T observed. The reported 2D giant magneto-optic devices work with the latest magnetic 2D objects, such as $Cr_2Ge_2Te_6$, Fe_3GeTe_2, and CrI_3, as well as opening up new 2D objects based on magneto-optic writing and also laying the groundwork for further deep-diving on the search for open source magneto-optic effects.

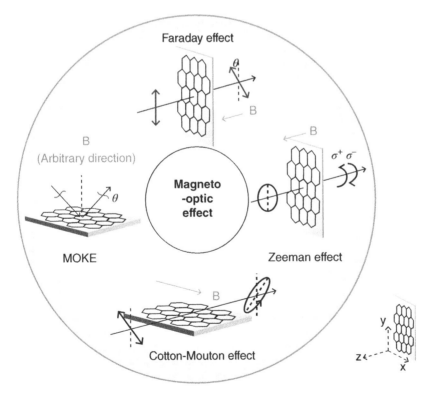

Figure 4.8 Typical magneto-optic effects of two-dimensional (2D) materials: Faraday effect, magneto-optic Kerr effect (MOKE), Cotton–Mouton effect, and Zeeman effect.

4.4 Faraday Rotation

The inception of Faraday rotation was explained by two possible models – the phenomenological model and the atomic model – by Balanis and Zvezdin respectively [9, 10].

4.4.1 Phenomenological Model

In this model magnetic field objects are replaced by a description of magnetic dipoles corresponding to each other creating the magnetic state of the object. When the outer field is used the dipoles are rotated by a field path. It is similar to the direction of the top of the magnetic field in a magnetic field, a fixed magnetic field that rotates with a constant frequency, known as the Larmor precession frequency, in front of the external magnetic field. As the way that torque exerted is by the gravitational field on a spinning top, single magnetic dipole moment m exerted by an external magnetic field exerts a torque T given as

$$T = \mu_0 m \times H_0 = m \times B_0$$

Where m= nIds, H_0 is the applied magnetic field and B_0 is the applied magnetic flux density.

4.4.2 Atomic Model

The atomic model describes magneto-optic (MO) features, so free-ion theory is useful in explaining the magneto-optical effects of quantum mechanics [4]. The Hamiltonian of a free ion can be written as

$$H = H_0 + H_{ee} + H_{so}.$$

H_0 is the sum of the one-half Hamiltonians with the force corresponding to the equal center-field projection, H_{ee} is the energy associated with the magnetic field and H_{so} is the Hamiltonian defining spin-orbit interaction. The state of the ion is determined by the distribution of electrons between single electrons, which is also determined by the Paul–Fermi principle and by the low power of the ion. Most of the magnetic-optical properties are controlled by ions touching incomplete 3d and 4f shells. For these Russel–Saunders coupling (LS coupling) means spin-spin and orbit-orbit coupling is the strongest. Here the orbital angular time of the electric field is so strong that it communicates with a different full L and/or different full S with a different power. Thus the ground configuration, defined by H_0, is divided into characteristics of the (Spin) S and (orbital) L time, and the state is defined by the waves of activity (k; S, L, Ms, M_L), where K is the configuration index.

If there is any solid or liquid is applied to the magnetic field of the field, and the field beam is polarized and passed through it in the direction of the magnetic lines of force (through holes in the pole of the electric field gate), they found that the transmitted light is still a fixed plane. This "optical rotation" is called the Faraday rotation (or Faraday effect) and contradicts significant aspects of a similar effect, called the optical phenomenon, occurring in a sugar solution [6]. In a sugar solution, the circular motion goes same way, whichever way the light beam is directed. In Faraday practice, however, the optical rotation, such as when looking inside a band, is restored when the light exceeds an object facing the magnetic field; that is, the rotation can be restored by changing the field direction or the direction of light. By setting an example between two pieces of Polaroid or two Nicol prisms, it can be arranged (with sufficient magnetic force) so that small light is passed

Figure 4.9 Polarization rotation due to the Faraday effect.

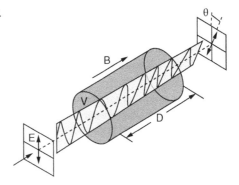

through the system in one way, while it can pass, eventually without exhaustion, in opposite directions. The result is that it allows the construction of an irreplaceable mirror object such that observer A can see observer B, while A cannot be detected by B.

Magneto-optic results are cases in which the optical properties of an object are altered by the presence of a quasi-static magnetic field (either externally applied, or due to the magnetization of that object). The first discovery of a magnetic-optical effect, now known as the Faraday effect, occurs when a light separated differs when passing through an object exposed to a magnetic field directly aligned to the light direction. Under these conditions, the polarization of plane appears to rotate; this cycle is called the Faraday rotation (Figure 4.9).

The strength of the polarization rotation due to the Faraday effect is defined according to the empirical relationship:

$$\theta = VBD$$

Here, V is the Verdet constant of the material, D the thickness of the sample, and B is the strength of the magnetic field. For most transparent materials, V is extremely small and is both temperature- and wavelength-dependent.

$$\epsilon = \begin{pmatrix} \epsilon_{xx} & i\epsilon_{xy} & 0 \\ -i\epsilon_{xy} & \epsilon_{yy} & 0 \\ 0 & 0 & \epsilon_{zz} \end{pmatrix}$$

The permitting tensor is the traditional first point of access for the Faraday Output (although the magneto-optic output can be the equivalent taken from the load tensor). To explain the Faraday effect, dielectric permittivity is treated as a non-diagonal tensor taking form (for propagation in the z direction). Here, the specification of the applied magnet (or magnetization) is taken as an axis. The diagonal properties represent the operation of objects in the absence of magnetic-optic results. The diagonal element ϵ_{xx} is the origin of the Faraday effect: it can be shown that non-zero ϵ_{xx} creates a phase lag between left-right and right-handed circularly polarized components of linearly polarized light, which results in rotation of the polarization plane.

One of the unique and important features of Faraday rotation is its non-reciprocal character; if the light makes two opposite faces passing through magnetic-optic objects, the Faraday rotation does not cancel (such as the polarization rotation due to the optical event) but rather doubles. This non-repetition allows the Faraday effect to be used as a operating principle for optical isolator and optical circulators, important components in optical telecommunications and other laser applications.

Bibliography

1 D. Papadimitriou et al. Optical multicast – a framework. OIF2001.093, April 2001.

2 Karl Heinz Bennemann. *Non-Linear Optics in Metals. No. 98.* Oxford, Oxford University Press, 1998.

3 R. Doverspike and J. Yates. *Challenges for MPLS in optical network restoration.* IEEE Communications, February 2001.

4 B. Mukherjee et al. *Light trees: optical multicasting for improved performance in wavelength router networks, IEEE Communications Magazine,* February 1999.

5 C. Leycuras, H. Le Gall, J. Desvignes, M. Guillot, and A. Marchand. Magnetic and magneto-optical properties of a cerium YIG single crystal. *IEEE Transactions on Magnetics,* 21(5): 1660–1662, 1984.

6 N. Ghani, J. Fu, D. Guo, X. Liu, Z. Zhang, P. Bonenfant, L. Zhang, A. Rodriguez Moral, M. Krishnaswamy, D. Papadimitriou, S. Dharanikota, and R. Jain. Architectural framework for automatic protection provisioning in dynamic optical rings. *OIF* 2001.041, January 2001.

7 D. Papadimitriou. Optical rings and optical hybrid mesh-rings topologies, Internet draft, work in progress, https://datatracker.ietf.org/doc/html/draft-papadimitriou-optical-rings-00, February 2001.

8 S. Ayandeh and P. Veitch. Dynamic protection and restoration in multilayer networks. *OIF* 2001.166, April 2001.

9 A.K. Zvezdin and Y.A. Kotov. *Modern Magnetooptics and Magnetooptical Materials.* Bristol, Institute of Physics Publishing, 1997.

10 C.A. Balanis. *Advanced Engineering Electromagnetics,* Chichester, John Wiley and Sons, 1989.

Further Reading

E.U. Condon and G.H. Shortley. *Theory of Atomic Spectra.* Cambridge, Cambridge University Press, 1959.

A. Cotton and H. Mouton. Influence of polydispersity on the phase behavior of colloidal goethite. *Annales de chimie et de physique,* 11:145, 1907.

M. Faraday. Experimental researches in electricity. Nineteenth Series. *Philosophical Transactions of the Royal Society London,* 136:1, 1846.

T. Hayakawa and M. Nogami. Dynamical Faraday rotation effects of sol–gel derived Al_2O_3–SiO_2 glass containing Eu^{2+} ions. *Solid State Communications,* 116:77, 2000.

T. Hayakawa, M. Nogami, N. Nishi, and N. Sawanobori. Faraday rotation effect of highly Tb_2O_3/Dy_2O_3-Concentrated B_2O_3–Ga_2O_3–SiO_2–P_2O_5 glasses. *Chemistry of Materials,* 14:3223, 2002.

T. Kato, S. Iwata, M. Yasui, K. Fukawa, and S. Tsunashima. Magneto-optical spectra of (Mn1-xMx)Pt3 (M-Fe, Co) ordered alloy films. *Journal of Magnetism and Magnetic Materials,* 177–181:1427, 1998.

J. Kerr. XLIII. On rotation of the plane of polarization by reflection from the pole of a magnet. *Philosophical Magazine Series 1,* 3:321, 1877.

J. Kerr. JXXIV. On reflection of polarized light from the equatorial surface of a magnet. *Philosophical Magazine Series 1,* 5:161, 1878.

M. Kucera, J. Bok, and K. Nitsch. Faraday rotation and MCD in Ce doped yig. *Solid State Communications,* 69:1117, 1989.

C. Leycuras, H. Le Gall, J.M. Desvignes, M. Guillot, and A. Marchand. Magneto-optic and magnetic properties of praseodymium substituted garnets. *Journal of Applied Physics,* 53(11): 7125–8425, 1982.

W. Liu, Y.H. Dai, Y.E. Yang, J.Y. Fan, L. Pi, L. Zhang, and Y.H. Zhang. Critical behavior of the single-crystalline van der Waals bonded ferromagnet $Cr_2Ge_2Te_6$. *Physical Review B*, 98, 214420, 2018.

K. Nassau. *Handbook of Laser Science and Technology, Supplement 2: Optical Materials*. Boca Raton, FL, CRC Press, 1995.

S. Novoselov, A.K. Geim, S.V. Morozov, D. Jiang, Y. Zhang, S.V. Dubonos, I.V. Grigorieva, and A.A. Firsov. Electric field effect in atomically thin carbon films. *Science*, 306:666, 2004.

M. Shamonin, T. Beuker, P. Rosen, M. Klank, O. Hagedorn, and H. Dotsch. Feasibility of magneto-optic flaw visualization using thin garnet films. *NDT and E International*, 33:547, 2000.

A. Vafafard and M. Sahrai. Tunable optical and magneto-optical Faraday and Kerr rotations in a dielectric slab doped with double-V type atoms. *Scientific Reports*, 10:8544, 2020.

W. Voigt. Doppelbrechung von im Magnetfelde befindlichem Natriumdampf in der Rischtung normal zu den Kraftlinien. *Mathematisch-Physikalische Klasse*, 1898:355, 1898.

Q.H. Wang, K. Kalantarzadeh, A. Kis, J.N. Coleman, and M.S. Strano. Electronics and optoelectronics of two-dimensional transition metal dichalcogenides. *Nature Nanotechnology*, 7:699, 2012.

Y. Xu, J.H. Yang, and X.J. Zhang. Quantum theory of the strong magneto-optical effect of Ce-substituted yttrium iron garnet. *Physical Review B*, 50:13428, 1994.

P. Zeeman. On the influence of magnetism on the nature of light emitted by a substance. *Astrophysical Journal* 5:332, 1897.

X. Zhang, Z. Lai, C. Tan, and H. Zhang. Solution-processed two-dimensional MoS_2 nanosheets: preparation, hybridization, and applications. *Angewandte Chemie*, 55:8816, 2016.

5

Acousto-Optic Switches

Sudipta Ghosh[1], Chandan Kumar Sarkar[1], and Manash Chanda[2]

[1] Department of Electronics and Telecommunication Engineering, Jadavpur University, Kolkata, West Bengal, India
[2] Department of Electronics and Communication Engineering, Meghnad Saha Institute of Technology, Kolkata, West Bengal, India

5.1 Introduction

Acoustic-optic switches are made based on the principles of diffraction of light through a propagating medium having periodical variation of refractive index. This index variation could formulate moving index grating or standing index grating, through the medium, caused by moving or standing acoustic wave respectively. The period and degree of modulation of index grating can be tuned by the frequency and amplitude of the acoustic wave through a transducer, controlled by electronic signals. A radio frequency (RF) electric field is applied across the electrodes of an acoustic transducer to generate an acoustic wave by piezoelectric effect. This acoustic wave induces a cyclical strain in the desired medium either on its surface to create surface acoustic wave (SAW) or in the bulk of that medium to create a bulk acoustic wave. Optical switches, couplers, frequency shifters, beam deflectors, and modulators have widely emerged as applications of acoustic-optic devices in recent times.

Compound semiconductors using III-V materials [1], like indium gallium arsenide or gallium phosphide, are desired candidates to construct acousto-optic devices as these are popular in manufacturing monolithic ICs. These are most suitable materials for producing optical wave guides, lasers, and photodetectors. However, the driving power, diffraction efficiency, and piezoelectric properties of these materials are not good enough to build acousto-optic devices. In a mean time lithium niobate (LN or LiNb03) [2] has emerged as a potential candidate in this domain for its excellent piezoelectric property. A well defined fabrication process with LN makes this material quite suitable for future AO switches and includes low-loss waveguides, frequency selective optical switches, wavelength division multiplexing switches (WDM), and networks.

5.2 Fundamentals of Acousto-Optic Effect

Photo-elastic effect is defined as the change of optical properties of the supportive medium on appliance of mechanical strain. The principle of acousto-optics is based on the phenomenon of the time-dependent periodic variations of the optical property of the medium affected through a

Optical Switching: Device Technology and Applications in Networks, First Edition. Edited by Dalia Nandi, Sandip Nandi, Angsuman Sarkar, and Chandan Kumar Sarkar.
© 2022 John Wiley & Sons, Inc. Published 2022 by John Wiley & Sons, Inc.

mechanical strain delivered by acoustic wave, i.e., photo-elastic effect is the main cause of acousto-optic phenomenon [3]. The strain causes the variation of refractive index of the medium, which can be expressed by the following equation:

$$\Delta n = \left(\frac{n^6 p^2 P_a}{2\rho v_a^3 A} \right)^{1/2},$$
(5.1)

where n = refraction index of the supportive medium
p = suitable component of photo-elastic tensor
P_a = total acoustic power (in Watts)
ρ = mass density
v_a = acoustic velocity
A = cross-sectional area of the supportive medium through which the wave propagates;
Equation 5.1 can be rewritten as

$$\Delta n = \left(\frac{M_2 P_a}{2A} \right)^{1/2},$$
(5.2)

where M_2 is known as "figure of merit of Acousto-Optics" and is expressed as [4]

$$M_2 = \frac{n^6 p^2}{\rho v_a^3}.$$
(5.3)

The constituent parameters of M_2 are subject to mode operations and the figure of merit as well. Both are dependent upon the polarizing mode and the direction of propagation of the acoustic wave and the polarization of the optical wave as well. The unit of figure of merit is cubic seconds per kilogram, which is equal to square meters per watt. The property of acoustic attenuation of material, due to acoustic absorption, is a considerable factor for many practical cases. The degree of attenuation increases with the square of sound frequencies. This phenomenon limits the use of materials like silica glass, TeO2, PbMoO4, and Ge etc. in any application below 1 GHz (typically between 10 and 500 MHz). Gallium phosphide (GaP) has a better use in mid-acoustic frequency range, typically between 500 MHz and 1 GHz. $LiNbO_3$ is very popular for the high-acoustic frequency range, typically up to 5 GHz.

Many crystals, for example $LiNbO_3$ and TeO_2, are interested in acousto-optic applications for their piezoelectric property. The coefficients of such elasto-optic crystals are interchangeable with a piezoelectric effect due to the fact that the acoustic wave can create an electric field in the crystal, which also causes the conversion of the index to the crystal by electro-optic effect. Correction for this second result can be very important for certain crystals. This conversion depends on the direction of acoustic wave propagation. In addition, they can rotate the ellipsoid index as well, creating decay, as unthinkable from previous experience and the effect of Pockels [5]. A complete description of the photo-elastic effect has to include the contributions from both strain and rotation. Therefore, to treat the acousto-optic diffraction through coupled-wave theory [6] one needs to consider the photo-electric effect in a medium, due to strain and rotation, both in terms of change in the permittivity of the medium.

5.3 Acousto-Optic Diffraction

The space and time-dependent acoustic wave equation can be represented by

$$S(r,t) = S \sin(K.r - \Omega t),$$
(5.4)

where K is the propagation constant and Ω is the angular frequency in radian. A standing wave is the combination of two similar waves propagating in opposite directions. Mathematically it can be expressed as follows:

$$S(r,t) = S\sin(Kr)\cos(\Omega t),$$ (5.5)

where $K = \dfrac{\Omega}{V} = \dfrac{2\pi}{\Lambda}.$ (5.6)

The time- and space-dependent permittivity of the medium changes introduced by the acoustic wave is given by

$$\Delta\varepsilon(r,t) = \Delta\acute{\varepsilon}\,\sin(K.r - \Omega t)$$ (5.7)

where the factor K counts on the propagation direction and polarization property of acoustic wave. Generally, $\Delta\acute{\varepsilon}$, that is, the change of permittivity, is a function of strain and rotation, caused by the acoustic wave, the photo-elastic coefficient of the medium, and the direction of the acoustic wave. It also depends on the polarization and frequency of the optical wave. However, it is independent of the factors, K and Ω. The interaction between the optical wave of frequency ω, and the incident medium having periodic variation of permittivity, given in (5.7), results in diffracted optical waves with frequencies $\omega \pm \Omega$. If the process keeps occurring this will end up with an array of diffracted waves with frequencies $\omega \pm n\Omega$, where n is a positive or negative integer and is referred as the "order of acousto-optic diffraction." Therefore, the frequency of the diffracted beam can be expressed as

$$\omega_{+1} = \omega_0 + \Omega$$ (5.8)

$$\omega_{-1} = \omega_0 - \Omega,$$ (5.9)

where Equations 5.8 and 5.9 represent mathematically the diffracted beam with upshifted and downshifted frequency respectively, as shown in Figure 5.1.

The total electric field can be expressed as a linear combination of all interacting components as follows

$$E(r,t) = \sum_n E_n(r)e^{-i\omega_n t} = \sum_n E_n(r)e^{ik_n.r - i\omega_n t},$$ (5.10)

where all the wave components, having frequencies $\omega_n = \omega + n\Omega$, are associated with a single wave vector k_n. Therefore, the coupled efficiency among the optical wave components of varying frequencies can be attributed as follows:

a) Neighboring optical frequency components are separated by an amount of $\pm\omega$ value.
b) The efficiency of the coupling relies on the direction of propagation and the polarization of the optical waves, being coupled, and the acoustic wave as well.
c) Coupling is possible between different polarized optical wave components in anisotropic and isotropic medium as well. The nature of $\Delta\acute{\varepsilon}$ is anisotropic itself. For an isotropic medium, coupling is possible provided $p_{11} \neq p_{12}$, i.e., the independent elements of the said medium should not be matched.
d) The amount of phase mismatch between two wave components is the governing factor of the coupling efficiency in any coupling process. The phase-match conditions between E_n, E_{n-1} and E_{n+1} are: $k_{n-1} = k_n - K$ and $k_{n+1} = k_n + K$ respectively.

Therefore, different diffraction phenomena are observed under different experimental condition, which are the main driving force for different application fields.

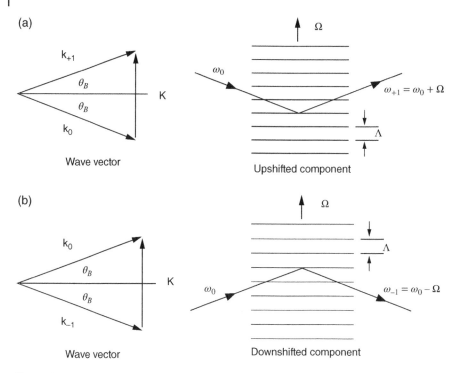

Figure 5.1 (a) Upshifted diffraction phenomenon. (b) Downshifted diffraction phenomenon.

5.4 Raman–Nath Diffraction

The Raman–Nath diffraction phenomenon is demonstrated in Figure 5.2(a) [7], where a plane optical wave of frequency ω gets diffracted by an acoustic column while travelling through a isotropic medium. It is considered that the propagation direction of the acoustic wave along the

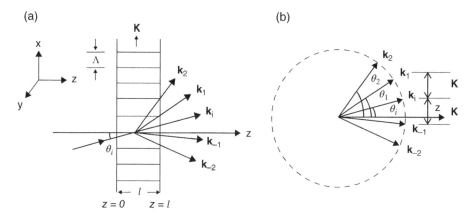

Figure 5.2 (a) Raman–Nath diffraction phenomenon and (b) wave vectors.

x-axis, as shown in the figure, and the propagation direction of the incident optical wave is normal or near normal to the acoustic wave propagation path, i.e., along the z-axis. θ_i is a small angle of diffraction with respect to the Z direction. The acoustic wave is considered to have a finite width in the Z direction and an infinite length in the X direction. It is also considered that there is no component of interaction along Y-direction, i.e. the interaction between two waves is purely two-dimensional. Here θ_n is the directional angle of $\mathbf{k}_n$, the wave vector, with respect to the Z-axis. If l be the interaction length along the Z direction, and its value be so small as to satisfy the condition

$$n\frac{K^2l}{k} \ll 1,$$

then one can neglect the cumulative phase mismatch effect over the interaction length. This situation allows a number of diffraction orders through acousto-optic coupling. This is known as Raman–Nath diffraction regime. The condition for Raman–Nath diffraction is given as

$$Q = \frac{K^2l}{k} = \frac{2\pi\lambda l}{n\Lambda^2} = \frac{2\pi\lambda f^2 l}{nv_a^2} \ll 1,$$

where the typical value of Q be lesser than equal to 0.3 for the Raman–Nath diffraction regime.

5.5 Bragg Diffraction

Optical diffraction by acoustic column causes either an array of diffracted beams or a single one, depending on refractive index, wavelength and the acousto-optic interaction length [8]. Raman–Nath diffraction usually causes an array of beams, whereas the Bragg diffraction phenomenon generates a single beam of diffraction. Owing to higher diffraction efficiency, the Bragg diffraction is more popular for a wide range of applications.

The condition for Bragg's refraction is such that the interaction length is large enough to satisfy the condition below:

$$Q = \frac{K^2l}{k} = \frac{2\pi\lambda l}{n\Lambda^2} = \frac{2\pi\lambda f^2 l}{nv_a^2} \gg 1.$$

Here, one cannot neglect the cumulative phase mismatch effect over the interaction length. In order to have higher "diffraction efficiency", the coupled wave components should have perfect or near-perfect phase matching and this is the condition for Bragg diffraction. The typical vale of Q is more than equal to 4π for Bragg diffraction phenomena. The condition for phase matching is given as

$$k_d = k_i \pm K$$

$$\omega_d = \omega_i \pm \Omega,$$

where d and i stand for diffracted wave and incident wave respectively. The ±sign refers to frequency upshift and downshift phenomena respectively. The vector diagram of the Bragg diffraction is shown in Figure 5.3. The Bragg angle is given by

$$\theta_B = \sin^{-1}\left(\frac{1}{2n}\left(\frac{\lambda}{\Lambda}\right)\right)$$

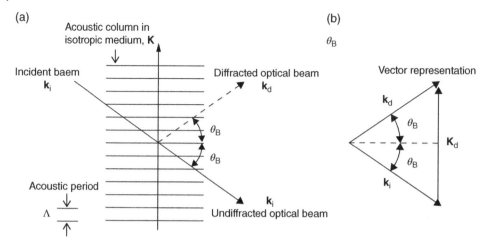

Figure 5.3 (a) Bragg diffraction phenomenon and (b) wave vectors.

5.6   Principle of Operation of AO Switches

The operation of AO switches is based on the principle of optical beam deflection or the collinear mode conversion. Figure 5.4 represents the basic operation of a 2×2 switch based on the principle of acousto-optic beam deflector (AOBD) in cross and bar states. In the cross state, the output beam is obtained from the output port O_1 (port O_2) with the input beam entered through port I_2 (port I_1), without any deflection since no radio frequency signal is applied. In bar state, the input beam, entered through input port I_1 (port I_2), is deflected and routed through output port O_1 (port O_2), following the principle of Bragg deflection. Such guided-wave acousto-optic switches are usually made of gallium arsenide, indium phosphide, and lithium niobate etc. [9].

The block diagram of a collinear polarization principle based acousto-optic switch is depicted in Figure 5.5. An acousto-optic polarization converter (AOPC) and a pair of identical polarizing beam splitters (PBS) are used to construct the switch. The system functions according to the polarization diversity concept, similar to the LN-based polarization-independent scheme applicable for tunable filters [10]. The first splitter dissolves the incident polarized monochromatic light beams into two orthogonal components, known as vertical (V) and horizontal components. Therefore, the switch routed the H polarized beam from upper (lower) input towards lower (upper) output and V polarized beam from upper (lower) input to upper (lower) output port. Here a bar-state switch and a cross-state switch is depicted in Figure 5.6 (a) and (b) respectively. For the bar-state switch,

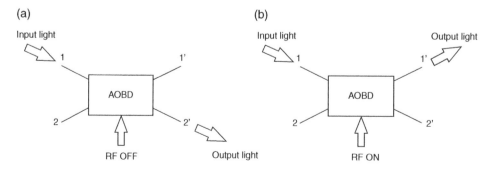

Figure 5.4   Acousto-optic beam deflector based optical switch (a) cross state and (b) bar state.

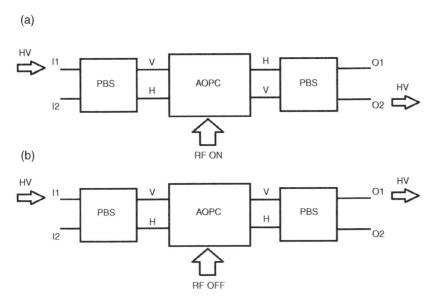

Figure 5.5 Block diagram of polarization-independent acousto-optic switch (a) cross state and (b) bar state.

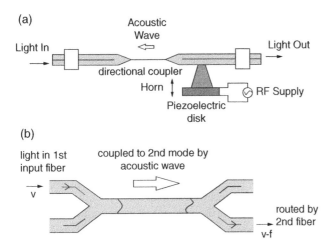

Figure 5.6 Making of (a) all-fiber switch and (b) acousto-optic null coupler.

polarization conversion doesn't take place as radio frequency control is absent. Therefore, signal is routed from input port I_1 to output port O_1 and from input port I_2 to output port O_2. For a cross-state switch, on a contrary, the applied radio frequency assists the AOPC to rotate the linear polarization orthogonally, followed by a second PBS at the output end, acts as a beam combiner. Therefore, the switching system guides the signal from I_1 to O_2 and from I_2 to O_1 respectively.

The quest for low-power, low insertion loss, and low-cost acousto-optic switch ends up with the all-fiber-made single-mode fused taper coupler. The coupler is known as null coupler [11, 12]. The coupler is constructed with two optical fibers with diameters, having difference to a such extent, that not to couple any light beam in the resultant coupler. It is implemented by pre-tapering one of the two fibers through a short length before both the fibers get fused and extended. Therefore, a 2×2 switch is constructed with identical single-mode ports. In the waist of the coupler, the fundamental mode is exited with one input beam, while the second mode is exited with the second input beam. However, an acoustic wave attunes the refractive index of the waist, resulting the input beam is guided to the second fiber due to the mode conversion, taken place at the waist of the coupler and this happens when the resonance occurs and is known as cross-state. In bar state, the beam from one fiber is guided to the output of the same fiber as no acoustic wave is there to influence it. An all-fiber acousto-optic switch with insignificant polarization sensitivity, improved drive power, lower switching time, and insertion loss has achieved this by twisting the waist of a taper-null-coupler.

5.7 Acousto-Optic Modulator

An acousto-optic modulator is an electronically controlled amplitude modulator which modulates the intensity of an acoustic wave. The diffraction efficiency of an acousto-optic modulator depends on the intensity of the acoustic wave. Therefore, it is an amplitude modulation of optical beam by acoustic wave. An acousto-optic modulator (AOM) operates in a Raman–Nath regime or in a Bragg regime. The first-order diffraction efficiency of a Raman–Nath modulator is equal to the diffraction efficiency of a Bragg cell, in low efficiency range.

There are certain disadvantages of a Raman–Nath-modulator over a Bragg-type modulator. First, the diffraction efficiency of the Raman–Nath type modulator is not more than 34%, even with the highest order of diffraction phenomena. On the contrary, the Bragg cell can attain 100% diffraction efficiency in a phase-matched operational condition. The interaction length, l, in a Raman–Nath type modulator is quadratically related with the acoustic frequency and linearly related with the optical wavelength. Hence, a very high acoustic frequency and a fairly high optical wavelength can only result in a significant interaction length. Therefore, a Raman–Nath type modulator is restricted to small bandwidth operations. Bragg cells are free from these limitations.

Acousto-optic modulators (AOM) have a wide range of applications. Previously, AO modulators were used in laser printers. The simple, low-cost AOMs were the popular choice for building the external modulator for laser printers, before the implementations of organic photo-resistors. AO devices have a unique application in combined operations as beam deflector and modulator in dither scanners.

Laser modulation in the infrared range has been another application domain of AOM. One important application is the use of AOM in laser communication as external modulator. Effective infrared AO materials cause a variety of laser modulation techniques with a large bandwidth, ranging from 1.06 μm to 10.6 μm. AO devices are preferred more than EO modulators in lower modulation bandwidth range, due to their low insertion loss. AOM are also used in CO_2 lasers for the applications of precision matching, range finding, and communications.

AO modulators are extensively used inside laser cavities, which is applicable for Q-switch, mode locking, and cavity dumping. Q-switching of YAG lasers has been a critical role in industrial laser applications. An intracavity AO modulator is required in Q-switching to maintain an insertion loss suitable for laser, to be kept below a threshold. Currently, UV grade fused silica is the most popular choice as an interaction medium for AO-based Q-switches, for their low optical absorption coefficient, good homogeneity, and lower strain-free materialistic properties.

A standing wave AOM is required inside the cavity to provide a loss modulation at a frequency equivalent to the longitudinal mode spacing, in mode-locking applications. The loss modulation causes a phase locking in longitudinal mode to generate short optical pulses. The cavity dumping is used to increase the repetition rates, which is constricted by the build-up time of the population inversion in a Q-switch. Appreciable research efforts have been delivered recently to AO devices, performing simultaneous operation of Q-switching with mode locking or cavity damping.

5.7.1   Acousto-Optic Q-Switching

Q-switching is a technique of generating high energy pulses of nanosecond order within solid-state lasers. The switch is placed inside a laser resonator. A high-power laser pulse is generated when the RF input of a laser resonator is turned off. A solid-state laser required high switching speed and/or high loss modulators. The brief operational principle [13] is demonstrated in the following literature.

The transducer, made of molten quartz, is activated with the RF source. The incident light is diffracted from the axis of laser optical resonator with low loss and suppressed oscillation. Meanwhile, Nd:YAG has been pumped continuously. Consequently, the accumulated energy is released in a form of high-power Q-switched pulses (i.e., the status of Q-value is high), as depicted in Figure 5.7, when the RF input is withdrawn.

Q-switched Tm^{+3} silica fiber laser is reported [14] to have a wavelength of 2 μm, when pumped with a Nd:YAG laser, operating at 1.319 μm frequency. The Q-switched laser is operated in the range of 2.9 m to 0.5 m and have the average laser output of 60 mW with a repetition frequency of Q-switching at 100–500 Hz. The shortest pulse width of 150 ns is obtained with 4.1 kW of maximum peak power, which is the highest value, reported to date [15–19] for a similar kind of device operating in low-order mode. An acousto-optic Q-switched Nd:YAG laser, operating at 1319 nm is recorded [20] to have highest peak power of 95 kW achieved with pulse duration of 150 ns and pulse repetition frequency of 5 kHz and 12.8% optical conversion efficiency. The conversion efficiency got even better, reached 17%, its highest value, assisted with pumping power of 555 W. Average output power is recorded as 94 W at 50 kHz PRF. Gold nano-rod (GNR) has been exploited as a saturated absorber (SA) for the first time [21] in a Q-switched Nd:YAG laser to obtain maximum pulse energy of 19 μJ at pulse repetition rate of 20 kHz.

AO Q-switched TEM_{00} grazing angle laser with ultra-high repetition frequency is realized [22] to have 2.1 MHz switching speed at average output power of 8.6 W and 2.2 MHz switching speed at average power of 10 W. The Nd:YVO_4 crystal is 3 at.% neodymium doped. The tolerance limit of the

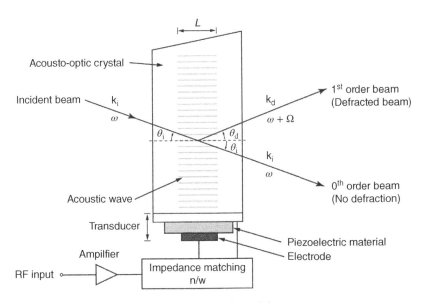

Figure 5.7   Schematic view of an acousto-optic modulator.

energy pulses is less than +/−6.7 % and +/−6.9 % at PRF of 2 MHz. Self-Raman multi-stokes Q-switched laser, working around 1.2 to 1.3 µm, is demonstrated in a c-cut Nd:YVO$_4$ crystal. Stokes emission is observed from 1215 nm to 1316 nm range in sequences. 17.1 W of pumping power is obtained at 10 kHz PRF. Average output is registered as 1.02 W for multi-stokes operation, resulting a slope efficiency of 9.1%. To ascertain high peak power, an AO-Q-switched HO:Y$_2$O$_3$ ceramic laser, operating at 2117 nm, pumped with fiber laser source at 1931 nm, is realized [23]. Average output power of more than 20 W is recorded at PRF of 10 kHz. Corresponding pulse energy is 2.0 mJ with a pulse duration of 33.1 ns and highest peak power of 60 kW. Search for further improvements on output performance is still under process.

5.7.2 Telecommunication Network

Cross-bar optical switch architecture using multichannel Bragg cell on GaAs substrate exhibits polarization-insensitive switching with low crosstalk, low insertion loss, and minimum access time [24]. A typical acousto-optic Bragg cell is shown in Figure 5.8. Multiple electrodes are constructed in a close proximity, on a common acoustic substrate. The design must consider the constraint of thermal budget, crosstalk limit, and figure of merit of the acousto-optic materials. In Bragg cell-based acousto-optic switches, interactive photons have to travel through an array of such cells in the system.

Wave division multiplexing (WDM) technique based optical telecommunication network has been a diverse domain of research for the last couple of decades. A WDM-based telecom switching system is investigated through add-drop and equalization functionalities, implemented in TeO$_2$, where trade-off has been made between spectral resolution of the device and AO figure of merit of the material for the sake of better filter performance [25]. The functionalities are implemented for all-fiber AO devices and TeO$_2$ is considered as AO material for its high value of figure of merit (M$_2$). Studies reveal that 40 mW average power is desirable for 100% selective deflection of light or intensity attenuation of 30 dB. Sidelobes have been suppressed for 1.55 µm wavelength. Three signals, to be multiplexed, are spaced by 1, 2, and 4 nm in input fiber for experimental setup of bar state and cross state as well. The system is proved to be completely insensitive to polarization. TeO$_2$ deflector based 2×2 optical switch/ multi-transducer is implemented through couple-mode arrangement of two, phase grating crystals [26]. The switch is attributed with 50-dB crosstalk, polarization sensitivity less than 0.5 dB and response time of less than 200 ns.

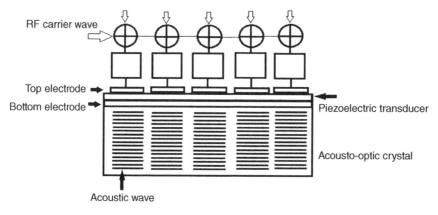

Figure 5.8 Schematic view of a multichannel Bragg cell.

A dynamic optical fiber add-drop multiplexer (OADM) is demonstrated based on principle of Bragg grating and AO effect [27]. The device possesses the virtue of minimum response time about 95 μs with a future technological projection of minimum leakage about −25 dB and response time about 50 μs and even less. Recent advancement on optical fiber laser is registered, that emits 17 wavelengths simultaneously, covering the entire C band [28]. Chromatic dispersion of the filter in optical network is measured using time flight method to justify the reliability of the laser source.

Mode division multiplexing (MDM) is one of the leading research trends in telecommunication network [29]. A highly efficient AO generator is realized for selective mode conversion from lower to higher-order modes for a few-mode fiber (FMF). The converter possesses 90% coupling efficiency and more than 10 dB extinction ratio for all higher-order modes, consisting of LP_{01}, LP_{11}, LP_{21}, and LP_{02} in a four-mode fiber. It proclaims a consistent response for step-index fiber. A novel Brillouin fiber sensor is reported [30] based on M-shaped single-mode fiber for performance analysis regarding temperature and strain sensing. Results revealed high accuracy, owing temperature error of 0.4 °C & strain error of 12.3 με. The sensing technique has utmost potential to be implemented in diversified fields. A fiber-optic array of accelerometer has been implemented using dual heterodyne phase sensitive optical time domain reflection (OTDR) concept [31]. Sensitivity about 36 rad/g is achieved subject to the system architecture design, consisting of three accelerometers spaced by 20 m in a single optical fiber in multiplexing mode. In this way, common mode noise is suppressed by 35 dB at a frequency of 100 Hz, which is a significant achievement in telecom domain.

5.8 Recent Trends and Applications

5.8.1 Emerging Spatial Mode Conversion in Few-Mode Fibers

Acousto-optic conversions in few-mode fibers have been emerged as a possible solution for the constraints of the conventional spatial mode switching devices. It enhances the switching capabilities and fast mode tuning in special mode through microwave signal modulated acousto-optic mode conversions. This effect endorses novel applications of dynamic mode controlled high-order mode fiber lasers, which finds diverse applications in the domain of optical communications, optical tweezers, laser manufacturing, structured light imaging etc. Spatial modes, consisting of cylindrical vector mode (CVM), linear polarization, and optical vortex mode, enhances the stability in optical fiber by a distributed electric field. For an ever-growing network traffic, this dynamic mode operation holds the sustainability of future communication system through spatial division multiplexing (SDM).

The quality of optical sources typically lies on the dynamic mode controlled spatial higher-order modes (HOM). Long-period grating (LPG), Fiber Bragg grating (FBG), photonic lanterns, and mode selective couplers (MSC) are the end product as applications of the spatial mode through static method in fiber-optics. MSC exhibits mode coupling through tapering fused fibers, where a single-mode fiber (SMF) is combined with a few-mode fiber (FMF) to construct MSC in order to obtain the HOMs from Gaussian-like fundamental mode [32].

Photonic lanterns are integral part of some specific HOMs, offering mode switching with low crosstalk and quite suitable for SDM technology [33]. The process of refractive index grating in fiber-optic generates mode conversions from forward propagating mode to either forward-propagating HOMs or backward-propagating HOMs in LPGs and FBGs respectively. But these applications are limited to HOMs unless a dynamic mode of switching is applied for spatial higher-order modes.

Acoustically modulated fiber-optic grating in optical fiber generates a high switching speed spatial mode in acousto-optic mode converters (AOMC). This is first reported and demonstrated by Kim et al. in 1986, with an AOMC based on a couple of mode fibers. Later, based on this unique concept, acoustic-optic interaction (AOI), various components have been derived. All-fiber acousto-optic tunable filters with low loss, optimum bandwidth, and reduced crosstalk problem is reported in the literatures [34–38]. More researches have been carried out on optimization the polarization sensitivity, utilization of tapered, and micro-tapered optical fiber and several structural modifications in filter design [39–47].

Different superlattice modulators based on acousto-optic Bragg grating phenomena have been reported in many literatures [48–52] as another application domain. The heterodyne detection/demodulation during the mode conversion of the frequency shifter [53–56] provides the vibration information under study. Photonic crystal fiber-based AOI devices [57–60], wavelength tunable lasers [61–65], AOCM-based ultrafast fiber lasers [66–75], and vector mode generators [76–80] have been potential research areas for the last couple of decades while exploring the device characterizations based on acousto-optic interactions. The acousto-optical birefringence with novel mechanisms, adopted in elliptic core FMF in recent times, have been supported to implement successfully the dynamic mode switching via FSK. This phenomenon has emerged as a potential technique in dynamic optical tunning for HOM lasers [81–87] and spatial mode-locking mechanism research [88].

However, in future, this research area will explores the time domain and spatial domain response of the nonlinear interaction between optical solitons and microwave signals. Multiple spatial mode inside the ML laser cavity introduces a new dynamic switching of mode conversion, which results in evolutionary spatial mode applications in HOM ML lasers. Advance studies have made it possible for AOMCs to integrate stimulate and depletion sources and super diffraction imaging. OAM mode operation in optical tweezers introduces the particle control to implement optical moving, catching, and rotation as well. The constraint regarding the controllability of the proportion of multi-HOMs has been mitigated through dynamic efficiency control of AOMCs. AOMC is a prospective approach in fabricating high-power lasers.

5.8.2 Lithium Niobate Thin Films

In the mid-90s, lithium niobate has been evolved as a prospective substrate material, best suited to integrated optic fabrication, for its acousto-optic and piezoelectric figure of merit [11]. An integrated acousto-optic switch with LiNbO$_3$ was reported, based on the principle of polarization conversion [89]. Like TE mode and TM mode, for two orthogonally polarized beams, two parallel Ti waveguides were fabricated. An unpolarized input beam is fragmented into two orthogonal components. At the output, the components are reunified and guided towards a particular output through a X-junction. A single-mode Ti waveguide is formed by indiffusion process for 10 hrs at 1050 °C. Acoustic barriers are fabricated by the same process, stated above, for 20 hrs of diffusion. The figure depicts a single-mode SAW waveguide with 100 μm spanning. Thermal diffusion is a well-known process to fabricate high-quality Ti:LiNbO$_3$ waveguides for acousto-optic switching. Most of the integrated acousto-optic switches are fabricated using x-cut, y-propagation with optical axis towards the z direction.

Lithium niobate has emerged as a potential candidate in the field of optoelectronics for the last couple of decades. It is popularly known as the "silicon of photonics" due to its novel acousto-optic, piezoelectric, and nonlinear (NL) properties. The nonlinear property has a natural dependency on the crystal orientation, i.e., poling orientation or domain. Frequency conversion, micro-phonon polariton excitation, and polarization rotation have been extensively studied through periodically-poled lithum niobate (PPLN). Introducing different physical process with domain engineering (DI) results in polarization-independent frequency conversion. Recently, commercially available LN thin film on Insulator (LNOI) made it possible to boost the packaging density of photonic devices and circuits. LNOI is a thin film LN over SiO_2/Ln substrate, like silicon-on-insulator (SOI) structure. The top LN film is x-cut or z-cut, having dimension of nm order. Two domain poling techniques has been experimented over micro-thick (~28 um) and sub-micro-thick (~540 nm) LNOI structure. Micron-thick PPLNOI was fabricated successfully and no domain back switching was experienced even after 35 days. For a submicron-thick sample the period is only for 25 hours and 50% of negative domain area could be sustained.

DI has a wide range of applications regarding quasi-phase-matched harmonic generators, optical parametric oscillators, and wavelength division multiplexing (WDM) frequency conversions. More recently, applications of DI have been exploited for large BW and low-voltage driving modulators, single sideband modulators, and bulk and surface acoustic waves with uniform electrodes.

5.8.3 Optical Fiber Communication and Networking

Light wave synthesized frequency sweeper (LSFS), an acousto-optic switch, is realized for wavelength division multiplexing (WDM), based on a chromatic dispersion method. Introducing lock-in detection and phase diversion technique in LSFS while measuring group delay results in fast measurement time, brilliant wavelength resolution, and high wavelength accuracy. The accuracy of frequency and the resolution of group delay are found about +/− 100 MHz and 8 ps respectively under experimental optical frequency range of 800 GHz, which could have further extended up to 4 THz by controlling the centre frequency of the BPF [90].

A novel all-fiber AO switch, constructed with AO tunable filters and mode selective couplers, is demonstrated with a two-mode fiber design [91]. The switch exhibits 2 dB or less operational loss in an operating bandwidth of 50 nm and more. The 3 dB bandwidth of the switched signal could be varied from 2.5 nm to 35 nm and even more. The extended work is a realization of optical add-drop multiplexer by this AO switch, which enhances the merit of this work to a great extent.

All-fiber acoustic tunable filters (AOTFs) are widely applicable for narrowband optical add-drop multiplexer (AODM) system, tunable bandpass filters (BPF), and WDM applications in optical communication field. A typical narrowband filter is designed [92] to have very low polarization sensitivity, owing to 3 dB bandwidth of 0.8 nm and interaction length of 10 cm.

A programable RF filter, designed for broadband range, is constructed with digital micro-mirror devices (DMD), acousto-optic tunable filter (AOTF), and a chirped fiber Bragg grating (CFBG). The filter is attributed to select higher number of taps, compared to the conventional ones. The filter nulls [93], registered at 6.90, 6.945, and 6.99 GHz for a three-tap design, proves its precision tunability property, which can enhance the stop-band and passband characteristics of the filter.

Research on acoustic sensor-based applications is carried out by exploring the distributed acoustic sensing (DAS) and distributed temperature sensing (DTS) simultaneously [94]. The DAS based on FDM and time-gated digital OFDR [95] resolved the trade-off between spatial resolution and maximum measurement distance and vibration response bandwidth and measurement distance as well. Two parallel vibration frequencies up to 9 kHz are recorded over 24.7 km long fiber, having spatial resolution of 10 m and SNR of 30 dB in one experiment whereas in second one [96] the measurement distance is 108 km with spatial distance of 5 m is achieved by a DAS system architected with distributed Raman amplifier.

A novel all-fiber mode converter frequency shifter (MCFS), based on mode selective coupler (MSC) and high-order mode converter (HOMC), is realized [97] to successfully implement the mode conversion from LP11 to LP01 and parallelly acts as a frequency shifter. Vibration frequencies range from 1 kHz to 300 kHz with minimum detectable amplitude of 0.019 nm. Conversion of modes from LP01 to higher orders like LP11a/b and LP21a/b is also achieved [98] at the same resonant frequency, attributed to broadband tunability in mode conversion.

Bibliography

1 C.S. Tsai. Integrated acousto-optic and magneto-optic devices for optical information processing. *Proceedings of IEEE*, 84(6):853–869, 1996.

2 D.A. Smith, R.S. Chakravarthy, Z. Bao, J.E. Baran, J.J. Jackel, A. d'Alessandro, D.J. Fritz, S.H. Huang, X.Y. Zou, S.M. Hwang, A. Willner, and K.D. Li. Evolution of the acousto-optic wavelength routing switch. *Journal of Lightwave Technology*, 14(6):1005–1019, 1996.

3 A. Yariv and P. Yeh. *Optical Waves in Crystals*. New York, John Wiley & Sons, 1984.

4 R.G. Hunsperger. Acousto-optic modulators. In: R.G. Hunsperger, *Integrated Optics: Theory and Technology*. Springer Series in Optical Sciences, vol. 33, 144–157. Berlin, Heidelberg, Springer, 1984. doi: 10.1007/978-3-662-13565-5_9.

5 W.P. Mason. Optical properties and the electro-optic and photo-elastic effects in crystals expressed in tensor form. *The Bell System Technical Journal*, 29(2):161–188, 1950. doi: 10.1002/j.1538-7305.1950.tb00464.x.

6 S.E. Miller. Coupled wave theory and waveguide applications. *The Bell System Technical Journal*, 33(3):661–719, 1954. doi: 10.1002/j.1538-7305.1954.tb02359.x.

7 D.T. Pierce and R.L. Byer. Experiments on the interaction of light and sound for advanced laboratory. *American Journal of Physics*, 41:314, 1973. doi: 10.1119/1.1987217

8 M. Ahmed and G. Wade. Bragg-diffraction imaging. *Proceedings of the IEEE*, 67(4):587–603, 1979. doi: 10.1109/PROC.1979.11285.

9 C.S. Tsai. Integrated acousto-optic and magneto-optic devices for optical information processing. *Proceedings of IEEE*, 84(6):853–869, 1996.

10 D.A. Smith, J.E. Baran, K.W. Cheung, and J.J. Johnson. Polarization independent acoustically tunable optical filters. *Applied Physics Letters*, 56(3):209–211, 1990.

11 T.A. Birks, D.O. Culverhouse, S.G. Farwell, and P. St. J. Russell. 2 x 2 Single-mode fiber routing switch. *Optics Letters*, 21(10):722–724, 1996.

12 D.O. Culverhouse, T.A. Birks, S.G. Farwell, and P. St. J. Russell. 3 x 3 All-fiber routing switch. *IEEE Photonics Technology Letters*, 9(3):333–335, 1997.

13 http://www.sintec.sg/products/ao/1286.html

14 A.F. El-Sherif and T.A. King. Analysis and optimization of Q-switched operation of a Tm^{3+}-doped silica fiber laser operating at 2 μm. *IEEE Journal of Quantum Electronics*, 39(6):759–765, 2003. doi: 10.1109/JQE.2003.811597.

15 P.R. Morkel, K.P. Jedrzejewski, E.R. Taylor, and D.N. Payne. Short pulse, high-power Q-switched fiber laser. *IEEE Photonics Technology Letters*, 4:545–547, 1992

16 Z.J. Chen, A.B. Grudinin, J. Porta, and J.D. Minelly. Enhanced Q-switching in double-clad fiber lasers. *Optics Letters*, 23:454–456, 1998.

17 C.C. Renaud, R.J. Selvas-Aguilar, J. Nilsson, P.W. Turner, and A.B. Grudinin. Compact high-energy Q-switched cladding-pumped fiber laser with a tuning range over 40 nm. *IEEE Photonics Technology Letters*, 11:976–978, 1999.

18 D.J. Richardson, P. Britton, and D. Taverner. Diode-pumped, high energy, single transverse mode Q-switch fiber laser. *Electronics Letters*, 33:1955–1956, 1997.

19 G.P. Lees, D. Taverner, D.J. Richardson, L. Dong, and T.P. Newson. Q-switched erbium doped fiber laser utilising a novel large mode area fiber. *Electronics Letters*, 33:393–394, 1997.

20 H. Zhu, G. Zhang, C. Huang, Y. Wei, L. Huang, and Z. Chen. Diode-side-pumped acoustooptic Q-switched 1319-nm Nd:YAG Laser. *IEEE Journal of Quantum Electronics*, 44(5):480–484, 2008. doi: 10.1109/JQE.2008.916698.

21 H.T. Huang, M. Li, L. Wang, X. Liu, D.Y. Shen, and D.Y. Tang. Gold nanorods as single and combined saturable absorbers for a high-energy q-switched Nd:YAG solid-state laser. *IEEE Photonics Journal*, 7(4):4501210. doi: 10.1109/JPHOT.2015.2460552.

22 X. Yan et al. 2 MHz AO Q-switched TEM_{00} grazing incidence laser with 3 at. % neodymium doped $Nd:YVO_4$. *IEEE Journal of Quantum Electronics*, 44(12):1164–1170, 2008. doi: 10.1109/JQE.2008.2003141.

23 E. Li, J. Tang, Y. Shen, F. Wang, J. Wang, D. Tang, and D.Shen. High peak power acousto-optically Q-switched $Ho:Y_2O_3$ ceramic laser at 2117 nm. *IEEE Photonics Technology Letters*, 32(8):492–495, 2020. doi: 10.1109/LPT.2020.2981642.

24 J. Sapriel, Vladimir Ya. Molchanov, G. Aubin, and S. Gosselin. Acousto-optic switch for telecommunication networks. *Proc. SPIE 5828, Acousto-optics and Applications* 3. doi: 10.1117/12.612841.

25 J. Sapriel, D. Charissoux, V. Voloshinov and V. Molchanov. Tunable acousto-optic filters and equalizers for WDM applications. *Journal of Lightwave Technology*, 20(5):892–899, 2002. doi: 10.1109/JLT.2002.1007946.

26 V. Quintard, A. Perennou, and J. Aboujeib. Characterization of a 2×2 optical switch based on a multitransducers acousto-optic deflector. *IEEE Photonics Technology Letters*, 21(24):1825–1827, 2009. doi: 10.1109/LPT.2009.2034538.

27 A. Diez, M. Delgado-Pinar, J. Mora, J.L. Cruz, and M.V. Andres. Dynamic fiber-optic add-drop multiplexer using Bragg gratings and acousto-optic-induced coupling. *IEEE Photonics Technology Letters*, 15(1):84–86, 2003. doi: 10.1109/LPT.2002.805867.

28 J. Maran, R. Slavik, S. LaRochelle, and M. Karasek. Chromatic dispersion measurement using a multiwavelength frequency-shifted feedback fiber laser. *IEEE Transactions on Instrumentation and Measurement*, 53(1):67–71, 2004. doi: 10.1109/TIM.2003.822008

29 D. Song, H. Su Park, B.Y. Kim, and K.Y. Song. Acousto-optic generation and characterization of the higher order modes in a four-mode fiber for mode-division multiplexed transmission. *Journal of Lightwave Technology*, 32(23):4534–4538, 2014. doi: 10.1109/JLT.2014.2360936.

30 Y. Dong, G. Ren, H. Xiao, Y. Gao, H. Li, S. Xiao, and Shuisheng Jia. Simultaneous temperature and strain sensing based on m-shaped single mode fiber. *IEEE Photonics Technology Letters*, 29(22):1955–1958, 2017. doi: 10.1109/LPT.2017.2757933.

31 X. He, M. Zhang, S. Xie, F. Liu, L. Gu, and D. Yi. Self-referenced accelerometer array multiplexed on a single fiber using a dual-pulse heterodyne phase-sensitive OTDR. *Journal of Lightwave Technology*, 36(14):2973–2979, 2018. doi: 10.1109/JLT.2018.2830114.

32 H. Yao, F. Shi, Z. Wu, X. Xu, T. Wang, X. Liu, P. Xi, F. Pang, and Xianglong Zeng. A mode generator and multiplexer at visible wavelength based on all-fiber mode selective coupler. *Nanophotonics*, 9(4):973–981, 2020.

33 X. Sai, Y. Li, C. Yang, Wei Li, J, Qiu, X. Hong, Y. Zuo, H. Guo, W. Tong, and J. Wu. Design of elliptical-core mode selective photonic lanterns with six modes for MIMO-free mode division multiplexing systems. *Optics Letters*, 42(21):4355–4358, 2017

34 D. Ostling and H.E. Engan. Spectral flattening by an all-fiber acousto-optic tunable filter. *IEEE International Ultrasonics Symposium*, 2(2):837–840, 1995.

35 D. Ostling and H.E. Engan. Acousto-optic tunable filters in two mode fibers. *Optical Fiber Technology*, 3(2):177–183, 1997.

36 H.S. Kim, S.H. Yun, I.K. Kwang, and B.Y. Kim. All-fiber acousto-optic tunable notch filter with electronically controllable spectral profile. *Optics Letters*, 22(19):1476–1478, 1997.

37 D.O. Culverhouse, S.H. Yun, D.J. Richardson, T.A. Birks, S.G. Farwell, and P.St.J. Russell. Low-loss all-fiber acousto-optic tunable filter. *Optics Letters*, 22(2):96–98, 1997.

38 I.K. Hwang, S.H. Yun, and B.Y. Kim. All-fiber tunable comb filter with nonreciprocal transmission. *IEEE Photonics Technology Letters*, 10(10):1437–1439, 1998.

39 R. Feced, C. Alegria, M.N. Zervas, and R.I. Laming. Acousto-optic attenuation filters based on tapered optical fibers. *IEEE Journal of Selected Topics in Quantum Electronics*, 5(5):1278–1288, 1999.

40 A. Diez, G. Kakarantzas, T.A. Birks, and P.S.J. Russell. Acoustic stop-bands in periodically microtapered optical fibers. *Applied Physics Letters*, 76(23):3481–3483, 2000.

41 D.A. Satorius, T.E. Dimmick, and G.L. Burdge. Double-pass acousto-optic tunable bandpass filter with zero frequency shift and reduced polarization sensitivity. *IEEE Photonics Technology Letters*, 14(9):1324–1326, 2002.

42 W. Zhang, F. Gao, F. Bo, Q. Wu, G. Zhang, and J. Xu. All-fiber acousto-optic tunable notch filter with a fiber winding driven by a cuneal acoustic transducer. *Optics Letters*, 36(2):271–273, 2011.

43 W. Zhang, L. Huang, F. Gao, F. Bo, G. Zhang, and J. Xu. All-fiber tunable Mach-Zehnder interferometer based on an acousto-optic tunable filter cascaded with a tapered fiber. *Optics Communications*, 292:46–48, 2013.

44 W. Zhang, L. Huang, F. Gao, F. Bo, G. Zhang, and J. Xu. Tunable broadband light coupler based on two parallel all-fiber acousto-optic tunable filters. *Optics Express*, 21(14):16621–16628, 2013.

45 H. Zhang, S. Kang, B. Liu, H. Dong, and Y. Miao. All-fiber acousto-optic tunable bandpass filter based on a lateral offset fiber splicing structure. *IEEE Photonics Journal*, 7(1):2700312, 2015.

46 G.R. Melendez, M. Bello-Jimenez, O. Pottiez, and M. Andres. Improved all-fiber acousto-optic tunable bandpass filter. *IEEE Photonics Technology Letters*, 29(12):1015–1018, 2017.

47 L. Huang, W. Zhang, Y. Li, H. Han, X. Li, P. Chang, F. Gao, G. Zhang, L. Gao, and T. Zhu. Acousto-optic tunable bandpass filter based on acoustic-flexural-wave-induced fiber birefringence. *Optics Letters*, 43(21):5431–5434, 2018.

48 W.F. Liu, P.S.J. Russell, and L. Dong. Acousto-optic superlattice modulator using a fiber Bragg grating. *Optics Letters*, 22(19):1515–1517, 1997.

49 W.F. Liu, P.S.J. Russell, and L. Dong. 100% efficient narrow-band acousto-optic tunable reflector using fiber Bragg grating. *Journal of Lightwave Technology*, 16(11):2006–2009, 1998.

50 P.S.J. Russell and W.F. Liu. Acousto-optic superlattice modulation in fiber Bragg gratings. *Journal of the Optical Society of America A*, 17(8):1421–1429, 2000.

51 W. Liu, I. Liu, L. Chung, D. Huang, and C.C. Yang. Acoustic induced switching of the reflection wavelength in a fiber Bragg grating. *Optics Letters*, 25(18):1319–1321, 2000.

52 N. Sun, C. Chou, M. Chang, and C.-N. Lin. Analysis of phase-matching conditions in flexural-wave modulated fiber Bragg grating. *Journal of Lightwave Technology*, 20(2):311–315, 2002.

53 W.J. Lee, B.K. Kim, K.H. Han, and B.Y. Kim. Dual heterodyne polarization diversity demodulation for fiber-optic interferometers. *IEEE Photonics Technology Letters*, 11(9):1156–1158, 1999.

54 H.M. Chan, R. Huang, F. Alhassen, O. Finch, I.V. Tomov, C.-S. Park, and H.P. Lee. A compact all-fiber LPG-AOTF frequency shifter on single-mode fiber and its application to vibration measurement. *IEEE Photonics Technology Letters*, 20(18):1572–1574, 2008.

55 W. Zhang, W. Gao, L. Huang, D. Mao, B. Jiang, F. Gao, D. Yang, G. Zhang, J. Xu, and J. Zhao. Optical heterodyne microvibration measurement based on all-fiber acousto-optic frequency shifter. *Optics Express*, 23(13):17576–17583, 2015.

56 W. Zhang, Z. Chen, B. Jiang, L. Huang, D. Mao, F. Gao, T. Mei, D. Yang, L. Zhang, and J. Zhao. Optical heterodyne microvibration detection based on all-fiber acousto-optic superlattice modulation. *Journal of Lightwave Technology*, 35(18):3821–3824, 2017.

57 A. Diez, T.A. Birks, W.H. Reeves, B.J. Mangan, and P.S.J. Russell. Excitation of cladding modes in photonic crystal fibers by flexural acoustic waves. *Optics Letters*, 25(20):1499–1501, 2000.

58 M.W. Haakestad and H.E. Engan. Acousto-optic properties of a weakly multimode solid core photonic crystal fiber. *Journal of Lightwave Technology*, 24(2):838–845, 2006.

59 M.W. Haakestad and H.E. Engan. Acousto-optic characterization of a birefringent two-mode photonic crystal fiber. *Optics Express*, 14(16):7319–7328, 2006.

60 K.S. Hong, H.C. Park, and B.Y. Kim. 1000 nm tunable acousto-optic filter based on photonic crystal fiber. *Applied Physics Letters*, 92(3):031110, 2008.

61 M.S. Kang, M.S. Lee, J.C. Yong, and B.Y. Kim. Characterization of wavelength-tunable single-frequency fiber laser employing acousto-optic tunable filter. *Journal of Lightwave Technology*, 24(4):1812–1823, 2006.

62 L. Huang, P. Chang, X. Song, W. Peng, W. Zhang, F. Gao, F. Bo, G. Zhang, and J. Xu. Tunable in-fiber Mach Zehnder interferometer driven by unique acoustic transducer and its application in tunable multi-wavelength laser. *Optics Express*, 24(3):2406–2412, 2016.

63 L. Huang, X. Song, P. Chang, W. Peng, W. Zhang, F. Gao, F. Bo, G. Zhang, and J. Xu. All-fiber tunable laser based on an acousto-optic tunable filter and a tapered fiber. *Optics Express*, 24(7):7449–7455, 2016.

64 N. Yan, X. Han, P. Chang, L. Huang, F. Gao, X. Yu, W. Zhang, Z. Zhang, G. Zhang, and J. Xu. Tunable dual-wavelength fiber laser with unique gain system based on in-fiber acousto-optic Mach-Zehnder interferometer. *Optics Express*, 25(22):27609–27615, 2017.

65 E.H. Escobar, M.B. Jimenéz, A.C. Avilés, R. López Estopier, O. Pottiez, M. Durán Sánchez, B. Ibarra Escamilla, and M.V. Andrés. Experimental study of an in-fiber acousto-optic tunable bandpass filter for single- and dual-wavelength operation in a thulium-doped fiber laser. *Optics Express*, 27(26):38602–38613, 2019.

66 D. Zalvidea, N.A. Russo, R. Duchowicz, M. Delgado-Pinar, A. Diez, J.L. Cruz, and M.V. André. High-repetition rate acoustic-induced q-switched all-fiber laser. *Optics Communications*, 244(1–6):315–319, 2005.

67 M. Delgado-Pinar, D. Zalvidea, A. Diez, P. Perez-Millan, and M.V. Andrés. Q-switching of an all-fiber laser by acousto-optic modulation of a fiber Bragg grating. *Optics Express*, 14(3):1106–1112, 2006.

68 C. Cuadrado-Laborde, M. Delgado-Pinar, S. Torres-Peiró, A. Díez, and M.V. Andrés. Q-switched all-fibre laser using a fibre-optic resonant acousto-optic modulator. *Optics Communications*, 274(2):407–411, 2007.

69 C. Cuadrado-Laborde, A. Diez, M. Delgado-Pinar, J.L. Cruz, and M.V. Andrés. Mode locking of an all-fiber laser by acousto-optic superlattice modulation. *Optics Letters*, 34(7):1111–1113, 2009.

70 C. Cuadrado-Laborde, A. Díez, J.L. Cruz, and M.V. Andrés. Experimental study of an all-fiber laser actively mode-locked by standing-wave acousto-optic modulation. *Applied Physics B*, 99(1–2):95–99, 2010.

71 M. Bello-Jimenéz, C. Cuadrado-Laborde, A. Diez, J.L. Cruz, and M.V. Andrés. Experimental study of an actively mode-locked fiber ring laser based on in-fiber amplitude modulation. *Applied Physics B*, 105(2):269–276, 2011.

72 C. Cuadrado-Laborde, A. Diez, J.L. Cruz, and M.V. Andrés. Q-switched and mode locked all-fiber lasers based on advanced acousto-optic devices. *Laser & Photonics Reviews*, 5(3):404–421, 2011.

73 M. Bello-Jimenez, C. Cuadrado-Laborde, A. Diez, J.L. Cruz, M.V. Andres, and A. Rodrıguez-Cobos. Mode-locked all-fiber ring laser based on broad bandwidth in-fiber acousto-optic modulator. *Applied Physics B*, 110:73–80, 2013.

74 W. Zhang, K. Wei, H. Wang, D. Mao, T. Mei, J. Zhao, F. Gao, and L. Huang. Tunable-wavelength picosecond vortex generation in fiber and its application in frequency-doubled vortex. *Journal of Optics*, 20(1):014004, 2018.

75 Y. Li, L. Huang, H. Han, L. Gao, Y. Cao, Y. Gong, W. Zhang, F. Gao, I.P. Ikechukwu, and T. Zhu. Acousto-optic tunable ultrafast laser with vector-mode-coupling-induced polarization conversion. *Photonics Research*, 7(7):798–805, 2019.

76 P.Z. Dashti, F. Alhassen, and H.P. Lee. Observation of orbital angular momentum transfer between acoustic and optical vortices in optical fiber. *Physical Review Letters*, 96(4):043604, 2006.

77 W. Zhang, L. Huang, K. Wei, P. Li, B. Jiang, D. Mao, F. Gao, T. Mei, G. Zhang, and Jianlin Zhao. Cylindrical vector beam generation in fiber with mode selectivity and wavelength tunability over broadband by acoustic flexural wave. *Optics Express*, 24(10):10376–10384, 2016.

78 W. Zhang, K. Wei, L. Huang, D. Mao, B. Jiang, F. Gao, G. Zhang, T. Mei, and J. Zhao. Optical vortex generation with wavelength tunability based on an acoustically-induced fiber grating. *Optics Express*, 24(17):19278–19285, 2016.

79 K. Wei, W. Zhang, L. Huang, D. Mao, F. Gao, T. Mei, and J. Zhao. Generation of cylindrical vector beams and optical vortex by two acoustically induced fiber gratings with orthogonal vibration directions. *Optics Express*, 25(3):2733–2741, 2017.

80 L. Carrión-Higueras, E.P. Alcusa-Sáez, A. Díez, and M.V. Andrés. All-fiber laser with intracavity acousto-optic dynamic mode converter for efficient generation of radially polarized cylindrical vector beams. *IEEE Photonics Journal*, 9(1):1500507, 2017.

81 J. Lu, L. Meng, F. Shi, X. Liu, Z. Luo, P. Yan, L. Huang, F. Pang, T. Wang, X. Zeng, and P. Zhou. Dynamic mode-switchable optical vortex beams using acousto-optic mode converter. *Optics Letters*, 43(23):5841–5844, 2018.

82 J. Lu, L. Meng, F. Shi, and X. Zeng. A mode-locked fiber laser with switchable high-order modes using intracavity acousto-optic mode converter. In *Optical Fiber Communication Conference*, paper W3C.3, 2019.

83 L. Meng, J. Lu, F. Shi, J. Xu, L. Zhang, H. Yao, and X. Zeng. Multi-orthogonal high-order mode converter based on acoustically induced fiber gratings. *IEEE Photonics Technology Letters*, 31(13):819–822, 2020.

84 F. Shi, J. Lu, L. Meng, P. Cheng, X. Liu, F. Pang, and X. Zeng. All-fiber method for real-time transverse-mode switching of ultrashort pulse. *IEEE Photonics Technology Letters*, 32(2):97–100, 2020.

85 H. Wu, J. Lu, L. Huang, X. Zeng, and P. Zhou. All-fiber laser with agile mode-switching capability through intra-cavity conversion. *IEEE Photonics Journal*, 12(2):1500709, 2020.

86 L. Meng, J. Lu, L. Zhang, F. Shi, and X. Zeng. Multi-orthogonal high-order modes converter. In *Conference on Lasers and Electro-Optics*, paper STh3L.4, 2019.

87 F. Shi, J. Lu, L. Meng, P. Cheng, and X. Zeng. Delivering transverse-mode switching based on all-fiber femtosecond Laser. In *Conference on Lasers and Electro-Optics Europe and European Quantum Electronics*, paper cj_11_2, 2019.

88 J. Lu, F. Shi, L. Meng, L. Zhang, L. Teng, Z. Luo, P. Yan, F. Pang, and X. Zeng. Real-time observation of vortex mode switching in a narrow-linewidth mode-locked fiber laser. *Photonics Research*, 8(7):1203–1212, 2020.

89 D.A. Smith, R.S. Chakravarthy, Z. Bao, J.E. Baran, J.L. Jackel, A. d'Alessandro, D.J. Fritz, S.H. Huang, X.Y. Zou, S.-M. Hwang, A.E. Willner, and K.D. Li. Evolution of the acousto-optic wavelength routing switch. *Journal of Lightwave Technology*, 14(6):1005–1019, 1996. doi: 10.1109/50.511601.

90 H. Takesue and T. Horiguchi. Chromatic dispersion measurement of optical components using lightwave synthesized frequency sweeper. *Journal of Lightwave Technology*, 20(4):625–633, 2002. doi: 10.1109/50.996582.

91 H.S. Park, K.Y. Song, S.H. Yun, and B.Y. Kim. All-fiber wavelength-tunable acousto-optic switches based on intermodal coupling in fibers. *Journal of Lightwave Technology*, 20(10):1864–1868, 2002. doi: 10.1109/JLT.2002.804035.

92 D.I. Yeom, H.S. Kim, M.S. Kang, H.S. Park and B.Y. Kim. Narrow-bandwidth all-fiber acoustooptic tunable filter with low polarization-sensitivity. *IEEE Photonics Technology Letters*, 17(12):2646–2648, 2005. doi: 10.1109/LPT.2005.859152.

93 N.A. Riza and F.N. Ghauri. High-resolution tunable microwave filter using hybrid analog-digital controls via an acousto-optic tunable filter and digital micromirror device. *Journal of Lightwave Technology*, 26(17):3056–3061, 2008. doi: 10.1109/JLT.2008.925045.

94 Z. Zhao, Y. Dang, M. Tang, L. Wang, L. Gan, S. Fu, C. Yang, W. Tong, and C. Lu. Enabling simultaneous DAS and DTS through space-division multiplexing based on multicore fiber. *Journal of Lightwave Technology*, 36(24):5707–5713, 2018. doi: 10.1109/JLT.2018.2878559.

95 D. Chen, Q. Liu, X. Fan, and Z. He. Distributed Fiber-Optic Acoustic Sensor with Enhanced Response Bandwidth and High Signal-to-Noise Ratio. *Journal of Lightwave Technology*, 35(10):2037–2043, 2017. doi: 10.1109/JLT.2017.2657640.

96 D. Chen, Q. Liu, and Z. He. 108-km Distributed acoustic sensor with 220-pε/$\sqrt{\text{Hz}}$ strain resolution and 5-m spatial resolution. *Journal of Lightwave Technology*, 37(18):4462–4468, 2019. doi: 10.1109/JLT.2019.2901276.

97 L. Zhang, J. Lu, L. Meng, P. Cheng, W. Li, J. Sun, T. Wang, and X. Zeng. A lower frequency shift based on mode conversion for optical heterodyne micro-vibration measurement. *Journal of Lightwave Technology*, 38(21):6057–6062, 2020. doi: 10.1109/JLT.2020.3004845.

98 L. Meng, J. Lu, F. Shi, J. Xu, L. Zhang, H. Yao, and X. Zeng. Multi-orthogonal high-order mode converter based on acoustically induced fiber gratings. *IEEE Photonics Technology Letters*, 32(13):819–822, 2020. doi: 10.1109/LPT.2020.2997364.

6

MEMS-based Optical Switches

Kalyan Biswas[1] and Angsuman Sarkar[2]

[1] Department of Electronics and Communication Engineering, MCKV Institute of Engineering, Howrah, West Bengal, India
[2] Department of Kalyani Government Engineering College, Kalyani, West Bengal, India

6.1 Introduction

Increasing demand for information carrying capacity in telecommunication networks because of the exponential growth in IP traffic may be met only with implementation of Optical Communication Networks. To implement an efficient all-optical network, several network components are required which can operate in the optical domain, for example, optical transmitters, optical receivers, optical fibers, optical amplifiers, filters, optical add-drop multiplexers (OADM), wavelength converters, optical switches, etc. [1]. An optical switch is a very important device that directs an optical signal from one optical path to a desired second optical path within a certain range [2–4]. A fiber-optic switch is one of the core devices for optical cross-connection, optical add/drop multiplexing, network monitoring, and automatic protection system. There are different implementation technologies for optical switches, including: mechanical optical switches, electro-optical switches, magneto-optical switches, thermo-optical switches, acousto-optic switches, liquid crystal optical switches, and MEMS-based optical switches. Opto-mechanical switches utilize moving parts to direct lights from one port to another. They apply different optical elements like a prism or micro-mirrors to direct the optical signal from one fiber into the desired output optical fiber [5, 6]. They may also utilize piezoelectric elements to manipulate optical collimators. Small stepper motors or piezoelectric actuators may be used for switching [7, 8]. In thermo-optic switches, actuation is done using thermo-optic effect where change of refractive index may be realized by changing temperature. Electro-optic switches uses the principle of change of material properties using electric field [9, 10]. Mainly refractive index of the material is changed using electric means for switching purpose. Very small switching time may be achieved by these electro-optic switches [9]. Nonlinear optical effects may also be utilized in efficient switching design [10]. By applying an electric field, refractive index gradient may be induced by the Kerr effect. The unpolarized input optical beam is instantaneously deflected from its initial direction and incident light beam can be moved to the required path by this device. There are also many coupler and resonator-based switches and also switches based on multiple-actuation techniques. Among these technologies, MEMS-based optical switches are extensively used because of their small size, good scalability, lower insertion loss, and low power consumption [4, 11].

Optical Switching: Device Technology and Applications in Networks, First Edition. Edited by Dalia Nandi, Sandip Nandi, Angsuman Sarkar, and Chandan Kumar Sarkar.
© 2022 John Wiley & Sons, Inc. Published 2022 by John Wiley & Sons, Inc.

Microsystems technology or micro electro-mechanical systems (MEMS) emerged as a new domain as it can co-fabricate sensors, actuators and control functions on silicon by using advanced integrated circuit (IC) fabrication process. One of the most exciting applications of MEMS-based technology is in the optical networking and particularly optical cross-connect switches. MEMS-based devices have both the static and movable components, having dimensions on the scale of microns. Technologies for optical switching are very critical for current generation and all-optical networks of the future. The cross-connect switches in present networks mostly rely on electronic cores. However, as the number of ports and data rates surge, these electronics-based switches become a bottleneck for the increasing speed of communication networks. This bottleneck has inspired researchers to have intensive research and develop all-optical switching technologies to replace the existing electronic cores. An optical switch also known as cross-connect (OXC) is a device having a number of input ports and output ports and that supports optical signals to be transferred from one port to another port. Networks having all-optical components offer many advantages in comparison to the traditional system, having requirements of electro-optics conversions. MEMS-based all-optical switches offer low insertion loss, transparency, immunity to electromagnetic interference, low crosstalk, and low power consumption. The impact of utilizing moving optical elements is much better than that of electro-optic effect or thermo-optic effects and these devices are also very efficient beam steering devices, which leads to their application as switches. Many different approaches have been in use for developing an optical switching technique [12–16], such as MEMS switching, electro-optic effect dependent switching, liquid crystal-based switching, and thermo-optic switching. In recent times, many works on free-space optical switch operating on MEMS technology have reported greater performance parameters. MEMS optical switches maintains the advantages of free-space optics such as having low losses and very low crosstalk and additionally include the benefits of small size, light weight, and very fast switching time. For free-space MEMS optical switching, mainly reflective MEMS elements are used. Alternatively, diffractive elements may also be used. Moreover, it is possible to integrate various micro-optics elements, different micro-actuators, micromechanical structures, and all the relevant electronics circuits on single silicon substrate using MEMS fabrication techniques.

This chapter reviews the state of the art of MEMS-based optical switches. In introduction, an overview of different optical switching technologies are given in section 6.1, Basic fabrication technology employed for MEMS optical switches is explained in section 6.2. Section 6.3 explains the switching architectures followed by mechanisms of actuations in section 6.4. Different optical switching parameters are explained in section 6.5 and challenges of the MEMS optical switches are highlighted in section 6.6. Finally, a conclusion is drawn in section 6.7.

6.2   Micromachining Techniques

Silicon is most preferred materials system for MEMS devices and different existing micromachining techniques are considered as the most appropriate fabrication process. There is no standard fabrication process technology exists for MEMS optical switches. Fabrication techniques depend on device structure and operating mechanism. Other than the steps used in conventional microelectronics fabrication process, special attention is given in proper etching technology to create 3D moving structures. There are two main types of micro fabrication techniques used in MEMS fabrication. These are bulk micromachining and surface micromachining. In bulk micromachining, material is removed from the bulk substrate using suitable etching process. In surface micromachining, thin-film structures are built on top of the substrate like standard IC technology.

Usually, polysilicon is considered as structural material whereas silicon dioxide is used as a sacrificial layer. Silicon is used as primary substrate because of its exceptional mechanical and optical properties.

6.2.1 Bulk Micromachining

To realize a 3D opto-mechanical structure, the bulk micromachining technique on silicon substrate has been employed for a long time. It is mainly used for aligning optical fibers or forming other optical MEMS devices [15]. As the name implies, bulk micromachining focuses on the creation of patterns or features within the bulk of some sort of starting material. Bulk micromachining is sometime called the subtraction process as silicon material is removed by etching process from the bulk silicon substrate selectively by application of suitable etchants. Using bulk micromachining techniques membranes, cantilever beams, different types of trenches, holes, and other types of structures may be formed by selectively removing silicon material using etching process from the bulk. In the etching process, there are two types of etching: (i) isotropic etching and (ii) anisotropic etching. Deep reactive ion etching (DRIE) has been developed as a special etching process and is extensively used in MEMS fabrication as a dry micromachining technique. Mostly dry etching technology is utilized in MEMS fabrication to obtain a microstructure with high aspect ratio. However, the actual etching process is much more complicated as various parameters influence the rate of etching, etch profile, and etch selectivity to materials of the mask used. Figure 6.1(a) shows a structure realized using bulk micro fabrication technology and multilayer structure by additional wafer bonding. As an example, a thin micro-mirror, one of the very important elements of MEMS optical switch, can be fabricated following a series of process steps. Starting with silicon substrate, silicon dioxide film is grown using thermal oxidation followed by a silicon nitride formation and polysilicon films deposition using LPCVD. Then Cr and Au metal films are deposited using sputtering process onto the front-side of the wafer substrate. From backside of the wafer, etching is done to remove silicon and expose silicon dioxide [12]. Figure 6.1(b) shows the structure of thin mirror fabricated using bulk micro machining technique.

6.2.2 Surface Micromachining

In the case of surface micromachining, a different approach is taken. Rather than etching into the bulk of the starting material, structures are built up on the substrate surface. These structures are created via the repetitive addition of layers selected for their various material properties, followed by the selective removal of these layers in a specific sequence. The basic need of surface micromachining is the proper choice of both the functional structural material and the removable sacrificial layer to fabricate the movable parts. Proper selection of materials is very important for the surface micromachining process. Polycrystalline silicon is mostly used as the structural layer and silicon dioxide is widely used for the sacrificial layer. Surface micromachining has attracted lots of interest during the last few years, for the fabrication of different optical subsystems like adaptive mirrors [17, 18] or optical cross-connects. The main drawbacks of this process are the time required, low yield, stress generated due to thermal mismatch in different components and stiction. A freestanding cantilever type structure fabricated using surface micromachining technique is depicted in Figure 6.2.

Apart from these traditional MEMS fabrication techniques, many other technologies have been developed and reported by the researchers to use for specific materials processing and for several other applications having their own merits and demerits. For example, the LIGA process is used

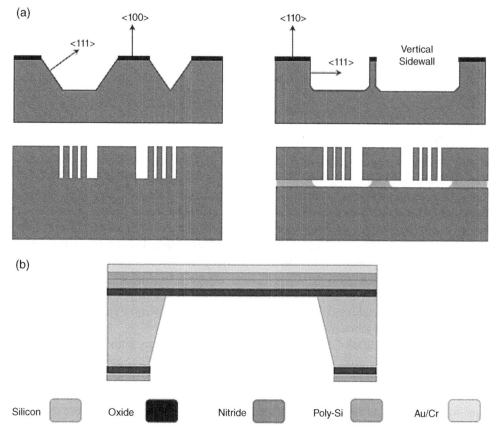

Figure 6.1 (a) Structures realized using anisotropic wet chemical etching and deep reactive ion etching. (b) Schematic diagram of thin micro-mirror. Source: Modified from [12].

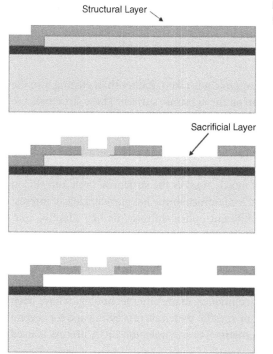

Figure 6.2 Surface micromachining process to release free-standing mechanical structures.

Figure 6.3 Cross-section of a silicon photonic MEMS switch.
Source: [22]/SPIE/CC BY-4.0.

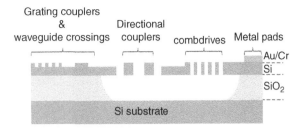

for very high aspect ratio microstructures [19]. Wet bulk micromachining may be used to fabricate fiber groove creation and fabrication of vertical smooth optical surfaces [20, 21]. Some other non-conventional machining techniques, such as ultrasonic and abrasive jet machining methods, are used by researchers for micromachining on glass or silica. A cross-section of a silicon photonic MEMS switch fabricated starting from SOI wafer is shown in Figure 6.3.

6.3 Switch Architectures

MEMS-based optical switches can be divided into two different categories such as free-space optical switches and optical waveguide-based switches depending on the light propagation method inside the switch. The optical signal propagates through free space during switching from one port to another for a free-space optical switch. This is realized by controlling the tiny movable micro-mirrors, having a diameter of the order of few hundreds of micrometers. On the contrary, the optical signal propagates through waveguides without travelling into free space in a waveguide optical switch. A waveguide or coupler is placed in a specific position to couple light from one waveguide to another waveguide. There are many popular approaches for efficient implementation of MEMS optical switch. Some of the common approaches are: (a) one-dimensional MEMS switch, (b) two-dimensional MEMS switches, and (c) three-dimensional MEMS switches [4].

6.3.1 One-Dimensional Switches

In this architecture, a dispersive element such as a grating is used to separate the input DWDM signal into its constituent wavelengths. Each different wavelength is directed to the required output fiber by an individual micro-mirror. A simple schematic of the 1D MEMS-based switch is shown in Figure 6.4(a) and (b).

6.3.2 Two-Dimensional MEMS Switches

A common approach to design 2-D MEMS switch is shown in Figure 6.5. In this approach, mirrors are arranged in cross-bar configuration. Each mirror is positioned in the light path between input and output ports of the switch in such a way that it can either reflect light (ON position) or may allow light to pass uninterrupted (OFF position). Simple control circuits based on TTL and other amplifiers may be used to control the mirror position. For an N×N switch design, total N^2 number of micro-mirrors are required. But the main drawback of this design is that insertion loss for all ports is not uniform. Insertion loss between different ports varies as the propagation distance of light path in free space varies for different ports. In a reflective-type optical switch, another important factor is the thickness of the mirror as it creates inevitable offset and insertion loss [12].

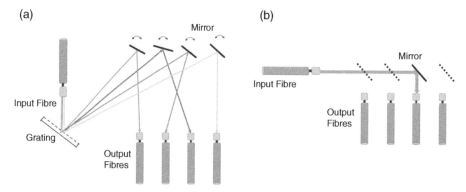

Figure 6.4 One-dimensional MEMS-based switches. (a) Basic principle of MEMS switching. (b) 1 x 4 switch using translating mirrors.

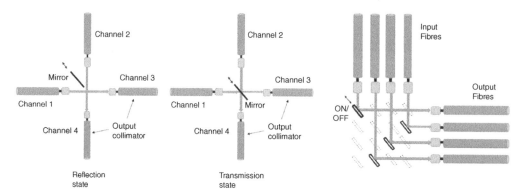

Figure 6.5 Two-dimensional MEMS-based switches.

Therefore, micro-mirrors with thinner dimensions are desirable. As an alternative approach, smaller 2D MEMS switch modules may be interconnected to have multi-stage network architecture. This approach may increase port count but increase the system complexity also.

6.3.3 Three-Dimensional MEMS Switches

The 3D optical switch architecture usually engages two separate arrays of mirrors, each aligned to a bundle of collimated input or output optical fibers. This arrangement has need of the use of 2N mirrors for a switch having N ports, which is significantly less than that of 2D architecture. In this architecture, as shown in Figure 6.6, mirrors are capable of rotating about two axes and light rays can be directed accurately in multiple angles as required by the port numbers. Light from each input fiber is directed to a particular mirror on the input array. The input micro-mirror then directs the optical beam to any of the output mirrors, which then directs the light to the required output fiber. This design may be scaled up to a number of ports with uniform insertion loss. The main advantage of this approach is that the differences in free-space propagation path along various ports-to-ports switching are very much less dependent on the port count. The design may be further improved for lower maximum angle requirement, reduction in micro-mirror size, improvement of tolerance in micro-mirror curvature, and switch crosstalk reduction by using Fourier transform lens between micro-mirror arrays [13]. The design of 3-D MEMS optical cross-connect switch is more complex as the requirement for optical design, the MEMS design along with mechanical design have to balance each other so that exceptional optical performance is achieved. Important parameters for selecting a mirror for MEMS switch are mirror size and fill factor, deflection angle, mirror reflectivity, resonance frequency, driving voltages, mirror flatness, and curvature along with mirror control system. Some typical configurations of reported MEMS-based optical switches are 1×2, 1×4. . .1×16, 2×2, 4×4. . ., 32×32, or even 128×128, 256×256 or 512×512. Most all-optical switches developed in recent times are using MEMS technology and a number of companies offer their solutions. Schematics of 3D MEMS switch is shown in Figure 6.6.

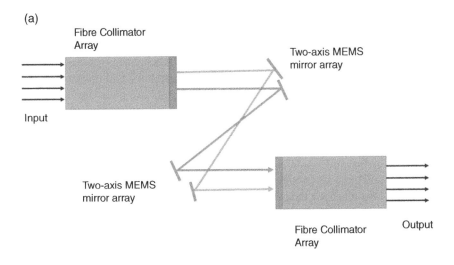

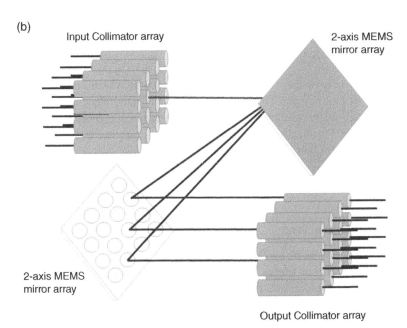

Figure 6.6 3D MEMS architecture for optical cross-connect switching. (a) Cross-sectional view. (b) 3D representation.

6.4 Mechanisms of Actuations

In all the optical switching technologies, the most important factor is the tilting of mirrors to alter light paths in free space to achieve switching functionality. The actuation mechanism to tilt the mirrors should be compatible with MEMS technology. There are different actuation mechanisms reported in the literatures. Among them, a few are described here.

6.4.1 Electrostatic Actuation

Actuation of MEMS optical switches may be accomplished using electrostatic forces. In this type of actuation mechanism, if a voltage is applied between the fixed electrode and the moving electrode, the moving electrode rotates about torsional axis. The rotation continues till the restoring torque becomes equal to the electrostatic force. There are two basic mechanisms for electrostatic actuation. The first one is based on the principle of parallel-plate capacitors, and the second one uses comb-drive capacitance. Comb-drive-based electrostatic actuators are more advantageous and provide a large angle of rotation and require lower voltage for actuation [23].

6.4.2 Magnetic Actuation

In magnetic actuation mechanism, a magnetic device interacts with an external magnetic field. Electrical current can induce the magnetic field and the induced magnetic field generates the force employed on the moving magnetic material [24, 25]. This mechanism is applicable when the fabricated device is in millimeter scale as the magnetic force depends on the volume of permanent magnet and on the total coil area for electromagnets. A magnetic actuation mechanism may help to reduce cost for device fabrication, reduce power consumption, and allow operation in a liquid environment. But this is not good for miniaturization of the device.

6.4.3 Thermal Actuation

Mismatch in coefficient of thermal expansion (CTE) of different materials may be utilized for generation of thermal actuation. Due to different CTE of two connected materials, they expand differently as temperature changes and a structural stress is generated which causes the structure to bend. Electrical current may be used for heating up the structure as required. The CTE difference for the materials involved for generating actuation should be high enough to obtain enough deflection [26]. The major advantage of this mechanism is that it may generate large deflection but temperature control and power consumption may be a concern.

6.4.4 Piezoelectric Actuation Mechanisms

Deformation of piezoelectric materials with applied electric field may be used for piezoelectric actuation mechanisms. Piezoelectric thin films of lead zirconate titanate (PZT) or aluminum nitride (AlN) can be used for large displacement vertical translational micro-actuator [27]. AlN-based devices are fully compatible with CMOS technology [28]. In MEMS devices, a thin piezoelectric material layer is deposited as a part of the MEMS beam between flexible layers. Within small footprint, large displacement and fast response of the actuators at less power consumption and lower actuation voltage make them very useful for MEMS optical switches.

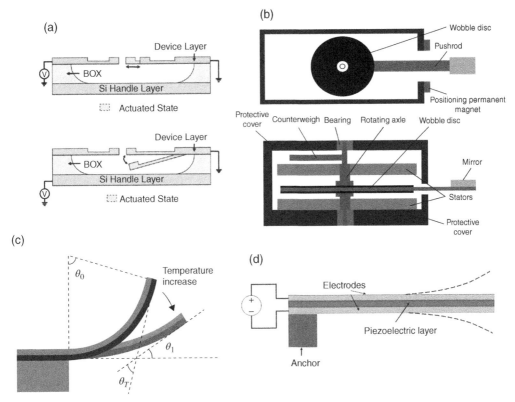

Figure 6.7 Different actuation mechanism for MEMS optical switches. (a) Electrostatic actuation. Source: [15]/IEEE. (b) Magnetic actuation. Source: [25]/MDPI/CC BY-4.0. (c) Thermal actuation. Source: [6]/IntechOpen/CC BY-3.0. (d) Piezoelectric actuation mechanisms.

6.4.5 Other Actuation Mechanisms

Ferromagnetic materials that change their shape when subjected to a magnetic field convert magnetic energy to mechanical energy. Therefore, magnetostrictive materials show a change in length per unit length when magnetized [29]. Vice versa, if an external force is used to produce a strain in a magnetostrictive material, the magnetic state of the materials change. This conversion of magnetic and mechanical forces is utilized for actuation of MEMS optical devices. The main advantage of this technology is the capability of remote actuation. Basic principles of different actuation mechanisms are illustrated in Figure 6.7. Different actuation mechanisms have several advantages and disadvantages. A comparative analysis of MEMS optical switches [30] with different actuation mechanism is given in Table 6.1.

6.5 Optical Switch Parameters

In order to consider the applicability of various technologies for all-optical switching, some important optical switch parameters are discussed here [31]. Switching times, insertion loss, crosstalk, wavelength, and power consumption need to be considered along with manufacturability and other reliability issues.

Table 6.1 Comparison of performance of different actuation mechanism.

Parameters	Actuation mechanism				
	Electrostatic	Magnetic	Piezoelectric	Electrothermal	Inertial type
Device size	Small	Large	Small	Small	Medium
Fabrication process used	Simple	Complex	Complex	Medium complexity	Simple
Power consumption	Very low	medium	Very low	High	Very low
Switching speed	Fast	Medium	Fast	Slow	Medium
Reliability	Good	Good	Good	Low	Low

6.5.1 Switching Time

Switching time is time required to change and set up a new optical path by a switch. To maximize the utilization of a switch, it is desirable to have smaller switching time. Ideally, a switch takes up a time for data exchange after the switching, leading to the total time period: $t_{switching} + t_{data}$. Thus, utilization is limited to $t_{data}/(t_{switching} + t_{data})$.

6.5.2 Insertion Loss

The loss of power incurred by an optical signal when it is traversed through a switch is known as Insertion loss. If P_{in} is optical signal power at the input end and P_{out} denotes the power at output port, the Insertion loss is given by Loss $= 10\log_{10}(P_{in}/P_{out})$. Insertion loss vary as optical path length in the switch vary and therefore depends on the various combinations of input-output ports.

6.5.3 Crosstalk

In reality, optical switch transfer a fraction of input signal to the undesired output end. It may be caused due to sufficient control of the structure or sensitivity of the device. Crosstalk is defined as the ratio (usually expressed in dB) of the total power from all the undesired inputs to the power at output from desired input. Since the out power from all the undesired input ports must be lower in comparison to the power at output port from the preferred input, crosstalk must be a negative number in dB.

6.5.4 Wavelength

The wavelength range describes the ability of an optical switch to operate over a certain desired wavelengths range. Some OxCs are capable of operating only at a certain range of wavelength that means those switches having very limited bandwidth. If a cross-connect can work over a wide wavelength range, that indicates the higher bandwidth of the switch. Larger wavelength of an optical switch is desirable.

6.5.5 Power Consumption

Power consumption of a switch is a very important characteristic. There are two main constituents of power consumption: static and dynamic power consumption. The power consumed while there is no switching activity is known as static power whereas the power required to create a new optical path, or paths, through the switch, is known as dynamic power. Total power consumption depends on the switching principle and fabrication technology used. Lower dynamic power consumption is desirable for fast operation of optical packet switching.

6.6 Challenges

In current scenario, MEMS-based optical switches are the front runner as the most suitable candidate for design of OXC for optical networking. But there remain some challenges which need to be considered in order to achieve widespread application in the core-transport network.

6.6.1 Optical Beam Divergence

There are excellent features of free-space optical switches such as low crosstalk, small polarization-dependence, little dispersion, and low wavelength-dependence. But loss performance of these devices has a tendency to be a little more as the insertion loss depends on the divergence of fundamental-mode Gaussian optical beams when propagates in free space. The divergence depends on beam waist and propagation distance. Though large optical beam diameter reduces the beam divergence, it is difficult to achieve in practice. To reduce the optical beam divergence, mode matching technique is widely used. But this approach is somewhat limited by beam diffraction. It is observed that complete mode matching can be accomplished only if the tilt angle of the mirror is sufficiently large. Therefore control of divergence of the optical beam is a big challenge for design of a compact MEMS optical switch [31].

6.6.2 Angular Control

In case of free-space optical switches, the most important switching parameters of concern is its insertion loss. The loss of an optical switch depends on angular misalignment [32, 33]. The requirement is more stringent in case of high port count 3D switches based on beam steering techniques as the optical path travelled by the light beam is higher than that of the 2D MEMS switches. Therefore the angular accuracy of the gimbal mirrors in analog fashion is considered the notable challenge for this technology.

6.6.3 Reliability of Optical MEMS

To analyze the reliability of MEMS-based optical devices, it is required not only to understand the technology-related variable but also external variables such as environmental and operational conditions. Because of the immense diversity of device designs, materials, and functions, reliability analysis of MEMS devices is very important to find out and understand the various causes of failure [34]. Failure mechanisms of optical MEMS are more complicated than the failure mechanisms of microelectronics devices. The basic difference between MEMS devices and microelectronic devices is the presence of movable parts in MEMS, which makes the reliability prediction of MEMS device a challenging task. Reliability testing of MEMS devices needs to be standardized.

One of the possible failure modes in MEMS-based optical switches is "stiction", which is caused in the surface micromachining process due to the surface adhesion in-between contacting interfaces exceeding the restoring force. Stiction between materials may restrict movements of the switch as the device dimensions are very small and gravity and other body forces does not play substantial role. Proper care may be taken during manufacturing process designs. Springs can be implemented on landing tips of micro-mirror to avoid stiction failure [35].

Friction is another crucial factor that limits the lifetime of MEMS-based devices. Friction occurs between contact surfaces that move against each other. This repeated frictional force leads to the increase of contacting stress and if that exceeds the yield strength, the material fails. Frictional force may be reduced by use of certain coatings or by eliminating rubbing surfaces during design of optical MEMS devices.

For MEMS optical switches, another reliability concern is vibration. Because of the fragile nature of optical MEMS, external vibrations can lead to terrible consequences. Extreme vibration may hamper either surface adhesion or break device support structures, which may lead to device failure [30]. Long-term and periodic vibration may also be a cause of fatigue. Mechanical shocks can also generate wire bond shearing, which is a common failure mode for most semiconductor devices.

Considering all the reliability challenges of MEMS optical switches, it is a big challenging task to design a proper packaging for the device [36, 37]. As MEMS optical switches may have very close contacts with the physical world through their mechanical components, proper care needs to be taken in packaging to minimize the effect of changing temperature, vibrations, humidity, and other environmental elements. The variation in packaging method may differ the device characteristics. Therefore, proper design of package should be considered to be of high importance during the initial design of the MEMS optical switching.

Other than these reliability challenges and Packaging issues, Optical MEMS have other challenges in terms of Manufacturability, Serviceability, Scalability, and Manufacturing automation. Considering all these issues, MEMS-based optical switches face challenges from other all-optical switching technologies and the emerging electronic switching systems.

6.7 Conclusion

In this chapter MEMS-based optical switching technology has been discussed. MEMS technology has been extensively and enthusiastically explored by the researchers over the past few years for possible application in optical communications systems. There are benefits to MEMS optical switching technologies and the limitations are significantly different from those of the conventional technology. Because of its unique IC-compatible fabrication process and free space nature, integration of optics, electronics, and mechanical structure on a single chip is easily achievable. Its compact nature, low cost means of implementation, and configurability with sustainable functionality adds to its merits lists. Conventional 3D MEMS-based switches provide low insertion loss, very good scalability (large number of input/output ports), an adequate switching time, and low crosstalk to be the most preferred switching technologies for future all-optical communication networks.

Bibliography

1 R. Ramaswami and K. Sivarajan *Optical Networks: A Practical Perspective*. New York, Morgan Kaufmann, 1998.

2 T.-W. Yeow, K.L.E. Law, and A. Goldenberg. MEMS optical switches. *IEEE Communications Magazine*, 39(11):158–163. doi: 10.1109/35.965375.

3 L.Y. Lin and E.L. Goldstein. Opportunities and challenges for MEMS in lightwave communications. *IEEE Journal of Selected Topics in Quantum Electronics*, 8(1):163–172. doi: 10.1109/2944.991412.

4 I. Plander and M. Stepanovsky. MEMS optical switch: Switching time reduction. *Open Computer Science*, 6:116–125, 2016.

5 O. Solgaard, A.A. Godil, R.T. Howe, L.P. Lee, Y. Peter, and H. Zappe. Optical MEMS: from micromirrors to complex systems. *Journal of Microelectromechanical Systems*, 23(3):517–538. doi: 10.1109/JMEMS.2014.2319266.

6 W. Piyawattanametha and Z. Qiu. Optical MEMS. In: N. Islam, editor, *Microelectromechanical Systems and Devices*, 2011. doi: 10.5772/27612.

7 G. Coppola, L. Sirleto, I. Rendina, and M. Iodice. Advance in thermooptical switches: Principles, materials, design, and device structure. *Optical Engineering*, 50(7):071112.

8 Á. Rosa, A. Gutiérrez, A. Brimont, A. Griol, and P. Sanchis. High performance silicon 2×2 optical switch based on a thermo-optically tunable multimode interference coupler and efficient electrodes. *Optics Express*, 24(1):191–198, 2016.

9 L. Qiao, Q. Ye, J. Gan, H. Cai, and R. Qu. Optical characteristics of transparent PMNT ceramic and its application at high speed electro-optic switch. *Optics Communications*, 284(16–17):3886–3890, 2011.

10 D. Gong et al. Electric-field-controlled optical switch using Kerr effect and gradient of the composition ratio Nb/(Ta+Nb). *Materials Research Bulletin*, 75:7–12, 2016.

11 V.A. Aksyuk et al. 238×238 surface micromachined optical crossconnect with 2dB maximum loss. In *Optical Fiber Communication Conference and Exhibit*, 2002, FB9. doi: 10.1109/OFC.2002.1036769.

12 K. Fan, W. Lin, L. Chiang, S. Chen, T. Chung, and Y. Yang. A 2×2 mechanical optical switch with a thin MEMS mirror. *Journal of Lightwave Technology*, 27(9):1155–1161, 2009. doi: 10.1109/JLT.2008.928955.

13 J. Kim, C.J. Nuzman, D.F. Lieuwen, J.S. Kraus, A. Weiss, C.P. Lichtenwalner, A.R. Papazian, R.E. Frahm, N.R. Basavanhally, D.A. Ramsey, V.A. Aksyuk, F. Pardo, M.E. Simon, V. Lifton, H.B. Chan, M. Haueis, A. Gasparyan, H.R. Shea, S. Arney, C.A. Bolle, P.R. Kolodner, R. Ryf, D.T. Neilson, and J.V. Gates. 1100×1100 port MEMS-based optical crossconnect with 4-dB maximum loss. *IEEE Photonics Technology Letters*, 2003, 15(11):1537–1539.

14 H. Hsieh, C. Chiu, T. Tsao, F. Jiang, and G.J. Su. Low-actuation-voltage MEMS for 2-D optical switches. *Journal of Lightwave Technology*, 24(11):4372–4379, 2006. doi: 10.1109/JLT.2006.883674.

15 N. Quack, A.Y. Takabayashi, Y. Zhang, P. Verheyen, W. Bogaerts, P. Edinger, C. Errando-Herranza, and K.B. Gylfason. MEMS-enabled silicon photonic integrated devices and circuits. *IEEE Journal of Quantum Electronics*, 56(1):8400210. doi: 10.1109/JQE.2019.2946841.

16 T.J. Seok, N. Quack, S. Han, R.S. Muller, and M.C. Wu. Highly scalable digital silicon photonic mems switches. *Journal of Lightwave Technology*, 34(2):365–371, 2016. doi: 10.1109/JLT.2015.2496321.

17 S.D. Senturia, *Microsystem Design*, 1st ed., Berlin, Springer US, 2001.

18 L. Qiao, W. Tang, and T. Chu. 32×32 silicon electrooptic switch with built-in monitors and balanced-status units. *Scientific Reports*, 7:42306, 2017.

19 S. Akkaraju, Y.M. Desta, B.Q. Li, and M.C. Murphy. A LIGA-based family of tips for scanning probe applications. In *SPIE, Microlighography and Metrology in Micromachining II*, Austin, TX, 1996, 191–198.

20 P. Helin, M. Mita, and H. Fujita. Self-aligned mirror and v-grooves in free space micro machined optical switches. *Electron. Lett.*, 36, 2000:563–564.

21 P. Dobbelaere, S. Gloeckner, S. Patra, L. Fan, C. King, and K. Falta. Design, manufacture and reliability of 2D MEMS optical switches. *Proc. SPIE 4945, MEMS/MOEMS: Advances in Photonic Communications, Sensing, Metrology, Packaging and Assembly*, 25 March 2003.

22 S. Han, J. Beguelin, L. Ochikubo, J. Jacobs, T.J. Seok, K. Yu, N. Quack, C.-K. Kim, R.S. Muller, and M.C. Wu. 32×32 silicon photonic MEMS switch with gap-adjustable directional couplers fabricated in commercial CMOS foundry. *Journal of Optical Microsystems,* 1(2):024003, 2021. doi: 10.1117/1. JOM.1.2.024003.

23 P.R. Patterson, D. Hah, H. Chang, H. Toshiyoshi, and M.C. Wu. An angular vertical comb drive for scanning micromirrors. In *IEEE/LEOS International Conference on Optical MEMS*, Sept. 25–28, 2001, Okinawa, Japan, 25.

24 I.-J. Cho and E. Yoon. A low-voltage three-axis electromagnetically actuated micromirror for fine alignment among optical devices. *Journal of Micromechanics and Microengineering*, 19, 2009. doi: 10.1088/0960-1317/19/8/085007.

25 S. Jia, J. Peng, J. Bian, S. Zhang, S. Xu, and B. Zhang. Design and fabrication of a MEMS electromagnetic swing-type actuator for optical switch. *Micromachines*, 12(2):221, 2021. doi: 10.3390/mi12020221.

26 L. Wu and H. Xie. A large vertical displacement electrothermal bimorph microactuator with very small lateral shift. *Sensors and Actuators A: Physical*, 145–146: 371–379, 2008.

27 Z. Qiu, J.S. Pulskamp, X. Lin, C.-H. Rhee, T. Wang, R.G. Polcawich, and K. Oldham. Large displacement vertical translational actuator based on piezoelectric thin films. *Journal of Micromechanics and Microengineering*, 20:075016, 2010.

28 H. Conrad, J.U. Schmidt, W. Pufe, F. Zimmer, T. Sandner, H. Schenk, and H. Lakner. Aluminum nitride: a promising and full CMOS compatible piezoelectric material for MOEMS applications. *Proc. SPIE* 7362, 73620J, 2009.

29 T. Bourouina et al. Integration of two degree-of-freedom magnetostrictive actuation and piezoresistive detection: application to a two-dimensional optical scanner. *Journal of Microelectromechanical Systems*, 11(4):355–361, 2002. doi: 10.1109/JMEMS.2002.800561.

30 T. Cao, H. Tengjiang, and Z. Yulong. Research status and development trend of MEMS switches: a review. *Micromachines* 11(7):694, 2020. doi: 10.3390/mi11070694.

31 M. Stepanovsky. A comparative review of MEMS-based optical cross-connects for all-optical networks from the past to the present day. *IEEE Communications Surveys & Tutorials*, 21(3):2928–2946, 2019. doi: 10.1109/COMST.2019.2895817.

32 L.Y. Lin, E.L. Goldstein, and R.W. Tkach. On the expandability of free-space micromachined optical crossconnects. *Journal of Lightwave Technology*, 18:482–489, 2000.

33 W.M. Mellette and J.E. Ford. Scaling limits of MEMS beam-steering switches for data center networks. *Journal of Lightwave Technology*, 33(15):3308–3318, 2015. doi: 10.1109/JLT.2015.2431231.

34 W. Merlijn van Spengen. MEMS reliability from a failure mechanisms perspective. *Microelectronics Reliability*, 43(7):1049–1060, 2003.

35 I. Stanimirović and Z. Stanimirović. Optical MEMS for telecommunications: some reliability issues. *Advances in Micro/Nano Electromechanical Systems and Fabrication Technologies*, IntechOpen, 2013. doi: 10.5772/55128.

36 T.R. Hsu. Introduction to reliability in MEMS packaging. In *Proceedings of International Symposium for Testing and Failure Analysis*, San Jose, California, November 5, 2007.

37 H.Y. Hwang et al. Flip chip packaging of digital silicon photonics MEMS switch for cloud computing and data centre. *IEEE Photonics Journal*, 9(3):2900210, 2017. doi: 10.1109/JPHOT.2017.2704097.

7

SOA-based Optical Switches

Xuwei Xue[1], Shanguo Huang[1], Bingli Guo[1], and Nicola Calabretta[2]

[1] State Key Laboratory of Information Photonics and Optical Communications (IPOC), Beijing University of Posts and Telecommunications, Beijing, China
[2] Eindhoven University of Technology, Eindhoven, Netherlands

7.1 Introduction

With the continuous advances in the experiments and development of the optical components and subsystems constituting the physical layer, capacity and speed in fiber-optic communication networks have increased over the past decades. Technologies of component miniaturization and integration that continue to improve the performance in the network while maintaining economic feasibility are of paramount importance. To achieve such extremely high data rates, utilizing semiconductor optical amplifiers (SOAs) as the key element in optical communication is a feasible scheme. Amplification of optical signals is not the only application SOAs are limited to. It is also validated performing as a significant element in high-speed optical switching, wavelength conversion, optical regeneration, demultiplexing, all-optical triodes, and all-optical logic doors. Benefits of SOAs are well-known including half-rack size, wide gain band, integration compared to the Erbium-doped Optical Fiber Amplifier (EDFA), especially their fast switching speed and the capability to modulate pulse signals with a high extinction ratio. However, the high coupling loss with optical fiber, mixed with the considerable noise and crosstalk which are sensitive to ambient temperature, is a crucial influence factor introduced by SOAs. The purpose of this chapter is to present metrics illustrations, design criteria, and performance enhancements for SOA-based switches. After our preface in this chapter, Section 7.2 will describe the structure of SOA. Subsequently, we will expound design criteria of SOA-based switches in Section 7.3. Moreover, improvements on SOA-based switches will be generalized in Section 7.4. In Section 7.5 and Section 7.6, the networks employing SOA-based switches, discussion, and future work will be summarized and proposed.

7.2 SOA Structure

The versatile structure of the SOA is comprised of an active region encompassed with passive cladding material, similar to the semiconductor laser. As depicted in Figure 7.1, in other words, the active region is the gain region, which is made of semiconductor materials such as INP. The main

Optical Switching: Device Technology and Applications in Networks, First Edition. Edited by Dalia Nandi, Sandip Nandi, Angsuman Sarkar, and Chandan Kumar Sarkar.

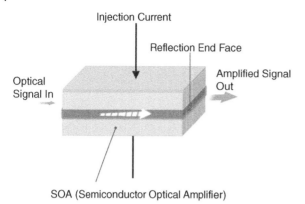

Figure 7.1 The amplifying process of SOA.

Injection Current

Reflection End Face

Optical Signal In

Amplified Signal Out

SOA (Semiconductor Optical Amplifier)

difference from semiconductor lasers is an anti-reflective coating to prevent reflection of the SOA end face and eliminate the resonator effect. A little clue, anti-reflective coating is arranged in the end face of a single or multilayer dielectric layer.

There are four types of SOAs: Fabry–Pérot SOA (FP-SOA), gain-clamped SOA (GC-SOA), traveling-wave SOA (TW-SOA), and quantum dot SOA (QD-SOA), where the third due to less or no internal reflection versus a large optical bandwidth and low polarization sensitivity is widely used [1].

7.2.1 Active Region

As it is described, the active region is the place where incident light is amplified. It is the result of the stimulated emission which occurred posteriorly the population inversion ($E_2 > E_1$). Actually, the process in the amplification begins with the injected electric currents. Meanwhile, these electrons in the heterojunction will be moved by the electric field to form a pair of electron holes, and when the incident light comes, the electrons will lose energy as photons and return to their ground state. Generally speaking, the active region is an attractive slide. Your children are raised by you (injection current), and they will slip when the wind (incident light) blows across the slide. Subsequently, following the children, it will generate the same frequency "wind" to enlarge the initial wind. The structure of polarization-insensitive SOA is shown in Figure 7.2.

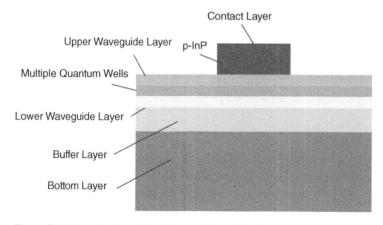

Contact Layer

Upper Waveguide Layer   p-InP

Multiple Quantum Wells

Lower Waveguide Layer

Buffer Layer

Bottom Layer

Figure 7.2 The structure of polarization-insensitive SOA.

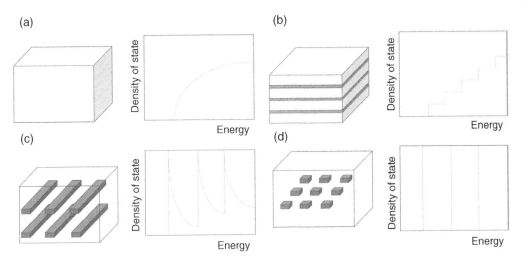

Figure 7.3 Different active region of SOAs: (a) bulk material, (b) quantum well, (c) quantum wire, (d) quantum dots.

In the SOA, the active region can be in form of bulk, quantum wells (QWs), quantum wire, and quantum dot [2]. As depicted in Figure 7.3(a), the bulk gain media density of state (DOS) increases as a parabola with the rising energy. With the improving doping technique and innovative structures, the performance of SOA keeps the momentum developing from quantum wells and quantum wires to quantum dots. The quantum well structures are constrained in one dimension, usually in the epitaxial growth direction. Figure 7.3(b) illustrates the correlation between the DOS and energy in a stepwise growing. Diverse with the quantum wells, carriers in quantum wires can only move freely in one direction due to the quantum constraints of two dimensions. Meanwhile, the delta-like correlativity of two factors is depicted in Figure 7.3(c). Furthermore, with the best property, the structure of quantum dots is similar to the atoms, where electrons and holes are quantum-restricted in three directions. The impulse line along with the energy level is shown in Figure 7.3(d).

7.2.2 Inter-Band Versus Intra-Band Transition

Based on the position of the electron transition, the excitation emission within the SOA is divided into two classes: inter-band and intra-band. Inter-band variation refers to the transfer of carriers in conduction and valence bands, as shown in Figure 7.4. When there is no injection current, most of the electrons are in the valence band, only a few electrons are transferred in it. Nevertheless, with the injection of current, the electrons drift to the conduction band while the electron density of the conduction band increases and holes in the valance band before the number of population inversion. They will spontaneously recombine with holes in sub-nanosecond time scales from the conduction band to the valence band, as well as releasing the energy by emitting photons. Moreover, the lifetime of the non-equilibrium free carrier is reduced to picosecond magnitude as a result of the stimulated emission process. Within the procedure of current injection, population inversion will occur between valence band and higher conduction band levels when the injection is beyond a threshold [3].

Because of the continuous intra-band energy levels, the injected hot carriers can quickly compensate for the reduced carriers due to stimulated emission by relaxation. The time of supplement

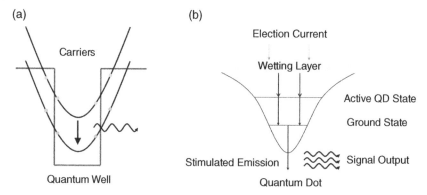

Figure 7.4 Intra-band and inter-band transition and stimulated emission of (a) QWs and (b) QDs.

is approximately 50 femtosecond or higher, much faster than the picosecond time scales of inter-band mechanisms. Hence, carriers in the whole active region are rapidly depleted while it ensues uniform gain saturation formed.

The aforementioned intra-band process is the ideal transition and recombination state of carriers, emitting phonons, however, there is an atypical non-radiation process called Auger recombination. It is a significant phenomenon with the recombination of electron-hole, where the excess energy from it is transferred to electrons or holes. Subsequently, the energy is excited to higher energy states within the same band instead of giving off photons (the radiative process). With high carrier or current density, the Auger recombination may play a crucial role in the resultant processing of gain droop.

7.2.3 Transparency Threshold

Transparency threshold in the SOA is the minimal current density of the gain media that the SOA with no amplification and absorption. Hence, if the injection current lies in or beyond the threshold current density, it can be described in a transparent state. The relationship between optical gain of SOA with the photon energy and the carrier density is quite complex. Whereas, we can simplify the whole relation in brief equations. I_{tA} is the threshold current in active region expressed in Equation 7.1:

$$I_{tA} \cong \frac{\alpha\left(\dfrac{1}{\tau_p \Gamma_\gamma} + N_{TR}\right)}{\tau_n},$$

(7.1)

where α is the electron charge multiplied by the volume of the active region, τ_p is the photon's lifetime, Γ is the confinement factor, γ is the differential gain coefficient multiplied with the group velocity, N_{TR} is the carrier density at the transparency condition, τ_n is the photon's lifetime. S_0 is the steady-state photon density above I_{TR} [4], as depicted in Equation 7.2.

$$S_0 \cong \frac{\Gamma \tau_p}{\alpha}\left(I_0 - I_{th}\right),$$

(7.2)

where I_0 depicts the current injection of the whole process into the SOA, meanwhile, I_{th} equals I_{tA} plus I_F, which is the leakage current. On the basis of the relationship between the density of states and the energy of the injected current in Figure 2.3, taking the quantum wells and the bulk materials as a comparison, it is effortlessly ascertained that there is a lower transparency current achieved in QW materials compared to bulk materials. In other words, it achieves higher concentration of carriers in the same current with a smaller volume by the enhanced confinement of the electrons/holes. Meanwhile, the size of conduction and valence band offset determine the efficacy of the constraint. To postpone the saturation point, a deeper QW is a convenient scheme, whereas the deeper wells fabricate other problems such as non-uniformity in carrier distribution across multiple wells. In the transparent pattern, moreover, several all-optical switching devices implement the on/off function by virtue of the transparency with nonlinearity. The concept and application of gain nonlinearity will be discussed in the next dot [5].

7.2.4 Gain Nonlinearity

The value of amplifier gain G is a significant parameter in elimination for the performance of SOAs. It is defined as the Equation 7.3:

$$G = \frac{P_{s,out}}{P_{s,in}} = \exp\left[\Gamma\left(g_m - \bar{\alpha}\right)L\right] = \exp\left[g(z)L\right], \tag{7.3}$$

G is increasing with device length, in turn with the reduction, when the internal gain is limited by gain saturation. Where g_m, α, and L are the material gain coefficient, the effective absorption coefficient of the material, and amplifier length respectively. In practice, the signal distortions are possibly introduced by SOA gain saturation, which manifests as the transmission quality degradation. In a feasible scheme, the adequate saturation output power is required at a high state to alleviate the influences of these distortions. The optical gain of SOA is defined in another form of the carrier density N by Equation 7.4:

$$G = \exp\left[\Gamma a\left(N - N_t\right)L\right], \tag{7.4}$$

where Γ is the confinement factor, a is the differential gain coefficient, N_t is the carrier density at transparency, and L denotes the SOA length. Therefore, the optical gain of SOA can be translated into the factor controlled by the carrier density generation.

With the nonlinearities of SOAs, the main types exploited in all-optical networks are cross-gain modulation (XGM), cross-phase modulation (XPM), self-phase modulation (SPM), and four-wave mixing (FWM) [1]. The saturation of SOAs is the basic requirement of XGM, which is, hence, converting the signal into another wavelength. In the comparison between XPM and XGM, the biggest difference may arise with the former where refractive index of the active region in SOA is not constant, which is relative to the carrier density and the material gain. The gain nonlinearity is always in used with the SOA-MZI configuration while the SPM compensates the pulse broadening after the transition in fibers. Additionally, FWM with the strong pump for the wavelength conversion can offer the phase and amplitude information of the signals whilst the weak probe signal is injected.

7.2.5 Polarization-Insensitive SOA

The gain polarization sensitivity of SOA can be explained in two ways: i) the different amplification for transverse electric (TE) polarization and for transverse magnetic (TM) polarization in waveguides even though the confinement to the active layer is comparable for the two states and ii) the anisotropy of the light-matter interaction in a quantum well [6]. The amplification of TE and TM modes is different, which is due to the rectangular shape and the crystal structure of the active region. It makes gain coefficient (g) and confinement factor (Γ) dependent on polarization.

There are many scenarios for fabricating polarization-insensitive SOAs. For example, the SOA may provide approximately polarization-insensitive characteristics with the intrinsic polarization-sensitive designs. It uses two identical amplifiers in cascades, where the difference of both amplifiers is the 90° included angle between them. Additionally, there are also two parallel amplifiers deploying for the two polarization directions. Furthermore, polarizing beam splitters (PBSs) are simultaneously deployed for these amplifiers front and back [7]. Besides, the polarization diversity circuit based (PDC-based) SOAs are several polarization-insensitive devices. One of the typical PDC-based SOA structures can be divided into three components: polarization-sensitive SOAs as mentioned before, half wave plates (HWPs), and PBSs, while the number of them is the same as mentioned before in two [8]. The PBS will split the incident signal into TE and TM components. Furthermore, another scheme is the use of a Faraday rotator between the device and the reflecting mirror with a bidirectional amplifier. Additionally, the balance of the gain between TE and TM mode is another feasible method. There is the introduction of tensile strain into the active layer to implement the envision. However, the requirement is an obstacle to the high-precision control of the thickness and the strain of the active layer. Meanwhile, the crystalline reliability of the active layer is also influenced with a mass of strain. Besides, the deployment of the polarization-dependent loss (PDL) is another approach. TE and TM polarized signal can be amplified with the application of one polarization-insensitive SOA. Nevertheless, inasmuch as the fiber has the predetermined length of the PDL unit, it introduces extra loss and the difficulty of integrating with other devices.

In the experiment of multiple-quantum-well structures of SOAs, the TE-polarized signal exceeds TM-polarized signal in the optical gain as observed in the use of typical self-assembled InAs/GaAs QDs, where the amplitude of output signal is affected with detrimental fluctuations [9]. Bulk and MQW achieve polarization insensitivity through choice of nearly square cross-sections of the active layers.

7.2.6 Noise in SOA

With the deficiency of analytical architecture of noise in the SOAs, we merely calculate the noise in detection or in our mathematical models [10]. To measure the noise, we will ascertain the definition of relative factors and relations. Firstly, Optical Signal to Noise Ratio (OSNR) is a key to represent the signal quality effected with the noise generated in long transmitted distance like amplified spontaneous emission (ASE) noise etc. This is expressed in a general formula in Equation 7.5:

$$OSNR = 10 * \log\left(S / N\right), \tag{7.5}$$

where S is the signal power and N is the noise power. Both of them are in watts or milliwatts. The higher ratio value of them manifests the better signal integrity after transforming the overall system, whose measurements are defined in IEC 61282-12 / b-IEC 61280-2-9 standards [11], illustrated in Equation 7.6:

$$OSNR_{\text{int}} = 10 \log\left(\frac{1}{B_r} \int_{\lambda_1}^{\lambda_2} \frac{s(\lambda)}{\rho(\lambda)} d\lambda\right), \tag{7.6}$$

where $s(\lambda)$ is time-averaged signal spectral power density without including ASE. Additionally, $\rho(\lambda)$ is the ASE spectral power density which is independent of polarization, both of which expressed in W/nm. B_r is the reference bandwidth expressed in nm with the default value 0.1 nm if there is no other statement [12]. To accommodate the amplification between the signal and noise, the OSNR degradation is expressed in terms of noise factor (NF) dB as Equation 7.7:

$$NF = \frac{OSNR_{INPUT}}{OSNR_{OUTPUT}} = \frac{G \times input\ noise\ level + added\ noise}{G \times input\ noise\ level}, \tag{7.7}$$

where G is the optical power gain of the device. Posteriorly, in general, with the higher driving current or pumping input power in the SOA, the noise will increase parabolically in the SOAs [13].

Moreover, electron density fluctuation in the SOA affects the noise characteristics in the lower frequency region than 100 MHz. Hence, corresponding to an inverse value of the S/N ratio, the relative intensity noise (RIN), which is supposed to be higher after passing through the SOAs, will reach a lower level than the RIN of the incident light when there is high optical input power beneath the 100MHz. Additionally, the RIN in the SOA including the ASE for operation is higher than RIN in other amplifiers like EDFAs with low optical input power, which synchronously get higher phase noise and frequency noise in the SOAs because of the larger optical absorption.

7.3 Design Criteria of SOA-Based Switch

Based on the different applications of SOA as the switch, differences should be highlighted in them. There are several SOA-based switches, for example, Terahertz Optical Asymmetric Demultiplexer (TOAD), the switches derived from the structure of Mach–Zehnder interferometer (MZI) like Colliding Pulse Mach–Zehnder (CPMZ) and the variation Symmetric Mach–Zehnder (SMZ) etc. The structure of them is illustrated in Figure 7.5. Furthermore, different features of SOA can be divided the switches in two categories, on the basis of the linear (gating) or nonlinear (wavelength conversion) operation task. Besides, all of them will make corresponding fine adjustments according to different functions. Briefly described, the temporal window of TOAD is controlled with the SOA relative position of the midpoint. Control pulses will introduce a differential phase shift in the Mach–Zehnder-based switches. Moreover, reduced current threshold with low confinement factors in the gating applications and the long active regions with high confinement factors for wavelength conversion are also considered.

7.3.1 Effect of Doping on Gain Dynamics

With the different doping techniques, it is categorized in p-doped (hole-doped) barriers and n-doped (electron-doped) barriers. In their applications, with the bulk of a certain level, the unsaturated optical gain can be greatly enhanced through p-type doping. Nevertheless, increasing p-type concentration will be associated with saturation density declining and an increase in the transparency current. In turn, the n-type doping used can possibly increase the saturation density and enhance the linearity of the SOA [14]. Not merely of these effects, the n-doping strongly influences the transparency level. Meanwhile, the amplifier noise enhancement factor n_{sp} is more effectively reduced by the n-doping than p-doping at a given excitation level [15], as illustrated in Figure 7.6 [16]. Meantime, there are various materials doped in the system. For instance, InAs/GaAs is the p-doping system, simultaneously, GaAs (AlGaAs) is the n-doping.

Above all, we will introduce the effect of p-doping on gain dynamics. Basically, the p-doping manifests the temperature insensitive threshold current, linewidth enhancement factor, high peak modal gain, and high modulation bandwidth [16]. Actually, the basic idea of p-doping is the way to fill the QD hole states. In a p-doped device, despite the occupation of electrons only

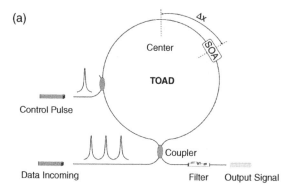

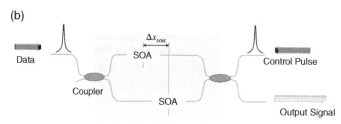

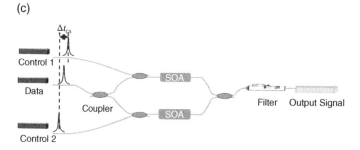

Figure 7.5 The structure of (a) TOAD, (b) CPMZ, (c) SMZ.

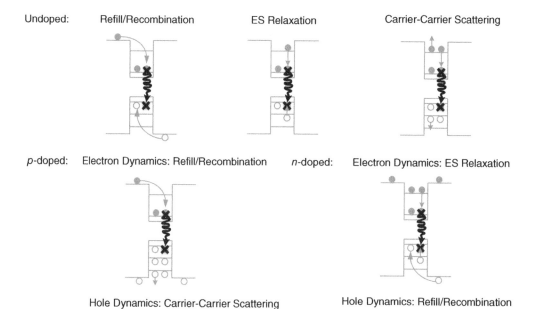

Figure 7.6 The differences between various doped material Source: 1–4: [16]/AIP Publishing; 5: modified from [16].

aggregated at the ground state (GS) due to the low injection of carriers, the built-in hole reservoir grants the occurrence of fast hole-hole scattering. Subsequently, the instant in the removed implementation of a GS electron-hole pair controlled by pump impulse, electrons will recover, while the unsaturated gain is enhanced. Nevertheless, in the p-doped InAs/GaAs QD amplifiers, such gain dynamics will be followed by slow electron dynamics on the 100 ps scale due to the lack of electron carriers in the higher energy states. In the experiment of Cesari et al. [17], it measures the ultrafast gain dynamics of the GS transition with the emission of 1.3 μm InAs/GaAs QD amplifiers in an electrical pump at room temperature. In fabrication under same conditions, the slower gain dynamics are observed in p-doped amplifiers compared with the undoped devices at the operation of the identical gain. In fact, the reduction of the excited-state electrons reservoir contributes to slower gain dynamics in p-doped quantum dot devices, which confines the recovery of the electron ground-state occupation mediated by intra-dot carrier-carrier scattering. Thus, it demonstrates that the benefits of p-doping are uncertain for the applications of ultrafast optoelectronics [16].

With the investigation of quantum dot vertical cavity semiconductor optical amplifier (QD-VCSOA), it is depicted that under the WL pumping operation, the gain and saturation threshold will increase with the acquisition of gain recovery dynamics [18].

7.3.2 Gain Dynamic for SOA

In the widest applications of high-speed systems, SOAs with faster gain dynamics are the best scheme in the expected future. To accurately describe the dynamic gain in SOA, we similarly use pump-probe spectroscopy technology to characterize the dynamic gain performance. In most experiments, the gain dynamic can be represented in the electron density, the recovery time, and the recombination time of the electron-hole pairs etc. On the basis of these characteristic analysis methods, the difference on gain dynamics between the configuration of co-propagating and counter-propagating for SOAs will be described briefly, and not merely this, but the different gain dynamics between the form of materials, the bulk, quantum well, and quantum dot amplifiers will be also organized in order subsequently.

The basic rate equation can be written in the same way as the propagation Equation 7.8:

$$\frac{dn}{dt} = \eta_I \frac{I}{qV} - R(n) - g(\lambda_{sig}, n) \frac{P}{\sigma} \frac{\lambda_{sig}}{hc} - g_{ase}(n) \frac{P_{ase}^+ + P_{ase}^-}{\sigma} \frac{\lambda_{ase}}{hc}, \tag{7.8}$$

where P_{ase}^+ and P_{ase}^- are the power of the forward- and backward-propagating ASE beam, respectively, λ_{ase} is a parameter only for the units to analyze, and σ is the cross-section of the gain region. I is the bias current, g_{ase} is modal gain of the ASE. Meanwhile, the effective value η_I is the efficiency coefficient of the injection of carriers, V is the volume of the active region, R(n) is the recombination rate totally, q is the constant of electron charge, and g is the modal gain. Meantime, P is the value of power and λ_{sig} is the wavelength of the input signal [19].

Based on this equation, according to experimental results, in the co-propagating configuration, it is observed that the gain for a bias current and input power before the threshold can be evaluated with the present overshoot in the gain recovery. Nevertheless, with the input power constantly increasing, another turning point brings the disappearance of the overshoot because of the recovery time rising. But with the differently injected wavelength, the performance and the effects will demonstrate distinctive properties. Moreover, in the counter-propagating configuration, the increase of injected power induces a considerable reduction in the recovery time. If the wavelength matches to the gain region, the improvement of counter-propagating will reduce as compared to the co-propagating scheme for high-injected power. Altogether, in the gain recovery, with the comparison to the two configurations in the same injected power, it is obvious that the counter-propagating with the overshoot is present, which is the cause of the carrier dynamics.

7.3.2.1 Bulk-Active Regions

In the experimental simulations, the bulk-active region should be able to achieve significantly 40 ps gain recovery time. Utilizing butt-joint growth (BJG) for bulk-active region definition, the bandgap of the bulk-active layer is 1565 nm and the layer thickness is 180 nm. The bulk-active region presents high confinement, which shows the excellent performance at 40 Gb/s [20]. Meantime, in order to use the SOA intra-band fast dynamics for high-speed applications, the slow gain compression should be kept to the smallest value possible for the slow dynamics of it. Furthermore, in the general of bulk-SOAs, the fast gain recovery time increases with increasing current and pulse energy [21]. Nevertheless, the time is slightly independent on the active region length. With varying the pulse energy, it is observed with the values of gain recovery time reducing. In fact, the different polarization waves of bulk SOAs are also containing the distinct gain recovery time [22]. In the TM axis, longer carrier recovery time is due to the fact that for a bulk tensile strained device there are more TM transitions than TE transitions because of the shifting of the sub-bands in the valence band. In this situation, the light holes giving rise to the TM transitions are favored.

7.3.2.2 Quantum Well/Multi-Quantum Well (MQW) Active Regions

The applications of MQW structures are related to the predictions of differential gain enhancement and of current threshold reduction with a high differential gain which has a positive impact on the performances on SOAs [23]. In MQW, there is also an overshoot occurring. At high current bias, in the region of the SOA where carriers are strongly depleted, as a sequence, the injection rate becomes slightly faster than the cumulative recombination rates. Moreover, another characteristic of MQW-SOAs will appear, the non-uniform distribution of carriers, under strong optical excitation. As a comparison, the tendency of results in experiment are similar. It indicates the QWs exchanges the necessary carriers to support stimulated recombination, creating a carrier flux toward the middle portion of the structure. The recombination procedures become faster as carrier density increasing to higher values, consequently, the smaller gain recovery time compared the MQWs, 170 ps. These recovery times are directly related to the maximum repetition rate at the performance of the amplifier in wavelength multiplexing.

7.3.2.3 Quantum Dots

In semiconductor quantum dot optical amplifiers, some researches have proposed some quantitative analysis methods. Furthermore, it is very different in the QD, and carrier dynamics in the QDs are critical for distinguishing these SOAs [24]. In these experiments, fast gain recovery (~100 fs) dynamics have been preciously observed in an InAs-InGaAs QD active waveguide [25]. To describe the relations in QD-SOA in details, the factors influenced the gain dynamics are exhibited below. The DOS for the QD, ρ_G, is extrapolated in Equation 7.9:

$$\rho_G = n_1 D_{3D} \varepsilon_G \frac{\gamma_H}{\eta_G^{FWHM}}, \tag{7.9}$$

where η_G^{FWHM} is a parameter of the influence of FWHM in the GS transition, the phenomenon manifests as inhomogeneous broadening. n_1 is the number of QD layers, decided after the fabrication in the device. D_{3D} is the density of 3D dot in a single dot layer versus the volume V. Moreover, ε_G is the degeneracy of the GS of a single dot, which is constant 2 because of spin. With the approximation to the GS transition, the majority of inhomogeneous broadening is arising from the same electron states in assumption.

The rate equations indicating the time revolution of the electrons are Equations 7.10 and 7.11:

$$\frac{dN_w}{dt} = \frac{1}{qV} - \frac{N_w}{\tau_w} - f_0^w \left(1 - f_e^G\right) \frac{\rho_G}{\tau_c} + \frac{f_e^G \left(1 - f_0^w\right) \rho_G}{\tau_{esc}}, \tag{7.10}$$

$$\frac{df_e^G}{dt} = \frac{f_0^W \left(1 - f_e^G\right)}{\tau_C} - \frac{f_e^G \left(1 - f_0^W\right)}{\tau_{esc}} - \frac{f_e^G}{\tau_G} - \frac{\hat{g}L}{V\rho_G}\left(f_e^G + f_h^G - 1\right)\frac{P_G}{\hbar\omega_0}, \tag{7.11}$$

where I is the injected current, and N_W is the carrier density of the wetting layer (WL). The latter is also standardized versus to the volume, meanwhile, τ_C is the capture time. τ_G and τ_W are, respectively, the spontaneous recombination time of the QD GS and WL. Furthermore, f_e^G and f_h^G are the electron and hole filling fractions of the QD states severally. $\hat{g}$ is the maximum modal gain, L is the length of one unit of the amplifier. P_G is the optical power of the signal injected at the center of the GS transition. f_0^W is the electron occupation probability at the edge of the WL, and the escape time is related to τ_C through considerations in a detailed balance.

For further illustration, the DOS of the QD holes ρ_h^D is validated as Equation 7.12:

$$\rho_h^D = \frac{1}{2\pi^2}\left(\frac{2m_h}{\hbar^2}\right)^{3/2}\sqrt{E - E_V^D}, \tag{7.12}$$

where m_h is the mass of the hole and E_V^D the location of the lowest QD hole state. ρ_h^{WL} is a traditional step function for a QW with a single bound state with the state located at E_V^{WL}. The Fermi energy of the valance band, $E_{F,V}$, is solved by following equation, while V_0^D is the volume of a single dot and V_W is the volume of the WL. The whole relationship between them is expressed in Equation 7.13.

$$f_e^G \rho_G + N_W = \int_{E_V^D}^{\infty} [V_0^D \rho_h^D \left(E\right)\frac{\rho_G}{\varepsilon_G} + \frac{V_W}{V} \rho_h^{WL}\left(E\right)\frac{1}{1 + \exp\left(\left(E - E_{F,V}\right)/k_B T\right)} dE \tag{7.13}$$

Measurements on QD devices have been observed with the consecutive varieties in the process of complete gain recovery in less than 1 ps. With analysis of results, it is demonstrated that the imbalanced states between the dot and wetting layer and the time scale of the equilibrium of electrons and holes decide the fast dynamics of quantum dot devices. Altogether, the gain dynamics on QD-SOA are better than the bulk and MQW. It is the major development tendency in the future.

7.3.3 Noise Suppression

There is a variety of noises in the SOA, like intensity noise, beat noise etc. The nonlinearity and saturation of optical gains in SOAs plays an important role for optical signal processing. To suppress the noise, there are two methods like using the gain-saturated SOAs or cascading the SOAs with the same level. But the bandwidth for the intensity noise reduction is limited by effective carrier lifetime, while it would still be favorable for few-Gbit/s optical communication systems [26].

In the suppression of intensity noise in SOA, the process is illustrated in Figure 7.7 [27]. For the beat noise, it exists in forms such as spontaneous-spontaneous beat noise, spontaneous emission

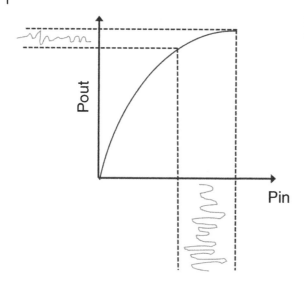

Figure 7.7 The noise suppression of SOA. Source: [27]/IEEE.

beat noise, and so on. At a high frequency, in other words beyond several GHz, increasing the SOA length is very useful for reducing the spontaneous emission beat noise in spectrum-sliced incoherent light and suppressing the intensity noise. The property of the aforementioned schemes has reached a 13 dB noise suppression ratio basically, which is enhanced by the cascading and lengthening the SOAs [26].

7.3.4 Scalability

The scalability of switches-based SOAs can be implemented in various schemes. One is the extension of a single stage, another is cascading more stages. Additionally, there are techniques to augment the capability of switching and expanding. For example, the enhanced extinction ratio is offered by the combination of SOA gates and an interferometer, which is a significant index for the scalable network. Moreover, the scaling of switched arrayed waveguide grating is also a feasible method.

Above all, it should first be pointed out: the potential capability for single-stage 8×8 switches at a data capacity of 10×10 Gbit/s is predicted to move switch performance to Tbit/s in a relatively simple single-stage switch fabric with a 1.6 dB power redundancy [28]. Nevertheless, the scalability to even higher levels of connectivity is confined by the incompatibility of existing waveguide crossing and waveguide bend techniques, which contributes to the lack of 8×8 switches and the focus on the effort of 1×8 monolithic connectivity [29].

Multi-stage in photonic networks is more complex than the implementation of broadcast-and-select architecture. As Figure 7.8 [29] illustrates, it is a switch network proposed to allow the scaling of four outputs per stage using the hybrid Clos/broadcast-and-select architecture [30]. The instance is interconnected with three stages to create the larger 16×16 network. It is noteworthy that an intermediate loss between each SOA gate in scalable stages limits the connectivity for Tbit/s.

In the experiment of White et al. [30], the power- and hardware-efficient scaling of switched optical networks has been demonstrated with the hybrid architectural approach combining Clos and broadcast-and-select routing. The Clos network balances the number of stages and complexity of each switch stage in a suitable case [30]. In the example depicted in Figure 7.9 [30], it is the universal architecture of a three-stage Clos network. Each twofold scaling in size for the n $\times$ n switch is implemented recursively with splitters around 2×2 switching elements.

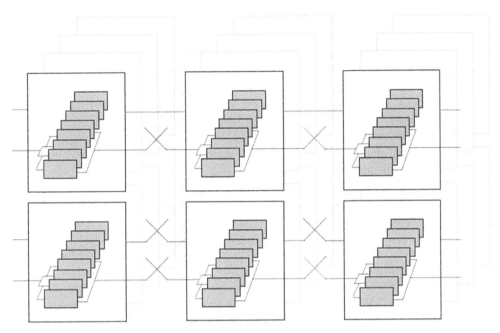

Figure 7.8 The scalable multi-stage switch architecture with SOAs. Source: Modified from [29].

Figure 7.9 The architecture of 32 × 32 Clos network. Source: Modified from [30].

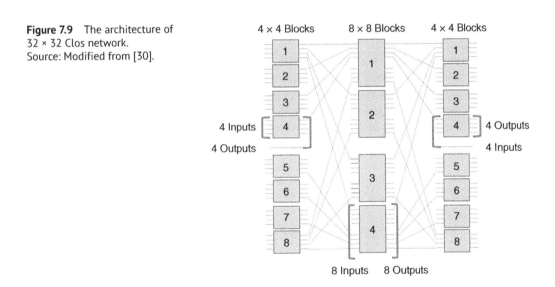

Physical layer scalability is another form of the network. Noteworthily, if we want to create multiport switch from SOA gates, it requires additional interconnecting passive circuit elements. Moreover, the impact of cumulative PDG on the physical layer scalability of space switch can be reduced through proper active region design for an expected SOA gain. However, it can be observed that though the measured noise figure (NF) is slightly smaller at higher gain, the SOA saturates more rapidly, leading to lower space switch scalability. There are many factors influencing the scalability. In our consideration of total power consumption, a larger number of cascaded SOAs at lower gain will not necessarily lead to better performance. While for the same number of ports, the architectures of a space switch determine the number of cascaded SOAs in the routing path and SOAs in the whole system.

Crosstalk will be taken into consideration with the disadvantages on scaling, which are due to non-ideal filtering and switching. The different order of crosstalk has various contributions on the scalability [31]. The interfering signal if suppressed once is called first-order crosstalk, of twice is called second-order crosstalk etc. To maintain the property of system after scaling, the crosstalk must eliminate or suppress in signal switching.

Not merely this, the feasibility demonstration of the scalability in the future is also a noteworthy topic. In the research of Cheng et al. [32], the 64 × 64 SOA-based optical switch has been theoretically substantiated with an experiment of an 8 × 8 SOA switch. In the simulation of the monolithically-integrated switch, the results depicted reaching the power penalty minimum as low as 1 dB with such a fully integrated switch.

7.4 Advancements on SOA-Based Switch

Before the chapter moves on to advanced SOA-based switches, it is worth concisely indicating the background to retain your perception of the theory of SOA switches. On the configuration of the electrical driving current, it is distinct of the control of the amplifier gain to deploy the on or off state, as shown in Figure 7.10. On the other hand, it is necessary to confirm our reference frequency, shown in Figure 7.11. The conception of switching rise time, which is the important figure of merit in the estimate of the optical space switches, is calculated on the 90–10% of steady-state output power. With the percentage of the exceeding steady optical signal, the overshoot is defined as the results of calculation. Moreover, the settling time is measured by the difference between the instant in steady level and the pulse start where the steady-level point is defined as the amplitude no longer higher than 5% of the steady-state level. Nevertheless, the stabilization point is correlated with the demand of different applications [33].

Afterward, techniques applied on the advancement and aspects are introduced respectively. Above all, the rise time is reduced by the preimpulse step-injected current (PISIC) in [34], which is based on the control of the injected current for on or off state in the SOA switch. The technique is deployed in the manipulation of carrier density and achieved by the injection of a narrow electric

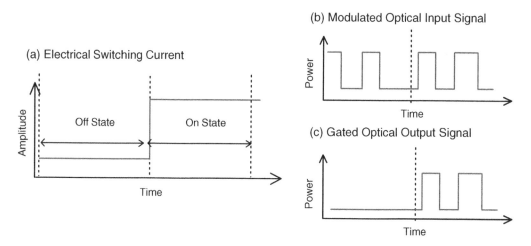

Figure 7.10 (a) The different electrical switching current drives SOA in on/off state. (b) The consecutive input of modulated optical signal. (c) The gated output signal comes after the pass-through of SOA biased by the switching current.

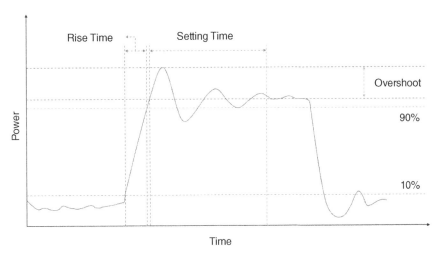

Figure 7.11 The definition of the rise time, settling time, and overshoot.

pulse before the valid switching signal impulse, which is implemented on the leading edge. The guarantee of sharp injection of electric carriers allows the increase of the carrier density in the active region of SOAs, which, in consequence, contributes to the faster transition of switching controls. The experimental results show an abrupt reduction of rise time from 2 ns to 200 ps in a bulk material of SOAs.

Nevertheless, to reduce the defect introduced by the PISIC, like the increase of the overshoot and a longer settling time etc., the multi-impulse step-injected current (MISIC) is proposed to possibly eliminate the practical impact of the potential results of SOA saturation and longer guard time [35]. The basic operation of the MISIC is the momentary injection of electrical carriers, which will compensate the different impairment in switches of various periods, for example, the undershoot, the gain oscillations, and the relaxation of carriers in SOAs. Altogether, the MISIC employs the extra carriers injected for the increase of the SOA optical gain. However, the stringent impulses synchronized and the evaluations of amplitude oscillation location and duration should be recognized. Moreover, the pulse intensity of every operated point and the higher energy consumption will be also the considerable effects on the implementation. Despite this, the applications of MISIC techniques are possibly waiting below the horizon.

Irrespective of which scheme can eventuate in the design on the basis of electrical switching, the signal is devoted to the compensation for oscillations of the electrical carriers in the SOA active region. Additionally, there are other scenarios like the enhancement of electro-optical performance through impedance matching and encapsulation optimization. The integration of the devices dramatically scales down while reducing the physical dimensions, improving the energy efficiency, and increasing the optical response bandwidth and stability. For example, the on-chip SOA-based optical space switch indicates it is competitive when used in practical applications in datacenters [36, 37].

To demonstrate this scheme, experimental results of the ultra-wideband in 2020 will be briefly described in this section. Renaudier et al. [38] developed the polarization diversity architecture, leveraging the single polarized SOAs to implement the 100+ Tb/s transmissions with tailor-made packaged devices. The schematic description of the structure is illustrated in Figure 7.12 [38]. Altogether, the complete integration of electro-optical parts is the most efficient manner for performance optimization in the application of SOA switches.

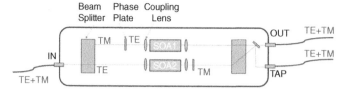

Figure 7.12 Schematic description for the polarization diversity architecture of SOAs. Source: [38]/IEEE.

7.5 Networks Employing SOA-Based Switch

Based on the commercial information, theses, and monographs, there are various architectures on the applications of SOAs. The classifications of networks may be based on the application scenario, or the role and position of SOA in the switches to separate them. In this section, the division will be on the basis of application scenarios and optical signal modulations. Moreover, each network deploying SOA-based switches will be depicted comprehensively.

7.5.1 Metro-Access Network

The work of Koenig et al. [39] demonstrated the cascade of at least four SOAs that can amplify the advanced-modulated optical signals in the format of BPSK, QPSK, 8PSK, and 16QAM in the convergence of optical metro networks and access networks. Furthermore, in the work of Calabretta et al. [40], the SOA-based photonic integrated wavelength division multiplexing (WDM) cross-connects for optical metro-access networks realized the switching speed by using an SOA. The main function of SOA enables the selection of any wavelength in the C band and compensates for the losses.

7.5.2 RF Network

Radio over fiber (RoF) is the technique to transmit the radio frequency signals with low dissipation. Shown in the paper of Qian et al. [41], 50 µs switching time has been attained while a proof-of-principle 10×10 switch is applied in a time-domain RoF network. It achieved a spurious-free dynamic range (SFDR) of 96.5dB·Hz$^{2/3}$ and an error vector magnitude (EVM) of 1.48%, moreover, based on which is scalable and cascaded enabled. Furthermore, the millimeter waves meet the definition of radio simultaneously, 30–300 GHz band. The nonlinear phenomena mentioned before like FWM, SPM, and XGM effects in SOA promote the performance improvement of switches in MMW-based 5G networks [42].

7.5.3 Silicon Photonic Switching

In the paper of Budd et al. [43], the design of a 4-channel SOA flip-chip based on the miscellaneous integration of III–V SOA is implemented in the silicon photonic switches. The photonic carrier with unique SiN waveguide coupling structures were arranged with individual SOAs using the precise flip-chip bonder (FCB), while a greater than 10 dB net SOA optical gain was recorded. Furthermore, the precise FCB is used in subsequent experiments like the thesis of Matsumoto et al. [44], in which the mode field on both the InP-SOAs and the Si waveguides was extended by spot-size converts (SSCs). The gain of the inline optical amplification based on the application of techniques mentioned and inline integration of SOAs with angled waveguides achieved 15 dB. Dramatically, the first demonstration of a lossless Si switch was achieved with the hybrid-integrated 4-ch SOA for the 4×4 Si switch. Not merely this, the experiment of Konoike et al. [45] is considerable, which fabricated even a four-stage transmission containing a bit error rate of the transmission signal below 20% forward-error-correction confinement with WDM channels.

7.5.4 Data Center Network

With the capacity demand of data center network (DCN), the applications of SOA-based switches are continuously increasing. In the concise introduction of SOA-based switches in Cheng et al. [46], the square principle increase of broadcast-and-select topology discourages scaling beyond 4×4 connectivity on the monolithic integration, which facilitates the application of stages. In the experiment, SOAs as on-chip gain in photonic WDM switches indicate the fast switching time and high contrast ratio of SOA, designed by Calabretta et al. [47]. Moreover, the SOA gates with nanoseconds switching completely compensated for the loss caused by the photonic integrated wavelength-selective switch (PIC-WSS), 8.1 dB, in the paper of Xue [48]. Not merely integrated in the switches, the SOA gain can take a role in the gain of compensating the loss of each buffer module in DCNs, as shown in the experiment of Singh et al. [49].

7.6 Discussion and Future Work

In view of the various researches on SOAs and SOA-based switches, this section will cover an outline of the matrix switch architectures, a variety of switches of different sorts, and SOAs used in spectrum-sliced systems. Altogether, the merits and impairments with the future work will be discussed.

The matrix architectures have been indicated by Stabile et al. [50] with space-division and wavelength division switching, which are classified into four types: (1) multi-stage, (2) cross-point, (3) broadcast-and-select (B&S) switch, and (4) wavelength-selective [51]. The basic point-to-point structure and point-to-multipoint switching architecture are both worth considering. Based on space division, the B&S has the potential to be implemented on the polarization division and mode division.

In the ideal SOA-based switch, the dynamic procedure is depended on the lifetime of carriers, which in the commercial devices are also influenced by the parasitic elements on switching velocity and overall property. Discussed in the research of Sutili et al. [33], the effects of parasitic elements are the quality of electrical switching pulses, the switch bandwidth reduction, and the dynamic stability, which are identified and quantified based the model presented by them. Moreover, a type of SOA-based polarization rotation switch on the basis of cross-polarization modulation in the communication of Raja et al. [52] is proposed and analyzed due to high extinction ratio and low switching energy. In the whole project, the presence or absence of light if polarized with a distinctive state are demonstrated as the feasible operation for the proposed designs. Additionally, on the basis of WSS technique, the optical network for the future adoption is studied by Martín González et al. [53]. The experimental results will be not cited particularly in this chapter, which manifest the performance of the SOA-based switch depending on the number of SOAs limits.

Furthermore, the thin-film facet integration is another lively research gate, which has the symmetrical thin-film resistive, thin-film microwave coupler etc. Take a brief description of both.

The former replaced the typical 47 Ω resistor in the thesis of Figueiredo et al. [54], which achieve the reduction of parasitic elements, 225 ps OFF-ON time, and 12% overshoot with the fast injection of PISIC, reaching the approximate Gaussian-response in the electro-optical spatial switching. Meantime, Sutili et al. [55] led the switch-time improvements and the reduction of settling time in a near 200 ps spatial switching time. Not merely this, coarse wavelength division multiplexing (CWDM) system with the impetus of capacity and distance extension requires the favorable performance of SOAs on the entire optical bandwidth (from 1260 nm to 1620 nm) meeting the demands. Additionally, the researches on SOA-based space switch of BER improvement and guard-time reduction through feed-forward filtering is also in the scope as Taglietti et al. [56].

Eventually, with the combination of SOAs, it may also be employed in spectrum-sliced systems in the future, manipulating as both noise suppression and other functions like wavelength conversion in WDM systems. Utilizing the optical nonlinearity properties of a saturated-SOA, it is a promising approach to reach the intensity noise suppression with the best suited performance as mentioned in the comparison of Jindal et al. [57]. The advantages of SOAs broadens the applications of SOA-based switches.

Altogether, with the development of hybrid architecture of SOAs and other techniques, the merits of integration, doping, nonlinearity, noise suppression, scalability, simplicity, high efficiency, and relative low cost will consecutively facilitate switching or other applications on the basis of SOAs.

Bibliography

1 D. Forsyth and F. Mahad. Semiconductor optical amplifiers: present and future applications. In N. Zulkifli, editor, *Recent Developments in Optical Communication and Networking*, Penerbit UTM Press, 2020.

2 T. Akiyama, M. Sugawara, and Y. Arakawa. Quantum-dot semiconductor optical amplifiers. *Proceedings of the IEEE*, 95:1757–1766, 2007.

3 A. Assadihaghi, H. Teimoori, and T.J. Hall. 6 – SOA-based optical switches. In: B. Li, S.J. Chua, editors. *Optical Switches*. Woodhead Publishing, 2010, 158–180.

4 R.C. Figueiredo, E.C. Magalhães, N.S. Ribeiro, C.M. Gallep, and E. Conforti, editors. Equivalent circuit of a semiconductor optical amplifier chip with the bias current influence. In *2011 SBMO/ IEEE MTT-S International Microwave and Optoelectronics Conference* (IMOC 2011, 29 Oct – 1 Nov 2011.

5 G.T. Kennedy, P.D. Roberts, W. Sibbett, D.A.O. Davies, M.A. Fisher, and M.J. Adams, editors. Intensity dependence of the transparency current in InGaAsP semiconductor optical amplifiers. *Summaries of papers presented at the Conference on Lasers and Electro-Optics*, 2–7 June 1996.

6 D.V. Taco, L. Daan, and B. Hans, editors. Polarization sensitivity of the amplification in semiconductor optical amplifiers. In *Proc. SPIE 2994, Physics and Simulation of Optoelectronic Devices V*, 6 June 1997. doi: 10.1117/12.275612.

7 D.R. Paschotta. Optical amplifiers based on semiconductor gain media. *Encyclopedia of Laser Physics and Technology*. www.rp-photonics.com/semiconductor_optical_amplifiers.html. 2008.

8 Z. Zhu, X. Li, and Y. Xi. A polarization insensitive semiconductor optical amplifier. *IEEE Photonics Technology Letters*, 28(17):1831–1834, 2016.

9 T. Kaizu, T. Kakutani, K. Akahane, and T. Kita. Polarization-insensitive fiber-to-fiber gain of semiconductor optical amplifier using closely stacked InAs/GaAs quantum dots. *Japanese Journal of Applied Physics*, 59(3):032002, 2020.

10 M. Yamada, editor. Noise in semiconductor optical amplifiers (SOA). In *2016 IEEE 6th International Conference on Photonics (ICP)*, 14–16 March 2016.

11 Vitextech. OSNR in Fiber Optic Communications. https://vitextech.com/osnr-meaning/.

12 OSNR: What does this mean; Why do we need and How to take care of it? 18 July, 2019. https://mapyourtech.com/entries/general/osnr-what-does-this-mean-why-do-we-need-and-how-to-take-care-of-it-.

13 M. Yamada. Analysis of intensity and frequency noises in semiconductor optical amplifier. *IEEE Journal of Quantum Electronics*, 48(8):980–990, 2012.

14 O. Qasaimeh. Effect of doping on the optical characteristics of quantum-dot semiconductor optical amplifiers. *Journal of Lightwave Technology*, 27(12):1978–1784, 2009.

15 K. Vahala and C. Zah. Effect of doping on the optical gain and the spontaneous noise enhancement factor in quantum well amplifiers and lasers studied by simple analytical expressions. *Applied Physics Letters*, 52:1945–1947, 1988.

16 V. Cesari, W. Langbein, and P. Borri. The role of p-doping in the gain dynamics of InAs/GaAs quantum dots at low temperature. *Applied Physics Letters*, 94:041110, 2009.

17 V. Cesari, W. Langbein, P. Borri, M. Rossetti, A. Fiore, S. Mikhrin et al. Ultrafast gain dynamics in 1.3μm InAs/GaAs quantum-dot optical amplifiers: The effect of p doping. *Applied Physics Letters*, 90:201103, 2007.

18 D. Razmjooei, A. Zarifkar, M.H. Sheikhi, editors. Effect of optical pumping to the wetting layer and excited state on the gain dynamics of QD-VCSOA: An equivalent circuit approach. In *2017 Iranian Conference on Electrical Engineering (ICEE)*, 2–4 May 2017.

19 G. Talli and M.J. Adams. Gain dynamics of semiconductor optical amplifiers and three-wavelength devices. *IEEE Journal of Quantum Electronics*, 39(10):1305–1313, 2003.

20 V. Lal, M.L. Masanovic, J.A. Summers, G. Fish, D.J. Blumenthal. Monolithic wavelength converters for high-speed packet-switched optical networks. *IEEE Journal of Selected Topics in Quantum Electronics*, 13(1):49–57, 2007.

21 L. Occhi, Y. Ito, H. Kawaguchi, L. Schares, J. Eckner, and G. Guekos. Intraband gain dynamics in bulk semiconductor optical amplifiers: measurements and simulations. *IEEE Journal of Quantum Electronics*, 38(1):54–60, 2002.

22 B.F. Kennedy, P. Landais, and A.L. Bradley, editors. Experimental analysis of polarization dependence of ultrafast gain dynamics in SOAs. In Proceedings of SPIE, 5825, Opto-Ireland 2005: Optoelectronics, Photonic Devices, and Optical Networks, 2005.

23 A. Reale, A.D. Carlo, and P. Lugli. Gain dynamics in traveling-wave semiconductor optical amplifiers. *IEEE Journal of Selected Topics in Quantum Electronics*, 7(2):293–299, 2001.

24 T.W. Berg, J. Mørk, and J.M. Hvam. Gain dynamics and saturation in semiconductor quantum dot amplifiers. *New Journal of Physics*, 6:178, 2004.

25 K. Kim, J. Urayama, T. Norris, J. Singh, J. Phillips, and P. Bhattacharya. Gain dynamics and ultrafast spectral hole burning in In(Ga)As self-organized quantum dots. *Applied Physics Letters*, 81:670–672, 2002.

26 T. Yamatoya, F. Koyama, and K. Iga, editors. Noise suppression and intensity modulation using gain-saturated semiconductor optical amplifier. In *Proceedings Optical Amplifiers and Their Applications*, 9 July 2000, Québec City: Optical Society of America.

27 F. Koyama, and H. Uenohara, editors. Noise suppression and optical ASE modulation in saturated semiconductor optical amplifiers. In *Conference Record of the Thirty-Eighth Asilomar Conference on Signals, Systems and Computers*, 7–10 November 2004.

28 T. Lin, K.A. Williams, R.V. Penty, I.H. White, M. Glick, and D. McAuley. Performance and scalability of a single-stage SOA switch for 10×10 Gbs wavelength striped packet routing. *IEEE Photonics Technology Letters*, 18(5):691–693, 2006.

29 R. Stabile and K.A. Williams. Photonic integrated semiconductor optical amplifier switch circuits. In P. Urquhart, editor. *Advances in Optical Amplifiers*. InTechOpen, 2011.

30 I.H. White, E.T. Aw, K.A. Williams, H. Wang, A. Wonfor, R.V. Penty. Scalable optical switches for computing applications. *Journal of Optical Networking*, 8:215–224, 2009.

31 L. Gillner, C.P. Larsen, M. Gustavsson. Scalability of optical multiwavelength switching networks: crosstalk analysis. *Journal of Lightwave Technology*, 17(1):58–67, 1999.

32 Q. Cheng, M. Ding, A. Wonfor, J. Wei, R.V. Penty, and I.H. White, editors. The feasibility of building a 64×64 port count SOA-based optical switch. *2015 International Conference on Photonics in Switching (PS)*, 22–25 September 2015.

33 T. Sutili, R.C. Figueiredo, B. Taglietti, C.M. Gallep, E. Conforti. Ultrafast electro-optical switches based on semiconductor optical amplifiers. In: A. Paradisi, R. Carvalho Figueiredo, A. Chiuchiarelli, and E. de Souza Rosa, editors. *Optical Communications: Advanced Systems and Devices for Next Generation Networks*. Cham, Springer International Publishing, 2019. p. 17–40.

34 C.M. Gallep and E. Conforti. Reduction of semiconductor optical amplifier switching times by preimpulse step-injected current technique. *IEEE Photonics Technology Letters*, 14(7):902–904, 2002.

35 R.C. Figueiredo, N.S. Ribeiro, A.M.O. Ribeiro, C.M. Gallep, and E. Conforti. Hundred-picoseconds electro-optical switching with semiconductor optical amplifiers using multi-impulse step injection current. *Journal of Lightwave Technology*, 33(1):69–77, 2015.

36 A. Wonfor, H. Wang, R. Penty, I. White. Large port count high-speed optical switch fabric for use within datacenters [Invited]. *IEEE/OSA Journal of Optical Communications and Networking*, 3(8):A32–A39, 2011.

37 R. Stabile, A. Albores-Mejia, and K.A. Williams. Monolithic active-passive 16×16 optoelectronic switch. *Optics Letters*, 37(22):4666–4668, 2012.

38 J. Renaudier, A. Arnould, A. Ghazisaeidi, D.L. Gac, P. Brindel, E. Awwad, M. Makhsiyan, K. Mekhazni, F. Blache, A. Boutin, L. Letteron, Y. Frignac, N. Fontaine, D. Neilson, and M. Achouche. Recent advances in 100+nm ultra-wideband fiber-optic transmission systems using semiconductor optical amplifiers. *Journal of Lightwave Technology*, 38(5):1071–1079, 2020.

39 S. Koenig, R. Bonk, H. Schmuck, W. Poehlmann, T. Pfeiffer, and C. Koos, et al. Amplification of advanced modulation formats with a semiconductor optical amplifier cascade. *Optics Express*, 22(15):17854–17871, 2014.

40 N. Calabretta, W. Miao, K. Mekonnen, and K. Prifti. SOA based photonic integrated WDM cross-connects for optical metro-access networks. *Applied Sciences*, 7(9):865, 2017.

41 Q. Xin, P. Hartmann, L. Sheng, R.V. Penty, and I.H. White, editors. Application of semiconductor optical amplifiers in scalable switched radio-over-fiber networks. In *2005 International Topical Meeting on Microwave Photonics*, 14 October 2005.

42 F. Saadaoui, M. Fathallah, A.M. Ragheb, M.I. Memon, H. Fathallah, S.A. Alshebeili. Optimizing OSSB generation using semiconductor optical amplifier (SOA) for 5G millimeter wave switching. *IEEE Access*, 5:6715–6723, 2017.

43 R.A. Budd, L. Schares, B.G. Lee, F.E. Doany, C. Baks, and D.M. Kuchta, et al., editors. Semiconductor optical amplifier (SOA) packaging for scalable and gain-integrated silicon photonic switching platforms. In *2015 IEEE 65th Electronic Components and Technology Conference (ECTC)*, 26–29 May 2015.

44 T. Matsumoto, T. Kurahashi, R. Konoike, K. Suzuki, K. Tanizawa, A. Uetake, et al. Hybrid-Integration of SOA on Silicon Photonics Platform Based on Flip-Chip Bonding. *Journal of Lightwave Technology,* 37(2):307–313, 2019.

45 R. Konoike, K. Suzuki, T. Inoue, T. Matsumoto, T. Kurahashi, A. Uetake, K. Takabayashi, S. Akiyama, S. Sekiguchi, S. Namiki, H. Kawashima, and K. Ikeda. SOA-integrated silicon photonics switch and its lossless multistage transmission of high-capacity WDM signals. *Journal of Lightwave Technology*, 37(1):123–130, 2019.

46 Q. Cheng, S. Rumley, M. Bahadori, and K. Bergman. Photonic switching in high performance datacenters [Invited]. *Optics Express*, 26(12):16022–16043, 2018.

47 N. Calabretta, K. Prifti, N. Tessema, X. Xue, B. Pan, and R. Stabile, editors. Photonic integrated WDM cross-connects for optical metro and data center networks. *ProcSPIE, Metro and Data Center Optical Networks and Short-Reach Links II*, 2019.

48 X. Xue, F. Nakamura, K. Prifti, B. Pan, F. Yan, and F. Wang, et al. SDN enabled flexible optical data center network with dynamic bandwidth allocation based on photonic integrated wavelength selective switch. *Optics Express*, 28(6):8949–8958, 2020.

49 P. Singh, J. Rai, and A. Sharma. Analysis of AWG-based optical data center switches. *Journal of Optical Communications*, 42, 2019. doi: 10.1515/joc-2019-0140.

50 R. Stabile, A. Albores-Mejia, A. Rohit, and K.A. Williams. Integrated optical switch matrices for packet data networks. *Microsystems & Nanoengineering*, 2(1):15042, 2016.

51 R. Soref. Tutorial: Integrated-photonic switching structures. *APL Photonics*, 3(2):021101, 2018.

52 A. Raja, K. Mukherjee, and J.N. Roy. Design analysis and applications of all-optical multifunctional logic using a semiconductor optical amplifier-based polarization rotation switch. *Journal of Computational Electronics*, 20(1):387–396, 2021.

53 L. Martín González, S. Van der Heide, X. Xue, J. Van Weerdenburg, N. Calabretta, and C. Okonkwo, et al. Programmable adaptive BVT for future optical metro networks adopting SOA-based switching nodes. *Photonics*, 5(3):24, 2018.

54 R.C. Figueiredo, T. Sutili, N.S. Ribeiro, C.M. Gallep, and E. Conforti. Semiconductor optical amplifier space switch with symmetrical thin-film resistive current injection. *Journal of Lightwave Technology*, 35(2):280–287, 2017.

55 T. Sutili, R. Figueiredo, N. Ribeiro, C.M. Gallep, and E. Conforti. Improvements evaluation of high-speed electro-optical integrated thin-film microwave coupler SOA-based space switch. *Journal of Microwaves, Optoelectronics and Electromagnetic Applications*, 17:477–485, 2018.

56 B. Taglietti, T. Sutili, R.C. Figueiredo, R. Ferrari, and E. Conforti. Semiconductor optical amplifier space switch BER improvement and guard-time reduction through feed-forward filtering. *Optics Communications*, 426:295–301, 2018.

57 J. Jindal, A. Kumar, and R. Kumar. Evaluation and analysis of different spectrum slicing techniques in free space optical systems. In *IOP Conference Series: Materials Science and Engineering*, 1033:012074, 2021.

8

Liquid Crystal Switches

Swarnil Roy[1,2] and Manash Chanda[2]

[1] IEEE SSCS Kolkata chapter, Kolkata, West Bengal, India
[2] Department of Electronics and Communication Engineering, Meghnad Saha Institute of Technology, Kolkata, West Bengal, India

8.1   Introduction

Bandwidth demand in modern communication systems calls for optical communication links and networks. Optical switches are an integral part of optical communication networks and optical sensing architecture because they can dynamically control the optical path connection. But if there is any optical to electrical conversion in the path that would crunch the bandwidth, all-optical switches are preferred in that context. Also, optical switches with no moving parts ensure high stability and reliability.

But it is also important to identify specifically where to apply liquid crystal (LC) technologies, as they have limitations as regards response time, which is not less than several microseconds. The main focus of this chapter is the purpose of LC optical switches in space switching (telecom and sensor applications), in protection and recovery applications, and optical add/drop multiplexing, which need fewer restrictions about switching time.

Networks that can detect a failure and have an additional path in order to maintain the transmission when failure is detected are called protection and recovery networks. Optical add/drop multiplexer (OADM) is an optical multiplexer which insert or extract optical wavelengths to or from the wavelength division multiplexing (WDM) optical transmission stream. Literatures have also referred to some reconfigured version of OADM (ROADM) [1]. These can be building blocks for an optical cross-connect (OXC); OXC is a switching matrix for incoming optical wavelength, where it provides a path from one input optical channel to any output channel. As an example, it is reported that the switching time of an OXC is within a few tens of milliseconds [2]. It should be mentioned also that the determination of the minimum response time required for WDM, transport network restoration, or flexible bandwidth allocation depends on several network management and service-related issues. On the other hand, LC-based switches are not suitable for packet switching applications which require a faster switching time (in the nanosecond range) [3]; hence, they will not be considered here.

Optical Switching: Device Technology and Applications in Networks, First Edition. Edited by Dalia Nandi, Sandip Nandi, Angsuman Sarkar, and Chandan Kumar Sarkar.

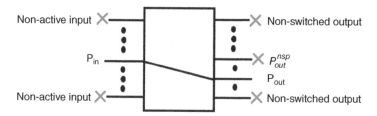

Figure 8.1 General optical switch block diagram.

Before getting into LC optical switching trend, it is important to have a brief understanding of the optical parameters [4]. For that purpose, we are considering only one active input with incoming power (Figure 8.1).

Insertion loss (IL): This is the ratio between the input signal power to output signal power, which is a measure of lost signal power between an input and an output-connected port of the switch. This loss is measured in decibels as given in the expression below and must be as small as possible.

$$IL = 10log\frac{P_{out}}{P_{in}} \tag{8.1}$$

Crosstalk: This is the fraction of the power leaked to the non-switched output port from the input port. It is a measure of the signal interference between channels. This ratio should be low. Here P_{out}^{nsp} is the power of the non-switched port.

$$CT = 10log\frac{P_{out}^{nsp}}{P_{in}} \tag{8.2}$$

Switching time: Switching time is generally expressed in terms of IL. It is the time required for IL of the switch path to achieve 90% of its final value from the moment the switching command is placed.

Polarization-dependent loss (PDL): This is measured as peak-to-peak difference between orthogonal states of polarization for transmitting light. Optical switches must have low PDL (typically <0.5 dB).

Power consumption: The electrical power required for the switching operation.

Bit Rate: The number of bits per second that the optical switch can manage to transmit.

Polarization Mode Dispersion (PMD): It occurs due to the fact that in a channel various states of polarization travel at slightly different speeds. When they pass through the switch these polarized lights create a lag between each other and that leads to PMD.

LC switches, in general, have the following advantages: they offer very low IL – less than 6 dB to access 40–80 channels. They also provide good CT (better than −40 dB) [5], low PDL and power consumption, and higher transmission capability (bi rate). They have no moving parts for switch reconfiguration and the technology has matured over time. What makes LC switches so attractive is that it provides all these at the same time, whereas other technologies could have one or two of these advantages but not all. Also, bearing in mind the recommendations of the Kyoto Protocol and Intergovernmental Panel on Climate Change, low power consumption has become essential and there is a need to reevaluate and reduce Information and Communication Technologies' Impact on the Energy Footprint; and LC in switching matrices might do just that.

In the next topics, we are going to review LC materials properties and principles, then a description of the main types of switches based on the mechanisms used for steering light, their parameters, and specific applications are presented.

8.2 Liquid Crystal and Its Properties

Liquid crystal (LC) has proved to be an important functional organic material in various hi-tech products surrounding us. The history of LCs can be traced back to the discovery of an optical property in cholesterol. Despite the appearance of a cloudy fluid, cholesterol displays an optical anisotropy (birefringence) like a crystal below a critical temperature T_c.

LCs essentially are substances whose molecular order, depending on the ambient temperature, is intermediate between that of crystalline solids and that of amorphous liquids. Some of their properties (like dielectric constant and refractive index etc.) are like crystalline solids. That is why the optical properties of LC can be modified using applied electric field.

Liquid crystal materials generally have several common characteristics. Among these are a rod-like molecular structure, rigidness of the long axis, and strong dipoles and/or easily polarizable substituents. These sharp anisotropic rod-like molecular structure (in some cases they are disc like) found in LC are called mesophases [6]. The orientation of these rod-like structures depends on temperature. As temperature is decreased to be low, the molecules show an orientational order with the molecular axis oriented in a particular direction, but they do not have a positional order (nematic phase, N). Decreasing the temperature further, the molecules show more and more ordered and layered structure, ultimately tending towards complete crystalline solid (Sematic-A, Sematic-C, Crystalline solid). On the other hand, this regular layered orientation can be achieved by applying electrical potential.

8.3 LC Structures for Optical Switching

As mentioned above, LC, having no movable part, is a promising candidate for optical switching in telecommunication. In this section we would discuss different LC structure for optical switching. Most popular structures are twisted nematic (TN) devices and surface-stabilized ferroelectric liquid crystals (SSFLC) [7–10]. Another structure based on polymer-dispersed liquid crystal (PDLC) has also been reported in recent literature [11–13].

8.3.1 Twisted Nematic (TN) cells

Though TN liquid crystals are the mostly used in the display systems, it can be used for optical switching as well. In these cells both planar and perpendicular alignment can be used. Because of its twisted nature, the LC molecules perform a 90° twist through the thickness of the LC cell. Thus, a linearly polarized light goes through a 90° phase shift provided it satisfies *Mauguin condition*: $d.\Delta n \gg \dfrac{\lambda}{2}$, where d is the cell thickness, Δn is the LC birefringence and λ is the light wavelength.

In display devices with TN cells, there are two linear polarizers in between LC layer. Depending on their orientation, TN cells can be made in either a normally black mode or normally white mode. In most display applications, two crossed polarizers are placed with the transmissive axis of each polarizer parallel to the rubbing direction of each alignment layer, operation of a TN device in this mode is normally white (NW) mode. In OFF state (no applied voltage) the incident light is transmitted, but when a voltage is applied molecules reorient themselves and block the light. A LC cell from an electrical point of view, acts as a capacitor with a non-ideal dielectric material whose electrical equivalent can be obtained using an experimental procedure based on the impedance spectroscopy technique reported in [14–16]. The equivalent circuit consists of voltage-dependent capacitor (C_{LC}) with series (R_S) and parallel (R_P) resistors as shown in Figure 8.2. The magnitude of C_{LC} for 1 cm^2 TN cell with of 5 µm thick LC is also depicted in Figure 8.2. The threshold (usually 1–2 V) and switching (usually 3–5 V) voltages can be derived from this electrical modelling.

C_{LC} varies with applied voltage due to the dielectric permittivity modifications as a result of molecular reorientations. At low voltages (below V_{th}) the capacitance is constant, but a nonlinear variation can be obtained if the voltage is increased after that. Finally, the capacitance remains almost constant for voltages greater than V_{sw}. Power consumption and response time of the TN device estimated from equivalent circuit simulation are of the order of nW and a few tens of ms respectively. The transmission factor for monochromatic polarized light can be calculated using Jones matrices [17] given as:

$$\frac{T}{T_0} = \frac{\sin^2\left(\frac{\pi}{2}.\sqrt{1+u^2}\right)}{1+u^2}, \qquad (8.3)$$

where $u = \dfrac{2\Delta nd}{\lambda}$ and T_0 = maximum transmission.

Typically, $u = \sqrt{3}$ is used in practical devices, since a smaller u gives a smaller cell gap and a faster response speed [18]. It should be mentioned that once contrast is maximized for a wavelength, optical transmission would vary for other λ. But this dependence can be reduced.

A simple polarization-independent liquid crystal Fresnel lens using the surface-mode switching of 90 twisted nematic liquid crystals (TN-LCs) is reported in recent literature [19]. This proposed structure reduces the transition time to a few milli-seconds under the surface-mode switching.

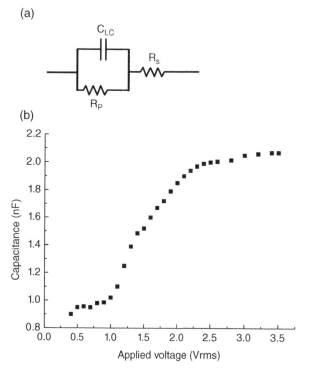

(a)

(b)

Figure 8.2 (a) Electrical equivalent of TN cell and (b) variation of capacitance with applied voltage.

Response time can be further improved by the use of low-viscosity LC materials and the transient nematic effect [20]. A hybrid-aligned nematic LC structure displays promising sub-millisecond response time [21]. Response time of such structures is reported as 0.75 mS experimentally. Another sub-millisecond response time has been reported for Bidirectional Field Switching Mode using inverse TN structure [22].

8.3.2 Surface-Stabilized Ferroelectric Liquid Crystal (SSFLC) Cells

In these devices, the ferroelectric LC material is sandwiched between two very thin (a few micrometers each) LC layers. A variety of molecular orientations exists for SSFLC [23], but the bistable bookshelf layer structure is most common one. In these materials, there is a tendency to create a unwanted helical formation due to macroscopic spontaneous polarization which must be stabilized [24]. The response time is a few microseconds but with a memory effect (bistability). In the presence of an electric field, the molecular orientation changes which is a rotation of the smectic cone driven by coupling between the polarization and the electric field, the device remains in this state until a reverse polarity voltage is applied. An electrical modeling of these devices that predicts switching voltage (a few volts), power consumption (nW), and response time (a few microseconds) in such devices as a function of fabrication parameters is reported in [25].

When an SSFLC device is placed between crossed polarizers, with one of them parallel to the molecular axis of one of the stables states, one of the two states will be black. Optical transmission can be calculated using the Jones calculus, and is given by:

$$T = T_0 sin^2\left(4\theta\right).sin^2\left(\frac{\pi\Delta nd}{\lambda}\right).$$

(8.4)

Here T_0 is the maximum transmission between parallel polarizers, θ is the cone angle of material (optimal value is 22.5°) and d is the thickness of the device which should be maintained at $d = \lambda/2\Delta n$ for optimal working. Like TN cells, the device performance is optimal for a given λ, but compared to TN cells optical transmission for other wavelengths varies much abruptly.

8.3.3 Spatial Light Modulator (SLM) Cells

A SLM modulates an optical beam's amplitude, phase, or polarization, using the birefringence properties of the LC cell in one or two dimensions. 2D LC SLMs are mostly electrical but optically addressed analog light valves are also proposed [26]. The spatial structure of an electrically addressed SLM is shown in Figure 8.3. The pixel pitch p is defined as spacing between two adjacent pixels and pixel gap i is the edge-to-edge spacing between adjacent pixels. Assuming square pixels, the geometrical fill factor F is defined as the ratio $(p/i)^2$ and this parameter puts an upper bound to the SLM optical efficiency. Most LC SLMs are panels consisting of a LC layer aligned between two glass sheets, with a thin-film transistor (TFT) control circuitry. Major drawbacks are rather large pixels and moderate 'fill factor' (<60%) due to large dead areas between TFTs. Lack of flatness is also a problem specially if the modulated beam is coherent.

More compact SLMs are obtained with the liquid crystal on silicon (LCOS) technology [27, 28]. The device structure consists of a LC layer sandwiched between a reflective silicon backplane and a transmissive counter-electrode shown in Figure 8.4.

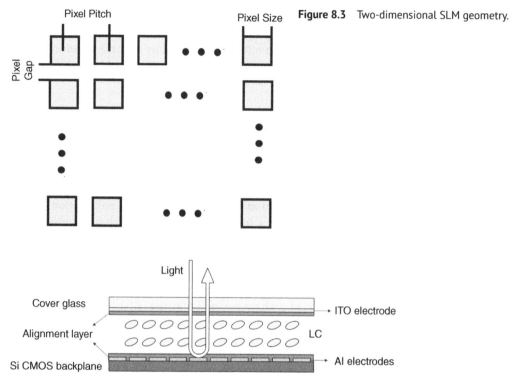

Figure 8.3 Two-dimensional SLM geometry.

Figure 8.4 LC on silicon SLM geometry. Source: [29]/MDPI/CC BY-4.0.

Using VLSI integration process, light and flat displays with high-definition (1920 × 1080), high-resolution (pixel pitch < 8 μm), and high fill factor (>95%) can be manufactured. Though main applications of LCoS SLMs are high-definition display, beam-steering for optical tweezers [30], optical switching matrices, or wavelength-selective switches (WSS) [31] can be implemented using this as well. But the characteristic of LCoS must be taken into consideration while designing optical fiber system [32].

8.4 Liquid Crystal Switches

Liquid crystal switches can be distinguished based on the light steering mechanisms with LCs, such as reflection, wave-guiding, polarization management, or beam-steering (planar or volume). Some of them are summarized in this section and compared against their switching parameters and applications.

8.4.1 Optical Crystal Switching Architectures

Optical space switches can be implemented in two architectures, namely, broadcast-and-select (BS) and space routing (SR). In BS switching, the incoming light information from a channel is split to all output channel array through an intermediate blocking stage. Switching the blocking stages ON and OFF would then select the desired output channel. In Figure 8.5 we can see a possible architecture for BS switches. We can see that from source, light information is going to channels 2, 3, and 4 via the blocking stage and due to the status of the blocking stage only channel 2

Figure 8.5 Broadcast-and-Select (BS) switch architecture.

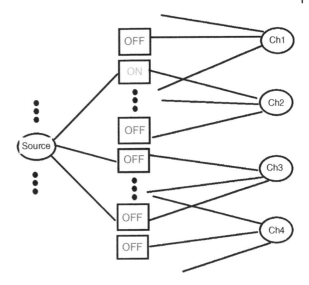

receives the signal. If the blocking stage is implemented by amplitude modulators, high contrast ratio needs to be maintained to avoid crosstalk. This scheme suffers from large complexity of the intermediate stage when the number of channel increases (2D input array require 4D selection mask) and a very poor power budget which forces optical amplification in this case.

The SR scheme is more suited for LC switching. In driving the information from an input to an output channel it exploits an orderly arranged SR intermediate element. If the number of switching elements is limited, this scheme can limit the IL as for the Benes or Banyan topologies [33].

Implementation of SR switches by individual light steering can be exercised in two ways: a multi-stage planar topology using arrays of 2 × 2 binary polarization switches [34] or single- and dual-stage schemes in which SR is performed by beam-steering in free space [35].

8.4.2 Switches Based on Polarization

The basic principle of TN displays using LC switches is polarization rotation configuration [36]. This switch works utilizing the property of LC cells that change the polarization state of the incident light when an electric field over the LC cell is applied. Along with the LC TN cell, a polarization beam splitter (PBS: generally implemented with calcite crystals) would allow us to implement an optical space router. Each polarization mode is considered parallelly to make the device polarization-insensitive and to minimize losses. It means that input signal is decomposed into TE and TM components, which are separately recombined at the switch output.

Figure 8.6 shows a possible architecture for space routing using LC switch and PBS [37]. Here light coming from source 1 falls on the first PBS and spilt into *P-S* polarizations. *P* polarization passes through to mirror 1 and *S* polarization is reflected and hots mirror 2. If LC cells are unbiased, again *P* passes through and *S* is reflected at the next PBS stage so they recombined and goes to channel 1. On the other hand, if LC cells are biased then *P-S* both rotates and they recombine at channel 2. But for source 2, if bias is applied information goes to channel 1 otherwise to channel 2.

A similar 1 × 2 space routing architecture (Figure 8.7) is reported in [38]. In this optic switch, there are two output states: the first one is straight state from port 1 to port 2 and a normal state from port 1 to port 3. For shifting from straight to normal states in the architecture, a 90° NLC polarization switch (PS) is used. There are three sets of lenses for collimating and focusing the light

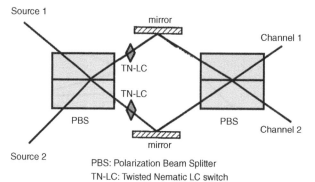

Figure 8.6 A possible architecture of 2 × 2 optical routing using beam splitter and TN-LC switch.

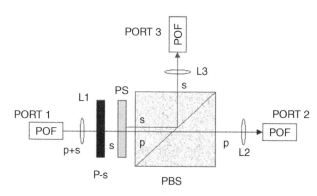

Figure 8.7 A simple diagram of 1 × 2 optical routing architecture using one beam splitter and TN-LC switch.

and a polarization beam splitter. When no voltage is applied to the LC cell information (*P* polarized light) goes to port 2 which is straight state. But by applying a voltage greater than V_{sw} to the LC cell, the polarization of light is reversed and light is forced to port3.

In these architectures transmission of information is controlled by the applied voltage of LC cells and lower voltages induce less polarization shifts. Thus, these switches can be employed as variable attenuators (VOAs). An applied voltage V between V_{th} and V_{sw} ($V_{th} < V < V_{sw}$) splits the input signal at both outputs with a variable ratio depending on the applied voltage.

8.4.2.1 Performance Analysis of Polarization-Based Switch Architecture

Based on the working principle discussed, different configurations can be created depending on the type and the number of elements used, like polarizing beam splitters and calcite plates, optional elements are: mirrors, half wave plates, quarter-wave plates, half-angle prisms, right-angle prisms, beam displacement prisms, total internal reflection prisms, birefringent crystals etc. Evolution of the state-of-the-art showing different implementations and their characteristics are shown in Table 8.1. Most of these consist of free-space optics elements, e.g. lenses for coupling light. Only a few of them use fiber-optic devices [39].

The table shows that FLC material switches produce the lowest response time which is in μS range. Using PLZT (lead lanthanum zirconate titanate EO material) ferroelectric material the switching response can be reduced to sub-microseconds (switching time of the LC material is ~100 ns) [50]. Zhang et al. [51] reports an integrated waveguide using LC material and can be operated as a binary switch or attenuator within 30–60 °C temperature range. NLC material produces response time in the

Table 8.1 Performance comparison of different architecture using LC switch.

Reference	Cell type	Architecture type	Wavelength (nm)	CT (dB)	IL (dB)	Response time	Voltage
[36]	TN-LC	1 × 2	633	−20	0.4	—	2.5V
[40]	TN-LC	2 × 2	632.8	−27	2.5	50/100mS	5V
[41]	TN-LC	2 × 2	633	−32	3	—	6V
[42]	NLC-FLC	2 × 2	—	−20	1.4	250μS	15V$_{rms}$
[37]	FLC	1 × 4	633	−21.6	3.5	50 μS	—
[43]	FLC	6 × 6	820/670	—	11.1	150 μS	—
[44]	TN-LC	2 × 2	1300	−43.3	2.2	—	—
[7]	FLC	2 × 2	1300	−34.1	6.94	35.5 μS	—
[45]	FLC	2 × 2	1550	−40	6.76	35.5 μS	—
[46]	NLC	1 × 2	650–850	−22	7	~mS	8V
[47]	NLC	3 × 1	650–850	−23	3	20–5mS	3V
[48]	NLC	3 × 2 (dual)	650–850	−20	—	13–5mS	5V
[49]	NLC/FLC	2 × 2	808	−36.2	2.5	60.6 μS / 35 μS	5V
[50]	FLC	1 × 6	850	—	—	< 1 μS	10V
[51]	FLC	—	1550	—	<4.3	<100 μS	—
[52]	NLC	1 × 2	1550	−20 (max)	8(max)	1mS	6.5V

order of ms. The table indicates usually it is in the range of mS in NLC-based optical cross-connect switches. A reduction of response time (60μS) can be obtained by combining NLC cells and the transient nematic effect (TNE) [49].

NLC switches operate in a wider wavelength range than FLC switches. Due to thickness of FLC cell there is a maximum wavelength in which the polarization shift is 90°. On the other hand, to create the polarization NLC switch cell thickness can be modified at first or second minimum for a multiband operation fulfilling Mauguin's regime. A broadband 3 × 1 reconfigurable optical multiplexer, from 650 to 850 nm, is presented in [47]. A 3 × 2 TN-LC multiplexer-based graded-index plastic optical fibers (GI-POF) has an operating range of wavelength from 850 to 1300 nm [48]. Another TN-LC based systems offer good performance in 1530–1560 nm range (C-band) [39].

Switch IL depends on the structure of the device. In that context, as simpler switches can manage only one type of polarization, so higher ILs are expected, 3.5 dB in [37] or higher. Using polarization diversity management complex switches exhibit low IL, 1.4 dB [42].

Optical switches with IL less than 1 dB, crosstalk from −20 to −45 dB, low PDL (around 0.1–0.2 dB) and low power consumption (~nWs) can be created, but it is not specifically reported in the literature.

8.4.3   LC Amplitude and Phase Modulator

The critical performances of the PolRot cells are contrast and bandwidth. The TN contrast is determined by the minima of Gooch and Tarry's law [17] as previously reported in Equation 8.3. The first minimum solution is $2\Delta nd/\lambda = \sqrt{3}$. The LC cell thickness of 3.8 μm (± .1 μm) generates 30 dB contrast at 1.5 μm wavelength. So, a clean cell assembly technology combined with the wide catalogue of nematic mixtures is perfectly suitable for C band telecom applications using PolRot switches [9].

It is an interesting and promising approach to use already existing *polymer-dispersed liquid crystals* (PDLC) as a variable switch between optical fibers [11, 53]. PDLC devices are a type of thin film that change their ability to pass the light through them in response to an electrical excitation. The PDLC film consists of an array of LC droplets (having a radius of the same size as the wavelength) filled polymer matrix. Liquid crystal molecules within each droplet have localized uniformity (meaning the nematic is uniformly aligned inside that droplet only); but from droplet to droplet, the nematic directors are random; thus, polarization is independent (Figure 8.8). Without the external voltage this structure scatters light and the switch is in the OFF state. With an applied voltage, nematic molecules align parallel to the electric field, and the structure becomes transparent (provided the refractive index of the polymer is close to the LC refractive index (n_0)) and the switch is in the ON state. An example of the most common PDLC mixture would be by weight 80% TL205 LC to 20% of PN393 monomer. The required contrast modification for a given wavelength is performed by sizing of the LC droplets as: $\Delta n.a/\lambda = 0.3$ where a is the radius of LC droplet [54]. The droplet size can be varied by changing the power of UV polymerization of the monomer.

For the phase modulation used in beam deflection gratings, the main issues are diffraction efficiency and response time. Phase modulation can be attained with parallel configuration of NLC cell with a 5° pre-tilt standard polyimide material. The desired phase shift of such a cell is related to parameters of the parallel NLC cell as: $\Delta\phi = 2\pi.\Delta n.d/\lambda$. To achieve a 2π phase shift at 1.5 μm wavelength LC cell size would be 7.5 μm with a 0.2 Δn. The diffraction efficiency can reach its theoretical limit for two-phase level grating which is 40%. Along with this, a reflective configuration would decrease the response time by a factor of 4, as it is proportional to viscosity, inversely proportional to the dielectric anisotropy and the square of the electric field. The birefringence Δn

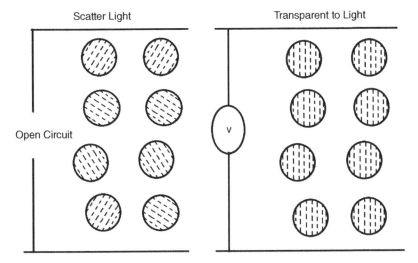

Figure 8.8 PDLC as switch without and with applied voltage.

typically decreases around 20% for telecom wavelengths compared to the visible range value because LC birefringence, irrespective of the material, reduces with increasing wavelength according to Cauchy law $n_\lambda = n_\infty + b/\lambda^2$ (b, specific for each mixture). For proper operation Δn should be as high as possible e.g., BL009 from Merck produce 0.29. Also, elastic splay constant (K_{11}) must be as low as possible.

In the arrangements above using NLC cells the response time is relatively large (typically around 20 ms). As discussed in previous section, FLC devices also display bi-level amplitude or phase gratings and the advantage is a symmetrical and fast response time (around 10–100 µs). But the demerit of using FLC material in this case is its inherent low Δn value (less than 0.15); so FLC material like SCE13 gives diffraction efficiency up to the range of 20%. Due to a thick cell size for telecom wavelengths, unstable LC alignment can be found with time. Special fiber arrays with both one- and two-dimensional non-uniform pitch are reported in literature to prevent the coupling of the parasitic diffraction orders [55, 56].

NLC cells capability of producing the gray scale permits a multi-phase grating and that results to high diffraction efficiencies [57]. For a N number of phase level which is blazed grating or saw-tooth profile, we can calculate the efficiency theoretically as η(N) in [58]:

$$\eta(N) = \left[\operatorname{sinc}(1/N) \right]^2. \tag{8.5}$$

This result gives η = 81% for N = 4 and η = 91% with N = 6, which means for smooth angle adaptations at each new routing configuration, numerous gray levels are needed. Experimentally measured diffraction efficiency with 1.55 µm wavelength reaches 80% for N = 6 [59]. Using an overdriving addressing scheme, the long response times can be reduced down to 50 ms rise time and 2 ms decay time [60]. Rise time can be further reduced by a factor of 4 by using reflective arrangement in LCOS implementation. NLC are better suited for LCOS application because of their inherent stable planar alignment [28].

The above discussion proves that LC modules can be used as amplitude and phase modulator. In binary grating ferroelectric LC material can be used for its high-speed response. As nematic LCs are effective for a wide range of wavelength (visible to IR), they are extensively used for Polarization

Rotation switches and high diffraction grating. Even with all these advantages and the fact that this technology is based on organic materials, LC material based optical switches are yet to be fully accepted in telecom industry.

8.4.4 LC-Based Wavelength-Selective Switches (WSS)

Devices like wavelength blocker [61] and wavelength-selective switches [62] made it possible to design efficient ROADM and OXC. Use of these devices are reported to be more flexible and less complex than optical space routing switches that has very intricate multiplexer and demultiplexer stages. Structurally a WSS subsystem can be viewed as an integrated 1 × N OXC stage (Figure 8.9). It allows a specific wavelength to move to one of the N output port from the input optical fiber. Power adjustment (mainly attenuation) for any possible power imbalance of the input WDM multiplex of individual routed channels can also be performed at that stage. A number of WSS-based implementations are present in the literature [63] but the most appreciated of them all are LCOS-based modulator [64] and MEMS-based system [65].

A generic WSS scheme consists of with an ultrashort optical pulse shaper [66], micro-beam displacement between a pair of high-resolution wavelengths dispersing diffraction gratings is reported. The first proposed solution for a WSS scheme, involving beam shifting by polarization rotation, is reported in [67]. The input polychromatic signals are split parallelly, then beamlets are angularly dispersed by 1st grating and individually focused by the lens as separate spots on the LC pixels of a TN array. It is then passed through a polarization selective deflector (Wollaston prism or calcite plate). As LC is ON, the wavelength is passes through the second lens without polarization rotation, after that when it falls to 2nd grating it recombines and goes to output channel 1. On the other hand, if LC is OFF, the same thing happens but with a polarization rotation which results in output going to output channel 2. The intermediate spatial shift is dependent on focal length of the lens and the distance between the output fibers. To achieve low CT and PDL values, polarization rotation and path length should be kept the same for the beamlets of the two polarizations of a given optical channel. One of the two eigen-polarizations of an input channel is rotated such a way that both beams have the same polarization and experience the same loss by crossing through the same optical components. It was first demonstrated in [67] for an eight-wavelength switch with 4 nm separation. The result shows −25 dB crosstalk and about 10 dB of IL. A 17-channel 1X2 WSS architecture with 2.8 mm × 6.5 mm of device size is also reported in the literature, where IL and crosstalk are 21 dB and −21 dB respectively [68]. Improvement of this architecture can lead up to 5 dB and −35 dB of IL and crosstalk for 80 channels. This architecture can be used for WSS with 10–50 Gbit/sec.

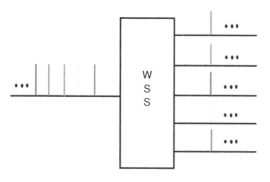

Figure 8.9 A conceptual diagram of 1 × 5 WSS system.

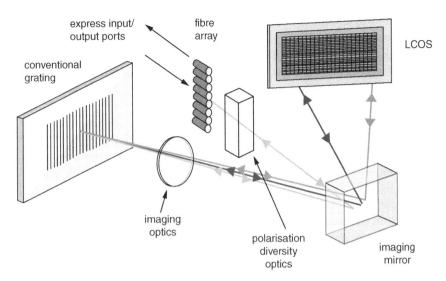

Figure 8.10 WSS architecture using LCOS beam steering. Source: [31]/IEEE.

8.4.4.1 WSS Based on LCOS

Figure 8.10 shows a wavelength-selective switch architecture using LCOS which provides integrated switching and optical power control in literature that include polarization diversity, control of mode [31]. In this system, the incident light from a single optical fiber is reflected from mirror and gets angularly dispersed by the grating array. This dispersed light is again reflected by the mirror and directed towards a specific portion of LCOS. This way the light of a different channel is pointed to a different section of LCOS. The same grating is used for multiplexing and demultiplexing around dynamic beam steering by LCOS, MEMS, and other elements. The wavelengths are dispersed along one dimension, and the orthogonal axis is used for port selection and possible amplitude control by beam shifts around the output fiber core center.

A multi-pole multi-throw (MPMT) WSS based on LCoS is presented in [69]. The MPMT function is implemented by the beam splitting with holographic phase modulation on the LCOS. It reports a 3–4.1 dB IL, maximum port crosstalk less than −20.0 dB. Both of these parameters can be further improved by the phase modulation technique. Xie et al. [70] report a *Finer-Grid* WSS based on LCoS for all-optical wavelength. To get a finer-grid WSS, the focal length is increased here rather than decreasing the pixel size of LCoS. Based on the proposed architecture a 1 × 9 WSS was designed and tested against DPSK, QOPSK, and 16QAM signals. Authors state that both the bandwidth setting resolution and grid granularity are improved from 12.5 GHz to 6.25 GHz. Yang et al. [71] demonstrates a 1 × 12 WSS module based on a 4k LCoS device with holographic 2D beam-steering capability and a laser-written 3D waveguide array. It achieves an average insertion loss of −8.4 dB, crosstalk of −26.9 dB with effective passband width of 43.4 GHz.

8.5 The Future of LC switches

8.5.1 Liquid Crystal Photonic Crystal Fibers

In recent years photonic crystal fibers (PCFs) caught the attention of scientific and commercial community as new class of optical waveguides. PCFs are microstructured waveguides (in silica or polymer) with air holes located in the cladding region. In PCFs, light can be guided either by effective index mechanism related to the modified total internal reflection (TIR) or through light confinement by the photonic band gap (PBG) phenomenon. By introducing different impurities into the air holes, the propagation parameters of PCF can be tuned; in this regard LCs seem to be particularly interesting substances to infiltrate PCFs since their refractive indices can be relatively easily modified either by temperature or by an external electric field. Such a new class of micro structured fibers is known as photonic liquid crystal fibers (PLCFs).

By using the field-dependent molecular orientation of LC, the refractive index contrast of the liquid and the fiber material can be modulated, which leads to light scattering (opaque state) or light propagation (transparent state). Thus, PLCFs allow for switching between a transparent and an opaque state. Tunable optical switches based on PLCF having a −60 dB of crosstalk, 1 dB of insertion loss with thermal tunability have been demonstrated [72]. On the other hand, PLCFs can be tuned electrically also, which allows for switching between two PBGs that depends on an ordinary refractive index of LC in OFF state and extraordinary index in ON state. This behavior offers tuning of PLCFs characterized by smooth changes in PBG positions [73, 74]. A wide and flat operating range from 600 to 1700 nm, optical switches based on PLCF with small power levels of 10 mW is reported [75], where IL is 3 dB and the CT is less than −20 dB.

8.5.2   Ring Resonators with LC

Micro-ring resonators (mRRs) are very compact devices that can be used as routers [76] and WSS [77]. The resonant frequencies of these switches are manipulated by changing the equivalent loop length by carrier injection [78], local heating and absorption [79]. To change refractive index thermally (by introducing a heater close to the resonator) may induce problems related to power dissipation when many resonators have to be integrated in a DWDM multiplexing system. This effect is reduced by the use of silicon-on-insulator (SOI) wafers with an electrical tuning and nematic LCs as the waveguide side cladding, achieving a tuning range of 0.22 nm [80]. The high refractive index contrast available in SOI allows compact mRRs with low loss and high Q.

Bibliography

1 K.G. Vlachos, F.M. Ferreira, and S.S. Sygletos. A reconfigurable OADM architecture for high-order regular and offset QAM based OFDM super-channels. *Journal of Lightwave Technology*, 37 (16):4008–4016, 2019. doi: 10.1109/JLT.2019.2905141.

2 R. MacDonald, L.P. Chen, C.X. Shi, and B. Faer. Requirements of optical layer network restoration. *Proceedings of Optical Fiber Communications Conference*, 3:68–70, 2000. doi: 10.1109/OFC.2000.868525.

3 Y. Liu, E. Tangdiongga, Z. Li, S. Zhang, M.T. Hill, J.H.C. van Zantvoort, F.M. Huijskens,de H. Waardt, M.K. Smit, A.M.J. Koonen, G.D. Khoe, and H.J.S. Dorren. Ultra-fast all-optical signal processing: towards optical packet switching. *Proceedings of SPIE: Optical Transmission, Switching and Subsystems IV*, 6353:635312, 2006. doi: 10.1117/12.687147.

4 G.I. Papadimitriou, C. Papazoglou, and A.S. Pomoportsis. Optical switching: switch fabrics, techniques and architectures. *Journal of Lightwave Technology*, 21(2):372–384, 2003. doi: 10.1109/JLT.2003.808766.

5 S. Hardy. Liquid-crystal technology vies for switching applications. *Journal of Lightwave Technology*, 44–46, 1999.

6 S. Chandrasekhar. *Liquid Crystals* (2nd edn). New York: Cambridge University Press, 1992.

7 N.A. Riza and S. Yuan. Low optical interchannel crosstalk, fast switching speed, polarization independent 2 × 2 fiber optic switch using ferroelectric liquid crystals. *Electronics Letters*, 34(13):1341–1342, 1998.

8 C. Vázquez, J.M.S. Pena, and A.L. Aranda. Broadband 1 × 2 polymer optical fiber switch using nematic liquid crystals. *Optics Communication*, 224(1–3):57–62, 2003.

9 F. Pain, R. Coquillé, B. Vinouze, N. Wolffer, and P. Gravey. Comparison of twisted and parallel nematic liquid crystal polarisation controllers. Application to 4 × 4 free space optical switch at 1.5 µm. *Optics Communications*, 139(4–6):199–204, 1997.

10 H. Yamazaki and M. Yamaguchi. 4 × 4 Free-space optical switching using real-time binary phase-only holograms generated by a liquid-crystal display. *Optics Letters*, 16(18):1415–1417, 1991.

11 P.C. Lallana, C. Vázquez, B. Vinouze, K. Heggarty, and D.S. Montero. Multiplexer and variable optical attenuator based on PDLC for polymer optical fiber networks. *Molecular Crystals and Liquid Crystals*, 502:130–142, 2008.

12 G.M. Zharkova, A.P. Petrov, V.N. Kovrizhina, and V.V. Syzrantsev, "Enhancing the luminophore emission of chiral polymer-dispersed liquid crystals. *Journal of Luminescence*, 194:480–484, 2018. doi: 10.1016/j.jlumin.2017.10.046.

13 V.N. John, S.N. Varanakkottu, and S. Varghese. Flexible, ferroelectric nanoparticle doped polymer dispersed liquid crystal devices for lower switching voltage and nanoenergy generation. *Optical Materials*, 80:233–240, 2018. doi: 10.1016/j.optmat.2018.05.003.

14 E. Barsoukov and J.R. Macdonald., *Impedance Spectroscopy: Theory, Experiment and Applications*. New Jersey: Wiley-Interscience, 2005.

15 J.M.S. Pena, I. Pérez, I. Rodríguez, C. Vázquez, V. Urruchi, X. Quintana, J. De Frutos, and J.M. Otón. Electrical model for thresholdless antiferroelectric liquid crystal cells. *Ferroelectrics*, 271:149–154, 2002.

16 I. Pérez, J.C. Torres, V. Urruchi, J.M.S. Pena, C. Vázquez, X. Quintana, and J.M. Otón. Voltage controlled square waveform generator based on a liquid crystal device. In *XVII Conference on Liquid Crystals*. Augustow, Poland, 2007.

17 C.H. Gooch and H.A. Tarry. The optical properties of twisted nematic liquid crystal structures with twist angles ≤90°. *Journal of Physics D: Applied Physics*, 8:1575–1584, 1975.

18 D.K. Yang and S.T. Wu. *Fundamentals of Liquid Crystal Devices*. Chichester, John Wiley, 2006.

19 C. Lin, H. Huang, and J. Wang. Polarization-independent liquid-crystal fresnel lenses based on surface-mode switching of 90° twisted-nematic liquid crystals. *IEEE Photonics Technology Letters* 22(3):137–139, 2010. doi: 10.1109/LPT.2009.2036738.

20 I.C. Khoo and S.T. Wu. *Optics and Nonlinear Optics of Liquid Crystals*. Singapore, World Scientific, 1993.

21 T. Choi, J. Kim, and T. Yoon, Sub-millisecond switching of hybrid-aligned nematic liquid crystals. *Journal of Display Technology*, 10(12):1088–1092, 2014. doi: 10.1109/JDT.2014.2346175.

22 S.P. Palto, M.I. Barnik, A.R. Geivandov, I.V. Kasyanova, and V.S. Palto. Submillisecond inverse TN bidirectional field switching mode. *Journal of Display Technology*, 12(10):992–999, 2016. doi: 10.1109/JDT.2016.2574929.

23 S.T. Lagerwall. *Ferroelectric and Antiferroelectric Liquid Crystals*. Weinheim: Wiley-VCH, 1999.

24 N.A. Clark and S.T. Lagerwall. Submicrosecond bistable electro-optic switching in liquid crystals. *Applied Physics Letters*, 36:899–901, 1980.

25 J.R. Moore and A.R.L. Travis. PSpice electronic model of a ferroelectric liquid crystal cell. *IEEE Proceedings – Optoelectronics*, 146(5):231–236, 1999.

26 G. Moddel, K.M. Johnson, W. Li, R.A. Rice, L.A. Pagano-Stauffer, and M.A. Handschy. High-speed binary optically addressed spatial light modulators. *Applied Physics Letters*, 55:537–639, 1989.

27 I. Underwood, D.G. Vass, and R.M. Sillitto. Evaluation of an nMOS VLSI array for an adaptive liquid-crystal spatial light modulator. *IEE Proceedings, Part J: Optoelectronics*, 133:77–83, 1986.

28 A. Lelah, B. Vinouze, G. Martel, T. Perez-Segovia, P. Geoffroy, J.P. Laval, P. Jayet, P. Senn, P. Gravey, N. Wolffer, W. Lever, and A. Tan. A CMOS VLSI pilot and support chip for a liquid crystal on silicon 8 × 8 optical cross-connect. *Proceedings of SPIE: Wave Optics and VLSI Photonic Devices for Information Processing*, 4435:173–183, 2001.

29 Huang Y, Liao E, Chen R, Wu S-T. Liquid-crystal-on-silicon for augmented reality displays. *Applied Sciences*, 8(12):2366, 2018. doi: 10.3390/app8122366.

30 W. Hossack, E. Theofanidou, J. Crain, K. Heggarty, and M. Birch. High-speed holographic optical tweezers using a ferroelectric liquid crystal microdisplay. *Optics Express*, 11:2053–2059, 2003.

31 G. Baxter et al. Highly programmable wavelength selective switch based on liquid crystal on silicon switching elements. *2006 Optical Fiber Communication Conference and the National Fiber-Optic Engineers Conference*, 2006. doi: 10.1109/OFC.2006.215365.

32 K. Heggarty, B. Fracasso, C. Letort, J.L. de Bougrenet de la Tocnaye, M. Birch, and D. Krüerke. Silicon backplane ferroelectric liquid crystal spatial light modulator for uses within an optical telecommunications environment. *Ferroelectrics*, 312:39–55, 2003.

33 C. Yu, X. Jiang, S. Horiguchi, and M. Quo. Overall blocking behavior analysis of general banyan-based optical switching networks. *IEEE Transactions on Parallel and Distributed Systems*, 17(9):1037–1047, 2006.

34 K. Hogari, K. Noguchi, and T. Matsumoto. Two-dimensional multichannel optical switch. *Applied Optics*, 30(23):3277–3278, 1991.

35 S. Fukushima, T. Kurokawa, and M. Ohno. Real-time hologram construction and reconstruction using a high resolution SLM. *Applied Physics Letters*, 58:787–789, 1991.

36 R.E. Wagner and J. Cheng. Electrically controlled optical switch for multimode fiber applications. *Applied Optics*, 19(17):2921–2925, 1980.

37 L.R. McAdams, R.N. McRuer, and J.W. Goodman. Liquid crystal optical routing switch. *Applied Optics*, 29(9):1304–1307, 1990.

38 C. Vázquez, J.M.S. Pena, and A.L. Aranda. Broadband 1 × 2 polymer optical fiber switches using nematic liquid crystals. *Optics Communications*, 224(1–3):57–62, 2003. doi: 10.1016/S0030-4018(03)01716-4.

39 S. Sumriddetchkajorn and N.A. Riza. Fiber-conectorized multiwavelength 2 × 2 switch structure using a fiber loop mirror. *Optics Communications*, 175:89–95, 2000.

40 R.A. Soref. Low-cross-talk 2 × 2 optical switch. *Optics Letters*, 6(6):275–277, 1981.

41 R.A. Soref and D.H. McMahon. Calcite 2 × 2 optical bypass switch controlled by liquid-crystal cells. *Optics Letters*, 7(4):186–188, 1982.

42 L.R. McAdams, R.N. McRuer, and J.W. Goodman. Liquid crystal optical routing switch. *Applied Optics*, 29(9):1304–1307, 1990.

43 G.J. Grimes, L.L. Blyler, A.L. Larson, and S.E. Farleigh. A plastic optical fiber based photonic switch. *Proceedings of SPIE: Plastic Optical Fibers*, 1592:139–149, 1991.

44 Y. Fujii. Low-crosstalk 2 × 2 optical switch composed of twisted nematic liquid crystal cells. *IEEE Photonics Technology Letters*, 5(6):715–718, 1993.

45 N.A. Riza and S. Yuan. Reconfigurable wavelength add-drop filtering based on a banyan network topology and ferroelectric liquid crystal fiber-optic switch. *Journal of Lightwave Technology*, 17(9):1575–1584, 1999.

46 C. Vázquez, J.M.S. Pena, and A.L. Aranda. Broadband 1 × 2 polymer optical fiber switch using nematic liquid crystals. *Optics Communications*, 224(1–3):57–62, 2003.

47 P.C. Lallana, C. Vázquez, J.M.S. Pena, and R. Vergaz. Reconfigurable optical multiplexer based on liquid crystals for polymer optical fiber networks. *Opto-Electronics Review*, 14(4):311–318, 2006.

48 P.C. Lallana, C. Vázquez, D.S. Montero, K. Heggarty, and B. Vinouze. Dual 3 × 1 multiplexer for POF networks. *Proceedings of International Conference on Plastic Optical Fibers*. Torino. 33–36, 2007.

49 J. Yang, X. Su, X. Liu, X. He, and J. Lan. Design of polarisation-independent bidirectional 2 × 2 optical switch. *Journal of Modern Optics*, 55(7):1051–1063, 2008.

50 X. Hu, O. Hadeler, and H.J. Coles. Ferroelectric liquid crystal mixture integrated into optical waveguides. *Journal of Lightwave Technology*, 30(7):938–943, 2012. doi: 10.1109/JLT.2012.2184738.

51 F. Zhang, H.-H. Chou, and W.A. Crossland. PLZT-based shutters for free-space optical fiber switching. *IEEE Photonics Journal*, 8(1):7800512, 2016. doi: 10.1109/JPHOT.2015.2511090.

52 T. Li, Q. Chen, and X. Zhang. Optofluidic planar optical cross-connect using nematic liquid-crystal waveguides. *IEEE Photonics Journal*, 10(4):6601417, 2018. doi: 10.1109/JPHOT.2018.2853759.

53 J. Qi, L. Li, and G. Crawford. Switchable infrared reflectors fabricated in polymer-dispersed liquid crystals. *OFC 2003 Optical Fiber Communications Conference*, 2003, 43–44. doi: 10.1109/OFC.2003.1247477.

54 D. Bosc, C. Trubert, B. Vinouze, and M. Guilbert. Validation of a scattering state model for liquid crystal polymer composites. *Applied Physics Letters*, 68:2489–2490, 1996.

55 B. Fracasso, J.L. de Bougrenet de la Tocnaye, L. Noirie, M. Razzak, and E. Danniel. Performance assessment of a liquid crystal multichannel photonic space-switch. *Photonics in Switching (Technical Digest)*, 24–26, 2001.

56 C. Letort, B. Vinouze, and B. Fracasso. Design and fabrication of a high-density 2D fiber array for holographic switching applications. *Optical Engineering*, 47(4):045401-1–045401-9, 2008.

57 W. Klaus, S. Shinomo, M. Ide, M. Tsuchiya, and T. Kamiya. Efficient beam deflector with a blazed liquid crystal phase grating. *Proc Opt Comp. Sendai.* 28–29, 1996.

58 J.W. Goodman and A.M. Silvestri. Some effects of Fourier-domain phase quantization. *IBM Journal of Research and Development*, 14:478, 1970.

59 N. Wolffer, B. Vinouze, R. Lever, and P. Gravey. 8 × 8 holographic liquid crystal switch. *Proceedings of SPIE: European Conference on Optical Communication*, III, 2000:275–276, 2000.

60 A. Tan, A. Bakoba, N. Wolffer, B. Vinouze, and P. Gravey. Improvement of response times of electrically addressed nematic liquid crystal blazed gratings. *Proceedings of SPIE: European Conference on Optical Communication*, 2000, 4089: 208, 2000.

61 M. Vasilyev, I. Tomkos, J.K. Rhee, M. Mehendale, B.S. Hallock, B.K. Szalabofka, M. Williams, S. Tsuda, and M. Sharma. Broadcast-and-select OADM in 80 × 10, 7 Gbit/s ultra-longhaul networks. *IEEE Photonics Technology Letters*, 15:332–334, 2003.

62 J.K. Rhee, F. Garcia, A. Ellis, B. Hallock, T. Kennedy, T. Lackey, R.G. Lindquist, J.P. Kondis, B.A. Scott, J.M. Harris, D. Wolf, and M. Dugan. Variable pass-band optical add-drop multiplexer using wavelength selective switch. *Proceedings of European Conference on Optical Communication*, 2001. Amsterdam. 550–551, 2001.

63 P. Bonenfant and M. Loyd. OFC 2003 workshop on wavelength selective switching based optical networks. *IEEE Journal of Lightwave Technology*, 22(1):305–309, 2004.

64 Yunshu Gao, Xiao Chen, Genxiang Chen, Ying Chen, Qiao Chen, Feng Xiao, Yiquan Wang, 1 × 25 LCOS-based wavelength selective switch with flexible passbands and channel selection, *Optical Fiber Technology*, 45:29–34, 2018. doi 10.1016/j.yofte.2018.05.008.

65 D.M. Marom, D.T. Neilson, D.S. Greywall, C.S. Pai, N.R. Basavanhally, V.A. Aksyuk, D.O. Lopez, F. Pardo, M.E. Simon, Y. Low, P. Koodner, and C.A. Bolle. Wavelength-selective 1 × K switches using free-space optics and MEMS micromirrors: theory, design and implementation. *IEEE Journal of Lightwave Technology*, 23(4):1620–1630, 2005.

66 J.P. Heritage, A.M. Weiner, and N.R. Thurston. Picosecond pulse shaping by spectral phase and amplitude manipulation. *Optics Letters*, 10:609–611, 1985.

67 J.S. Patel and Y. Silberberg. Liquid-crystal and grating-based multiple wavelength cross-connect switch. *IEEE Photonics Technology Letters*, 7(5):514–516, 1995.

68 H. Asakura et al. A 200-GHz spacing, 17-channel, 1 × 2 wavelength selective switch using a silicon arrayed-waveguide grating with loopback. In *2015 International Conference on Photonics in Switching (PS)*, 52–54, 2015. doi: 10.1109/PS.2015.7328950.

69 K. Yamaguchi, Y. Ikuma, M. Nakajima, K. Suzuki, M. Itoh, and T. Hashimoto. M × N wavelength selective switches using beam splitting by space light modulators. *IEEE Photonics Journal*, 8(2): 0600809, 2016. doi: 10.1109/JPHOT.2016.2527705.

70 D. Xie *et al.* LCoS-based wavelength-selective switch for future finer-grid elastic optical networks capable of all-optical wavelength conversion. *IEEE Photonics Journal*, 9(2):7101212, 2017. doi: 10.1109/JPHOT.2017.2671436.

71 H. Yang *et al.* 24 [1 × 12] wavelength selective switches integrated on a single 4k LCoS device. *Journal of Lightwave Technology* 39(4):1033–1039, 2021. doi: 10.1109/JLT.2020.3002716.

72 T.T. Larsen, A. Bjarklev, D.H. Sparre, and J. Broeng. Optical devices based on liquid crystal photonic bandgap fibres. *Optics Express*, 11(20):2589–2596, 2003.

73 T.R. Wolinski, K. Szaniawska, S. Ertman, P. Lesiak, and A.W. Domanski. Photonic liquid crystal fibers: new merging opportunities. In *Proceedings of the Symposium on Photonics Technologies for 7th Framework Program*, 95–99, 2006.

74 M.W. Haakestad, T.T. Alkeskjold, M.D. Nielsen, L. Scolari, J. Riishede, H.E. Engan, and A. Bjarklev. Electrically tunable photonic bandgap guidance in a liquid-crystal-filled photonic crystal fiber. *IEEE Photonics Technology Letters*, 17:819–821, 2005.

75 J. Tuominen, H. Hoffren, and H. Ludvigsen. All-optical switch based on liquid-crystal infiltrated photonic bandgap fiber in transverse configuration. *Journal of the European Optical Society, Rapid Publications*, 2, 2007.

76 C. Vázquez, S.E. Vargas, and J.M.S. Pena. Design and tolerance analysis of arouter using an amplified ring resonator and Bragg gratings. *Applied Optics*, 39(12):1934–1940, 2000.

77 J.E. Heebner and R.W. Boyd. Enhanced all-optical switching by use of a nonlinear fiber ring resonator. *Optics Letters*, 24(12):847–849, 1999.

78 K. Djordjev, S. Choi, S. Chou, and P. Dapkus. Microdisk tunable resonant filters and switches. *IEEE Photonics Technology Letters*, 14(6):828–830, 2002.

79 B.E. Little, H.A. Haus, J.S. Foresi, L.C. Kimerling, E.P. Ippen, and D.J. Ripin. Wavelength switching and routing using absorption and resonance. *IEEE Photonics Technology Letter*, 10(6):816–818, 1998.

80 B. Maune, R. Lawson, C. Gunn, A. Scherer, and L. Dalton. Electrically tunable ring resonators incorporating nematic liquid crystals as cladding layers. *Applied Physics Letters*, 83(23):4689–4691, 2003.

9

Photonic Crystal All-Optical Switches

Rashmi Kumari, Anjali Yadav, and Basudev Lahiri

Nano Bio Photonics Group, Department of Electronics and Electrical Communications Engineering, Indian Institute of Technology Kharagpur, Kharagpur, West Bengal, India

9.1 Idea of Photonics

Electronic devices for applications such as information processing and signal communication have been used in past decades. These devices possess the inherent limitation of operating at higher frequencies and bandwidths due to power dissipation and hardware heating at such high frequencies. In contrast, optical devices can work better at higher frequencies or bandwidths. To meet the demand of increasing bandwidths, researchers today are moving towards optical devices and minimizing the role of electronic devices. Although optics is a solution to meet the increasing demand for higher operating frequencies or bandwidths, it has its physical limitations too. Electronic devices operate at very low energy (a few femtojoules), whereas optical devices cannot operate at such low energy. Optical devices require very high driving energy for their operation and are very large in size for their integration with other networking components. A lot of effort is needed to substitute electronic devices with optics ones as they have the limitation of confining light in small volumes [1]. Any desired device should also be integrable with other devices to reduce power consumption and operating costs. The use of photonics offers a solution to this problem, as it can manipulate light at the nanoscale at very high frequencies and is also integrable with other optical components.

Photonic devices can beat the sub-diffraction limit and mold the flow of light at very high frequencies, which is much desired for various applications such as data processing at high bit rates, keeping the high-speed properties of optical signals with reduced power consumption [2]. Materials with periodic variation of refractive index possess a property of photonic band gap or a range of frequencies within which light cannot propagate [3]. In particular, these are called "photonic crystals", made up of periodically varying dielectric structures, and can mold the flow of electromagnetic waves in the same way as semiconductor materials can control the propagation of electrons [3]. The ability of photonic crystals to control the flow of light clearly shows the potential of a new generation of active and passive optical components for photonic integrated circuits.

Optical Switching: Device Technology and Applications in Networks, First Edition. Edited by Dalia Nandi, Sandip Nandi, Angsuman Sarkar, and Chandan Kumar Sarkar.
© 2022 John Wiley & Sons, Inc. Published 2022 by John Wiley & Sons, Inc.

Moreover, tuning the photonic bandgap led to the fabrication of photonic integrated devices such as tunable filters, optical switches and gates, channel drop filters, optical interconnects. This tuning is achieved by changing the refractive index of the medium which can be done by external mechanism either optical or electrical after the fabrication of the crystal.

9.2 Principles of Photonic Crystal All-Optical Switches (AOS)

To obtain optical transmission of signals, one needs optical-electrical-optical conversions, which are time and energy-consuming. Thus, the demand for all-optical switching using photonic crystal is increasing these days. Highly efficient switching applications involve quantum information processing, integrated all-optical signal processing, etc. which require strong light-matter interaction, preferably in a minimal volume. Thus, photonic crystal is a promising candidate for all-optical switching operations due to the requirement of low switching energy, high-quality factor, on-chip integration, small volume. The tuning of the photonic bandgap of the photonic crystals on a very short time scale is essential for ultrafast switching. This modification or tuning of the photonic bandgap happens due to a change of refractive index through optical Kerr nonlinearity [3]. The operating principle of all-optical switching is shown in Figure 9.1. The input signal (probe signal) and the pump signal are injected into the photonic crystal simultaneously. The pump signal is of high energy and causes a shift in the photonic bandgap by inducing a change in its refractive index nonlinearly via Kerr effect. This shift in the photonic bandgap is responsible for the transmission control of the input signal.

Scalora et al. in 1994 showed nonlinear propagation of ultrashort pulse in a one-dimension photonic bandgap structure consisting of a multilayer stack of alternating high (n_2) and low (n_1) indices material [4]. They showed the dynamic change in the medium response with the incident field, as the refractive index changes with the incident field intensity as given by Equation 9.1:

$$n_2^2 = n_0^2 + \chi_3 I \tag{9.1}$$

where χ_3 is the Kerr coefficient of the nonlinear medium, I is the intensity of the strong pump capable of altering the index of refraction of the χ_3 medium, and n_0 is the fixed background index. Here, the frequency of the probe light $\omega_0(I_0)$ is set in the band below the bandgap and the pump

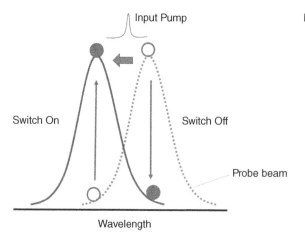

Input Pump

Switch On

Switch Off

Probe beam

Wavelength

Figure 9.1 Operating principle of switch.

Figure 9.2 (a) Both pump and probe beam initially lie in the passband and get transmitted. (b) Situation when refractive index changes the bandgap and probe beam I_0 lies in the bandgap and thus gets reflected back. Image courtesy Scalora et al.

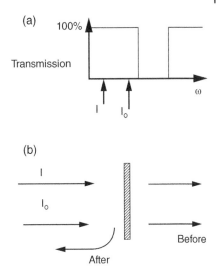

light frequency $\omega(I)$ is also located below the bandgap but is far below the gap. At first, the probe and pump light are getting completely transmitted and the switch is in "ON" state as shown in Figure 9.2 (a). When the pump light changes the refractive index of the nonlinear medium due to the Kerr effect, it results in the variation of the effective refractive index of the photonic crystal and the widening of the bandgap. This bandgap widening is such that the probe light frequency $\omega_0(I_0)$ now comes within the bandgap, and so is not allowed to transmit through the crystal. Thus, the switch is in an "OFF" state after the pump pulse changes the bandgap of the crystal, as shown in Figure 9.2(b).

The optical nonlinearity required for switching greatly depends on the electric field enhancement inside the structure. From the theory of electronic crystals, it is known that when a defect is introduced into the lattice it results in strong localization of energy around the defect. The analogue of this strong localization exists in the photonic crystal when we introduce a defect into the crystal. Tran in 1997 showed a switching mechanism in a one-dimensional photonic crystal with defect [5]. The defect was introduced in the middle of the quarter-wave stack, which provides a strong field enhancement near the band edge. The pump beam was used in the gap and near the band edge of the crystal, where the damping is not large and can change the effective refractive index. Initially, the probe beam gets transmitted through the crystal and the switch is in an "ON" state. After that when the pump beam is applied, the refractive index changes nonlinearly and shifts the photonic bandgap, which then makes the probe beam inside the gap, and the switch is in an "OFF" state. As the pump beam is applied in the bandgap, it does not propagate through the crystal and is thus easy to distinguish between the pump and a probe beam.

For high-speed, low-power all-optical switching, it is essential to obtain strong field confinement in a minimal volume. The first method to enhance optical nonlinearities is high-quality factor photonic crystal nanocavities in a very small volume. This can be achieved geometrically by designing the nanocavity carefully for efficient light-matter interaction [1]. This high Q/V ratio gives larger field enhancement inside the cavity. In practice, the choice of quality factor should be made on the basis of operating speed, as a quality factor also shows photon decay time inside the cavity. If the operating speed is more than 50 ps, then the quality factor should be less than 4×10^4, as photons should decay completely before it is switched on again. Another method to enhance optical nonlinearity in the photonic crystal is to exploit its material nonlinearities, i.e., to use ultraslow light (USL) medium with extremely large nonlinear optical response [6]. Ultrafast changes in refractive index happen either through optical Kerr nonlinearity related to Re(χ) or two-photon absorption related to Im(χ) [7]. Nonlinear absorption of free carriers in the cavity of the photonic crystal determines the shifting of the bandgap of the photonic crystal, and thus affects the switching mechanism. Material possessing high nonlinearity will function more effectively as a switch as compared to materials with less nonlinearity.

Optical switching based on defects introduced in the photonic crystal cavity shows another path to control the propagation of light waves within the crystal. Rapid photon flux-based switching was investigated by Wang et al. in 2004 in two-dimensional photonic crystal consisting of long dielectric nanorods [8]. They showed rapid switching based on out-of-plane light propagation through the entire crystal as the off state and light propagation through defect channels as the on state. A defect mode was created inside the photonic crystal by placing a dielectric rod of a larger diameter at the center. The probe beam is set around the central wavelength of the defect mode and is allowed to propagate within the crystal, i.e., on state. On exciting with the pump beam, the refractive index of the crystal changes, and so the defect mode central wavelength changes and the probe beam is not allowed to propagate through the crystal, i.e., off state.

9.3 Growth and Characterization of Optical Quantum Dots

9.3.1 Integration of PhCs-Based AOS with Optical Quantum Dots (QDs)

Federation of light and matter has helped to understand the properties of discrete particles and then utilize them for applications that could revolutionize our daily world. One such innovation is the development of nanostructures often known as Quantum Dots (QDs) that are confined in three-dimensional space. This structural bounding imparts them with properties that are exhibited by naturally occurring atoms and provide strong quantum confinement. When charged particles like electrons and holes are subjected to this quantum confinement they cannot move along the confined directions and the electronic density of states are strongly distorted, forming discrete energy levels [9]. The transition between these discrete levels will produce a distinct response in optical spectra [10]. Optical transitions between them are allowed only for discrete values, thereby producing a highly coherent and narrow linewidth response. The functioning of a quantum dot can be simply understood by a two-level system as shown in Figure 9.3. For transition 1, an incident photon is absorbed, causing the electron from the ground state to move towards the excited state. It remains in this state for some time, which is usually defined as the lifetime of carriers, and finally relaxes via intra- [11, 12] and inter-band [13] transitions. Relaxation of excited electrons in semiconductors results in radiative recombination and thus emits a photon with energy equal to the difference in the energy transition. It is defined as fluorescence.

Absorption and relaxation of excited electrons can occur at different resonance wavelengths depending on the size and the route taken by them to relax. Larger-size QDs emit and absorb at longer wavelength, i.e. are redder, whereas small-size QDs show their characteristics at a shorter wavelength (bluer). In addition, the properties of quantum dots show dependency not only on size and shape [14] but also on structural and material parameters [10]. Output characteristics like emission or absorption wavelength, relaxation time, and dephasing time are highly governed by the physical and chemical properties of QDs as well as the surrounding environment. These parameters can be controlled during synthesis. Thus, depending on the application of QDs, they

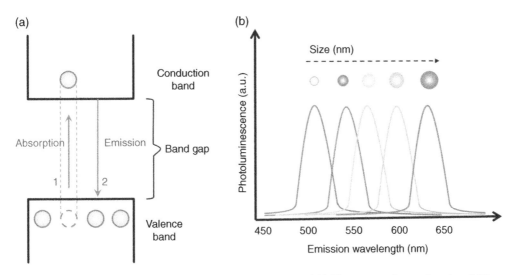

Figure 9.3 Optical response of QDs: (a) two-level diagram and (b) PL spectrum for varying size of QDs.

can be fabricated by using several organic and inorganic materials. For the present study, we will confine our discussion to optical quantum dots for application in all-optical switches based on photonic crystals.

As mentioned earlier, strong localization of the electromagnetic field is required to control and enhance light-matter interaction for various devices interconnected to form optical networks to provide on-chip photonic platforms. All-optical processing, with the additional advantage of its inherent capability to provide high-speed transmission and processing of data, holds out a promising and fast route for future developments in optical networks. One true way to achieve this is manipulating light with itself – i.e. LIGHT – and this is only possible with the use of *Optically Active Medium (OAM)*. Materials exhibiting nonlinear response impart dynamic tunability, and enhanced light-matter interaction in these OAMs can be further achieved (a) using materials with strong nonlinear effects, and (b) by using structures that boost interaction of light with the nonlinear medium. But all naturally occurring materials offer weak nonlinear properties. Thus integrating QDs that provide quantum confinement in all three directions can be one solution to the problem defined above.

All-optical switches, having the benefit of low power consumption and ultrafast response, have the capacity to integrate multiple channels [15]. The amount of switching energy required and response time are two main attributes of an optical switch. Several non-optical approaches to control switching operation are based on the thermo-optic effect [16] and nematic liquid crystals [17] have also been studied. Their large time scale response of hundreds of microseconds as compared to all-optical processes limits their use for applications requiring fast operational speed. One of the most attractive features of using PhCs for all-optical switching is its small energy requirement for switching function. Slow group velocity offered by photonic crystals results in enhancement of nonlinearity of the material used for obtaining switching applications [1, 18]. So, here we focus mainly on an all-optical operation where inclusion of OAM surpasses limitations offered by non-optical ways of tuning. One of the most common configurations of PhCs switches is the symmetric Mach–Zehnder (SMZ) type, where the two arms are decorated with material whose properties can be varied for ON/OFF switching [19]. One major disadvantage of SMZ- or MZI-based switches is their large footprint, which limits their integration for compact devices. Efforts have been made in the all-optical domain to reduce device size and increase interaction between light and switching material to make it more efficient. One way to achieve this is the formation of cavities inside PhCs. Micro- or nano-cavities provide extreme confinement of light and enhance its interaction with matter along with a reduction in device size [20]. Cavities made in PhCs support ultra-high quality (Q) factor modes and extremely small modal volume (V) [21]. Now the frequency shift required for switching is reduced by Q/V along with switching power by a factor of V/Q^2. The widely known advantage of quantum confinement and extremely high focusing of light of QDs can be further exploited for enhancing light-matter interaction. Thus the combination of quantum dots along with high-Q cavity photonic crystal is a great choice for all-optical devices. With the inclusion of nonlinear optical (NLO)-QDs, photonic band can be dynamically changed with light. This can be used for subsequent adjustment of transmission from PhCs from "on" (on-state) to "off" (off state) to simply function as switch.

Integration of QDs with PhCs to achieve switching response works in two basic ways – (a) NLO-induced phase shift and (b) saturable absorption. In the former case, SMZ-configured PhCs switches are made where the two arms contain QDs layers made up of materials that exhibit non-linear property. When light is incident it changes its refractive index depending on its intensity, which in turn provides a phase shift to the passing wave, thus adding destructively to produce null output [22]. In the latter case, assembled QDs interact strongly with light, giving rise to a saturable absorption phenomenon which plays a significant role at OFF state [23]. Quantum dots placed in high-Q cavities formed in PhCs strongly couples with the trapped photons and thus light-matter interaction can be greatly enhanced [24]. In 2007, Yoshimasa et al. [25] incorporated QDs with SMZ-based PhCs and sufficient nonlinear phase shift at extremely low optical control energy of 100fJ is observed. In 2011, Bose et al. [26] utilized strong coherent coupling between QDs and PhC cavity to get switching speed on the picosecond scale (~120 ps), and energy requirement is reduced up to 140 photons pulse energy. In another report in 2012, Volz et al. [27] reported switching time of ~50 ps for a single photon switch working in a strong coupling regime. Thereon, various efforts have been made for optimization of design as well as fabrication tools to increase the overall performance of switch in terms of enhanced Q, reduced switching speed, and minimal energy requirements.

9.3.2 Growth and Characterization of Quantum Dots

9.3.2.1 Growth of Quantum Dots

Q factor for experimental calculations shows dependence on structural design, and imperfections introduced during fabrication and absorption, where the first two parameters allow easy assessment of performance but the difficulty in the investigation of last one limits the performance. Although designed structures would have a Q factor as high as 10^5–10^9, these cannot be obtained experimentally even if the PhCs are precisely designed. The main reason behind this low output is the structural imperfections and absorption or scattering by QDs embedded in PhCs. To overcome such structural and experimental limitations efforts have been made by researchers to reduce losses and improve the performance of QDs. Here, we explain basic techniques for the synthesis of QDs along with various methods to resolve fabrication-based limitations.

9.3.2.2 Colloidal Solution Via Chemical Synthesis

This process is similar to any chemical synthesis process with the difference that the product remains suspended in the solvent. Heating the temperature of the solution in a controlled manner allows the precursors to form monomers via decomposition and then nucleate to form nanocrystals. Temperature and monomer concentration here plays an important role in crystal growth. QDs produced by this method are less practical for device fabrication as they are suspended in a solvent. Therefore, to produce them in a more usable form, QDs are embedded in a solid-state matrix usually made up of polymers [28]. In 2005, Pang et al. [29] used photosensitive material poly(methylmethacrylate) (PMMA) (positive resist) and SU-8 as UV sensitive (negative resist) for inclusion of CdSe/ZnS and PbSe colloidal QDs. Prepolymerization of precursors for preventing the agglomeration of QDs-polymer and sensitivity of resistance towards electron beams makes it an attractive approach in the fabrication of photonic devices. In 2008, Sun et al. [30] synthesized different-colored CdS QDs within the polymer matrix where the size is controlled by the amount of cross-linker introduced during growth. In 2015, Krini et al. [31] modified the surface of QDs by functionalizing them with silica shells to make them more stable in the polymer matrix. This method involves an additional step for converting solvent-suspended QDs to usable form i.e. matrix form. Therefore, another method that allows the formation of QDs along with the fabrication steps of PhCs is highly preferred and is discussed below.

9.3.2.3 Self-Assembly Technique

Quantum dots produced by this technique are usually 5–50 nm in size. The lattice mismatch between two materials induces strain at the surface sites and it is widely known that every material that exists in nature tries to get relieved to maintain an equilibrium state. Formation of QDs can be thus achieved by exploiting this relieving nature of materials. A common method named as Stranski–Krastanov (SK) is used for the growth of QDs [32]. It is one of the primary modes used for the growth of thin layers via epitaxy. Epitaxy allows layer-by-layer growth in a controlled manner where fabrication parameters like the temperature of the substrate, flow rate of adsorbates, lattice mismatch, and chemical potential between subsequently grown layers plays an important role. Control of these parameters enables monolayer film deposition as well as the growth of nucleated islands that finally behave as 3D quantum structures showing quantum confinement. Practical techniques such as Molecular Beam Epitaxy (MBE) and Metal-Organic Chemical-Vapor Deposition (MOCVD) are used for obtaining ultra-clean conditions for in-situ growth. Up to a certain thickness, the growth process continues to produce films of several monolayers thick and upon reaching a critical thickness, island growth and coalescence of nucleated debris further allow the formation of thin films [33]. It is then when QDs are produced. They are usually randomly nucleated and form a pyramidal shape which can be further altered by using different growth conditions [34]. Also, by controlling the nucleation of QDs on the surface, periodically arranged QDs can be obtained [35].

Various reports about improvement of quantum properties of QDs have been made, where control of substrate temperature, the strain of wetting layer, pressure of the chamber, and the inclusion of additional atoms in the spacer layer to achieve highly dense, uniform and homogenously distributed QDs are obtained for use in nano-photonic applications [36–40]. However, QDs produced by the above-described SK mode emit only a single wavelength and are not area-specific. Therefore, in 2007, Takata et al. [28] proposed the use of the metal mask (MM) method for selective area growth (SAG) of quantum dots made of InAs. In this method, multiple patches of QDs can be obtained in one single growth process. The inclusion of a strain-reducing layer to embed QDs allows one to obtain tunability of wavelength at which absorption occurs. This technique provides the advantage of monolithic growth along with different absorption wavelength of QDs. In a similar work by Ozaki et al. [41], highly dense and highly uniform QDs were obtained with a spectral width of 30meV and QD density ~4×10^{10}cm^{-2}. Also, a shift in wavelength from 1240nm to 1320 nm without additional optical degradation is achieved from various thicknesses of strain-reducing layer (SRL). In 2009, Ozaki et al. [42] modified the MM method for large-scale production of QDs on GaAs wafer and also studied the effect of thermal fluctuations from MM that produces variable linewidth output. Another factor that determines the performance of AOS is the decay rate of photons, where a slowly decaying further degrades switching performance. In typical QDs, excited carriers decay slowly at a rate of ~ns and this promotes a pulse pattern effect. In this effect, output signal intensity shows strong dependency on input pulse pattern for high-bitrate operation and is highly undesired for ultrafast optical switches. In 2009, Kitada et al. [21] proposed the use of SRL to embed QDs showing fast decay rate of ~18 ps for photo-generated carriers in In As QDs on GaAs (100) substrate. Extension in absorption wavelength 1.35–1.65 µm is also observed as a result of a reduction in a biaxial strain which otherwise would produce increased energy bandgap. Doping is another technique through which a reduction in decay rate can be achieved [43]. In 2017, Takashi et al. [44] illustrated the importance of cleaning the bottom side of the buried oxide layer used for obtaining cavities. A high Q factor of more than ten million is achieved experimentally in Si PhCs. Surface passivation of fabricated AOS is proposed by Kuruma et al. [45], whereby the passivating surface of fabricated PhC with sulfur produced three times increase in Q factor (~160 000) with respect to that obtained without passivation.

Thus, we can say that by incorporating techniques like the metal mask method, surface passivation, doping, strain-reducing layer, surface passivation and other fabrication parameters, one can obtain enhanced performance of AOS both in terms of high Q and fast switching speed.

9.3.2.4 Characterization of Quantum Dots

Optical characterization of QDs includes its absorption and emission properties, which are usually determined by Photoluminescence (PL) and UV-Vis Spectroscopy. Using these spectroscopic methods, one can easily determine the bandgap of QDs. These are explained briefly in the below text. Imaging of QDs or determination of size are carried by using techniques like Atomic Force Microscopy (AFM), Transmission Electron Spectroscopy (TEM), Scanning Electron Microscopy (SEM), and Dynamic Light Scattering (DLS). Since these techniques are common for imaging a wide variety of materials, they have not been defined here.

9.3.2.5 Photoluminescence Spectroscopy

Photoluminescence spectroscopy, often known as PL, is a non-destructive and non-contact method used for determining the optical properties of materials. In this method light from a laser source tuned close to an energy corresponding to the bandgap is directed onto the sample with the help of a focusing lens. These incoming photons are absorbed, causing electrons to transit from lower energy levels to higher energy levels. After some time, usually in the range of picoseconds, these exciting carriers release some energy in form of thermal vibrations and finally relax to the lowest energy state by emitting a photon. This emission is called photoluminescence, as it was assisted by photons. This emitted energy or light is collected via a collimating lens setup and is directed to the photodetector. The response so obtained is the emission spectrum of QDs under illumination. The emission spectra give valuable information about the optical properties of QDs. Another method includes Time-Resolved PL spectroscopy, which is similar to the above-described method with the only difference that it monitors output, i.e. emitted light intensity with respect to time, and is used to determine the lifetime of charge carriers in QDs.

9.3.2.6 UV-Vis Spectroscopy

QDs made up of semiconductors find applications in optical as well as in optoelectronics. Semiconducting QDs exhibit their optical response in the UV-Vis range. Thus UV-Vis Spectroscopy is used to extract bandgap information from its absorption properties. One major difference between the UV-Vis and PL spectroscopy is that the later measures transition from excited to the ground state while the former does its opposite. This method uses UV-Vis light to be incident on the sample and then obtain absorbance spectra plotting radiation versus wavelength. A peak in this absorption spectra indicates maximum absorption at this wavelength.

The methods of QDs characterization defined above allow one to get not only bandgap information but also allows to determine the particle size via spectrum peaks. Therefore, by careful designing and synthesis of QDs, they can be exploited for AOS based on PhCs integrated QDs.

9.4 Design and Fabrication

This section is intended to explain the fabrication procedure of a photonic crystal. The main steps involved in the fabrication of two-dimensional photonic crystal are as follows.

9.4.1 Sample Preparation

The first step is to clean the substrate on a spin coater with acetone, ethanol, and isopropanol, and it is finally blown dry with N_2. After this, a thick layer resist (positive or negative) which is a polymer sensitive to UV radiation or electron beam is spun on the top of the substrate. The resist is then baked to dry off the solvent and harden the resist.

9.4.2 Lithography

9.4.2.1 Electron Beam Lithography (EBL)

Lithography is used to transfer the pattern to the resist. For high resolution, research-purpose electron beam lithography (EBL) is used. The electron beam has few nanometers precision which can control feature sizes of sub-microns. This high precision cannot be obtained by optical lithography because of its sub-diffraction limit. In EBL, the pattern is transferred directly onto resist by focused electron beam, making an extremely small feature size pattern. However, due to its serial way of patterning the process is very slow and cannot be used for large volumes.

9.4.2.2 Optical UV Lithography

In order to transfer the pattern to a large volume for practical purposes, a parallel process is needed which can pattern entire resist simultaneously. Optical lithography is much needed for this reason but is limited in patterning small feature sizes in the nanometer range due to its sub-diffraction limit. Feature size comparable to illumination wavelength becomes fuzzy or does not print at all. To overcome this limitation deep UV lasers of illumination wavelength 193 nm and 248 nm are excited to fabricate structures with dimensions below 100 nm [46].

9.4.3 Etching

Etching is a process of removal of the mask/non-mask layer depending on negative resist and positive resist respectively to make active structures.

9.4.3.1 Wet Etching

The structure is immersed into the liquid acid solution to dissolve the mask/non-mask area.

9.4.3.2 Dry Etching

Dry etching is performed using RF-induced plasma instead of acid to remove the mask, also called reactive ion etching (RIE). The reactive species then chemically etch the material in its immediate proximity. It is the simplest process capable of highly directional etching with the energetic ion bombardment on the substrate. The reactive gases commonly used in RIE are CF_4, SF_6, O_2, Ar, and CHF_3.

9.5 Device Structure and Performance Analysis of Photonic Crystal All-Optical Switches

Photonic processing requires highly efficient all-optical switches which would require low energy and power along with high speed. Various all-optical switches have been reported using optical nonlinearity such as all-optical switches making use of the inter-sub-band transition in semiconductor quantum wells which can operate in few picoseconds but require several picojoules of energy [47]. All-optical switches based on semiconductor optical amplifiers (SOAs) and parametric processes also exist. These require very small switching energy but the total power is higher [48, 49]. Villeneuve in 1996 showed nonlinear optical switching in one-dimensional photonic crystal consisting of holes etched in a semiconductor with a defect in the center. They showed a large change in the refractive index by using the photorefractive effect based on the ionization of the deep donor levels known as the DX centers. Two such microcavities were placed in series with two distinct resonant frequencies. One cavity is triggered using photorefractive effect for switch on condition and the other cavity is triggered for the switch of condition.

Chen et al. in 2002 showed optical bistable switching in a one-dimensional nonlinear photonic crystal consisting of alternating layers of high and low index regions with a defect in the middle [50]. They achieved two resonant states depending on the thickness of the Kerr medium or the difference between the refractive indices of two alternating layers. The wavelength of the probe laser is set near the central wavelength of the defect mode. The transmittance of the probe laser increases with the increment of the intensity of the probe laser. The photon localization effect and the third-order nonlinearity can offer nonlinear feedback for the probe light. As a result, the transmittance spectrum of the probe light takes on the properties of optical bistability. When the intensity of the probe light is above the threshold value, the switch is turned on and when the intensity is below the threshold value, the switch is turned off. Hache et al. in 2000 showed the effect of two-photon absorption on the all-optical switching by one-dimensional photonic crystal consisting of a periodic lattice of amorphous silicon and silica [51]. They showed the generation of free carriers via two-photon absorption process which then alters the refractive index of the photonic crystal required for switching near the band edge with 12 ps switching time. Johnson et al. in 2002 reported ultrafast optical switching based on the photonic density of states (DOS) in a three-dimensional photonic crystal [52]. They showed a dramatic change in the photonic DOS in three-dimensional GaAs inverse opals via a two-photon absorption process on excitation of the laser pulse. Two-photon absorption process changes the effective refractive index of the crystal and hence the position and the width of the bandgap changes as mentioned above. Pump pulse is used to switch the DOS from a high to a nearly zero value i.e., off state or from zero to a very high value which is on state at a 100 fs timescale.

Soljacic et al. in 2005 model a hybrid system of photonic crystal microcavities incorporating ultraslow light medium as shown in Figure 9.4(a) for enhanced nonlinearity effect in low-power

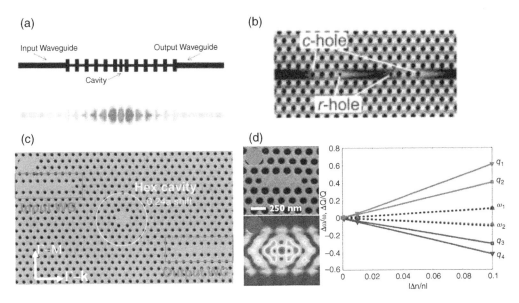

Figure 9.4 (a) Photonic crystal microcavity by introducing defect in the center by narrowing its width. Courtesy of [1]/AIP Publishing. (b) 4-point defect photonic crystal cavity. Courtesy of [21]/AIP Publishing. (c) Hexapole photonic crystal cavity coupled to input and output waveguides. Courtesy of [53]/AIP Publishing. (d) L3 type photonic crystal cavity showing change in Q factor with change in refractive index. Courtesy of [55]/AIP Publishing.

all-optical switching [6]. They combined both the two approaches to have strong and instantaneous light–matter interaction, one using a high Q photonic crystal cavity, and the other using an ultraslow light-medium which has larger Kerr coefficients. The ultraslow light atom which shows enormous dispersive behavior was inserted at the center of the microcavity to overcome the absorption at the cavity resonance and enhance nonlinearity. They reported the required input power to be 8.5 μW and with a switching time of 52 ps. Tanabe et al. in 2005 demonstrated all-optical switching in a two-dimensional high Q silicon photonic crystal nanocavities coupled to waveguides [21]. They made a three-point defect with high Q nanocavity by reducing the size of the holes in W1 waveguide as shown in Figure 9.4(b). As switching energy required scales as $V/_{Q^2}$, careful designing of the high Q nanocavity with very small modal volume led to very low energy requirement of 100 fJ with a switching time of 50 ps. In 2007, they showed all-optical switching using carrier plasma dispersion in an argon ion (Ar^+) implanted photonic crystal nanocavity as shown in Figure 9.4 (b) [53]. The schematic image of hexapole photonic crystal nanocavity is as shown in Figure 9.4(c). They increased the speed of all-optical photonic crystal switches by carrier killing technique by Ar^+ ion. Recombination rate was increased by creating traps for the carrier by annealing the waveguide and increasing the recombination of the carriers allowing fast switching recovery time. This resulted in drastic reduction in carrier lifetime of 150 ps with very low switching energy of 100 fJ and 70 ps switching time. In 2008, they numerically studied the carrier dynamics in silicon photonic crystal nanocavities [54]. They showed that even though silicon devices possess long bulk carrier lifetime, due to the small nanocavity, the switching speed achieved by two-photon absorption is significantly faster. They demonstrated the p-i-n structure based photonic crystal nanocavity to extract the carriers before they recombine, thus reducing the thermo-optic effect. Also, they showed that increased carrier diffusion is due to the small nanocavity which gives fast switching recovery time of 20 ps. In 2008, they used the same ion implantation method in two-dimensional photonic crystal cavity consisting of ring resonator coupled to two waveguides to enhance the two-photon absorption in the cavity and increase the switching speed to demonstrate bistable switching. They used the bistable switch further to show its memory application in flip flops.

Fushman et al. in 2007 introduced the defect in two-dimensional photonic crystal made of GaAs and containing InAs quantum dots [55]. They used the free carrier generation method, which is more prominent in GaAs for changing the effective refractive index of the photonic crystal. The cavity resonance shifts to a shorter wavelength on excitation of pump beam via free-carrier effect. They used a large surface area and small modal volume by removing three holes from the center as shown in Figure 9.4(d) for increasing the photon lifetime inside the cavity. Thus the free carrier can be swept easily away from the cavity by applying low potential, essential for ultrafast switching. They demonstrated thermal tuning of the cavity resonance on applying optical heating by pump laser which gets absorbed by the sample. To enhance the emission rate, a photonic crystal cavity was made to spectrally overlap with the QD emission. On applying the pump beam, the dielectric constant $\varepsilon = n^2$ of the cavity changes, which changes the refractive index n of the cavity. The change in refractive index Δn can be calculated from:

$$\frac{\Delta\omega}{\omega} \approx -\frac{1}{2}\frac{\int \Delta\varepsilon |E|^2 \, dV}{\int \varepsilon |E|^2 \, dV} \approx -\frac{\Delta n}{n}, \tag{9.2}$$

where ω is the resonance of the unshifted cavity, $\Delta\omega$ is the shift in resonance frequency, $|E|^2$ is the amplitude of the cavity mode, and the integration is done over full volume. Nozaki et al. in 2010 demonstrated an all-optical photonic crystal switch by introducing H_0 cavity defect [56]. They designed the H_0 nanocavity to have high Q with the smallest modal volume of 0.025 μm^3 to achieve a fast switching speed of 40 Gbps with reduced switching energy (<1 fJ) required and significantly short switching time (20 ps). High Q nanocavity also enhances the optical nonlinearity, and to enhance it further they used quaternary compound semiconductor InGaAsP because it has the largest carrier-induced nonlinearity.

Apart from photonic crystal nanocavities to function as a switch, many other configurations of photonic crystal were also used, such as waveguide-based directional coupler, and symmetric Mach–Zender interferometer as a switch. Sharkawy et al. in 2002 modeled a photonic crystal waveguide-based switch [57]. It consists of a silicon rod in the air and the two waveguides are brought in close proximity with each other to form a directional coupler. The switching mechanism was based on the change in the conductance in the coupling region between the two waveguides. The conductance of the coupling region was changed by doping it with charge carriers and according to that, the output is obtained at the two ports or single port. Nakamura et al. in 2004 demonstrated a symmetric Mach–Zender type photonic crystal configuration with InAs QDs to enhance the nonlinearity [58]. A pump-probe beam was used for switching purposes. Ideally, the switch is on, and when the pump beam is incident on it, then the refractive index of the coupling region changes and so the beam cannot pass through the waveguide and the switch is off. Though they possessed an efficient switching mechanism, due to large volume, the total energy required was greater, with reduced switching times and increased complexity in tuning.

9.6   Challenges and Recent Research Trends of Photonic Crystal All-Optical Switches

There has been tremendous development in the field of all-optical switching in the past twenty years due to the increasing demand for ultrafast processing systems. All-optical switching is an essential component in all-optical communication networks, as it prevents the losses incurred during electrical-optical-electrical signal conversion and also provides higher information processing rates. Various practical all-optical switching structures have been developed, having different physical structures and different optical behaviors as regards their spectral response, switching time, and the energy required. Among all the technologies developed to realize optical switches, photonic crystal all-optical switches have attracted greater attention since they support room temperature operation, compatibility with on-chip integration, large Q factor, and small nanobeam cavity [59]. The switching principle is mainly based on the change in the refractive index of the cavity with the incident pump beam via optical nonlinearity. Recently, Yu et al. in 2015 demonstrated photonic crystal Fano structures for all-optical switching [60]. They used InP-based two-dimensional photonic crystal with a nanocavity and a waveguide to have Fano resonance and demonstrated improved switching contrast and speed without adding any extra phase modulation. Fano resonance is determined by the constructive interference between the discrete resonance and the continuum. It shows sharp asymmetric resonance in the transmission spectrum within a narrow wavelength range. As the refractive index shift greatly depends on the sudden transition in the transmission spectrum which is small in this case, this enables low switching energy. Bekele in 2016 demonstrated InP photonic crystal with a defect between the directional couplers based Fano structure for all-optical switching [61]. The structure possessed bistable switching with the reduced power consumption of 60 fJ and high switching contrast. Silicon-on-insulator platforms have always been attractive to realize on-chip integration due to their low cost, compatibility with complementary metal-oxide-semiconductor (CMOS), small size, and high stability. Dong et al. in 2018 demonstrated a high-contrast and low-power all-optical switch using Fano resonance based on a silicon nanobeam cavity [62]. They improved the performance of the switch using Fano resonance; however, the switching contrast mainly depends on silicon carrier lifetime. They improved carrier lifetime and reduced switching recovery time with the help of the blue detuning filtering method [63] and showed a high extinction ratio of 40 dB with a switching contrast of 9.53 dB and low switching energy of 113 fJ.

The concept of slow light was introduced to provide enhanced nonlinearity required for switching. Slow light can be generated in two ways either by material change or by structural change. Slow light refers to the optical modes with small group velocity and high dispersion. Electromagnetic-induced transparency (EIT) is a phenomenon that can be used to achieve slow light. Due to high nonlinearity in the slow light region, the required switching power is greatly reduced. Granpayeh et al. in 2019 modeled an optical switch based on a photonic crystal directional coupler [64]. They have used the EIT phenomenon by modifying the photonic crystal structure by changing the refractive index of the central rod in the waveguide region to reduce the desired power to 23 mW/cm^3. They also showed the device to function as tunable multiplexer demultiplexer and beam splitter. In addition to achieving slow light by changing material properties, it can also be achieved by making structural changes such as in photonic crystal nanobeam cavity. Zhan et al. in 2020 demonstrated a one-dimensional high Q cavity photonic crystal waveguide of SiN on silica substrate [65]. They achieved very high Q to attain slow light in the waveguide with a group index of 24, which further reduced the power required for switching with high

Table 9.1 Performance analysis of various photonic crystal all-optical switches.

Structure	Time	Energy	Year	Reference
Ultrafast all-optical switching in a silicon-based photonic crystal	12 ps		2000	[51]
Ultrafast switching of photonic density of states in photonic crystals	100 fs		2002	[52]
Ultralow-power all-optical switching	52 ps	8.5 uW	2005	[6]
All-optical switches on a silicon chip realized using photonic crystal nanocavities	50 ps	100 fJ	2005	[21]
Ultrafast nonlinear optical tuning of photonic crystal cavities	60 fJ	3ps	2007	[55]
Fast all-optical switching using ion-implanted silicon photonic crystal nanocavities	100 fJ	70 ps	2007	[53]
Carrier diffusion and recombination in photonic crystal nanocavity optical switches		100 ps	2008	[54]
Sub-femtojoule all-optical switching using a photonic crystal nanocavity	<1 fJ	20 ps	2010	[56]
All-optical switching improvement using photonic crystal Fano structures	60 fJ		2015	[60]
Photonic crystal Fano resonances for realizing optical switches, lasers, and non-reciprocal elements	60 fJ		2016	[61]
High-contrast and low-power all-optical switch using Fano resonance based on a silicon nanobeam cavity	113 fJ		2018	[62]
All-optical ultrafast graphene-photonic crystal switch		0.3 ps	2019	[68]

contrast ratio (>50 dB). Recently the two-dimensional material graphene has shown its application in all-optical switching. Graphene possessed high nonlinear coefficients of 10^{-15} m^2/W [66]. Liu in 2017 showed electrically tunable switching based on photonic crystal waveguide loaded graphene stacks [67]. They designed photonic crystal waveguides consisting of graphene/Al$_2$O$_3$ stacks to achieve tunable switching. Azizpour in 2019 modeled an ultrafast graphene-photonic crystal switch, consisting of graphene–SiO$_2$ stacks surrounding the resonant ring [68]. They achieved a high contrast ratio of 82% with a switching time of 0.3 ps. The performance analysis of the abovementioned photonic crystal all-optical switches is summarized in Table 9.1.

Bibliography

1 M. Soljačić and J.D. Joannopoulos. Enhancement of nonlinear effects using photonic crystals. *Nature Materials*, 3(4):211–219, 2004.

2 M. Notomi, A. Shinya, K. Nozaki, T. Tanabe, S. Matsuo, E. Kuramochi, et al. Low-power nanophotonic devices based on photonic crystals towards dense photonic network on chip. *IET Circuits, Devices and Systems*, 5(2):84–93, 2011.

3 J.D. Joannopoulos, S.G. Johnson, J.N. Winn, and R.D. Meade. *Photonic Crystals: Molding the Light*. Princeton, NJ, Princeton University Press.

4 C.A. Chatzidimitriou-Dreismann, T. Abdul-Redah, B. Kolaric, and I. Juranic. Optical limiting and switching of short pulses by use of a nonlinear photonic bandgap structure with a defect. *Physical Review Letters*, 84(22):5237, 2000.

5 P. Tran. Optical limiting and switching of short pulses by use of a nonlinear photonic bandgap structure with a defect. *Journal of the Optical Society of America B*, 14(10):2589, 1997.

6 M. Soljačić, E. Lidorikis, J.D. Joannopoulos, and L.V. Hau. Ultralow-power all-optical switching. *Applied Physics Letters*, 86(17):1–3, 2005.

7 L. Vestergaard Hau, S.E. Harris, Z. Dutton, and C.H. Behroozi. Light speed reduction to 17 metres per second in an ultracold atomic gas. *Journal of Internet Technology.* 2011;12(6):887–897.

8 X. Wang, K. Kempa, Z.F. Ren, and B. Kimball. Rapid photon flux switching in two-dimensional photonic crystals. *Applied Physics Letters*, 84(11):1817–1819, 2004.

9 W.J. Parak, L. Manna, and T. Nann. Fundamental principles of quantum dots. In: Schmid, G, editor. *Nanotechnology. Volume 1: Principles and Fundamentals.* Weinheim, Wiley, 2008, 73–96.

10 G. Bastard and J. Brum. Electronic states in semiconductor heterostructures. *IEEE Journal of Quantum Electronics*, 22(9):1625–1644, 1986.

11 X.-Q. Li, H. Nakayama, and Y. Arakawa. Phonon bottleneck in quantum dots: Role of lifetime of the confined optical phonons. *Physical Review B*, 59(7):5069–5073, 1999.

12 U. Bockelmann and T. Egeler. Electron relaxation in quantum dots by means of Auger processes. *Physical Review B*, 46(23):15574–15577, 1992.

13 M.C. Tatham, J.F. Ryan, C.T. Foxon. Time-resolved Raman measurements of intersubband relaxation in GaAs quantum wells. *Physical Review Letters*, 63(15):1637–1640, 1989.

14 D.J. Norris, A.L. Efros, M. Rosen, and M.G. Bawendi. Size dependence of exciton fine structure in CdSe quantum dots. *Physical Review B*, 53(24):16347–16354, 1996.

15 M. Saruwatari. Multiple-channel output all-optical OTDM demultiplexer using XPM-induced chirp compensation (MOXIC). *Electronics Letters*, 34(6):575–576, 1998.

16 E.A. Camargo, H.M.H. Chong, and R.M.D. La Rue. 2D photonic crystal thermo-optic switch based on AlGaAs/GaAs epitaxial structure. *Optics Express*, 12(4):588–592, 2004.

17 H.M. van Driel, S.W. Leonard, H.-W. Tan, A. Birner, J. Schilling, S.L. Schweizer, et al. Tuning 2D photonic crystals. In: P.M. Fauchet and P.V. Braun, editors. *Tuning the Optical Response of Photonic Bandgap Structures.* Proceedings of SPIE 5926, 1–9, 2005.

18 M. Soljačić, S.G. Johnson, S. Fan, M. Ibanescu, E. Ippen, and J.D. Joannopoulos. Photonic-crystal slow-light enhancement of nonlinear phase sensitivity. *Journal of the Optical Society of America B*, 19(9):2052–2059, 2002.

19 A. Martínez, P. Sanchis, and J. Martí. Mach–Zehnder interferometers in photonic crystals. *Optical and Quantum Electronics*, 37(1):77–93, 2005.

20 F. Cuesta-Soto, A. Martínez, J. García, F. Ramos, P. Sanchis, J. Blasco, et al. All-optical switching structure based on a photonic crystal directional coupler. *Optics Express*, 12(1):161–167, 2004.

21 T. Tanabe, M. Notomi, S. Mitsugi, A. Shinya, and E. Kuramochi. All-optical switches on a silicon chip realized using photonic crystal nanocavities. *Applied Physics Letters*, 87(15):1–3, 2005.

22 H. Nakamura, K. Kanamoto, Y. Nakamura, S. Ohkouchi, H. Ishikawa, and K. Asakawa. Nonlinear optical phase shift in InAs quantum dots measured by a unique two-color pump/probe ellipsometric polarization analysis. *Journal of Applied Physics*, 96(3):1425–1434, 2004.

23 H. Nakamura, Y. Sugimoto, K. Kanamoto, N. Ikeda, Y. Tanaka, Y. Nakamura, et al. Ultra-fast photonic crystal/quantum dot all-optical switch for future photonic networks. *Optics Express*, 12(26):6606–6614, 2004.

24 T. Yoshie, A. Scherer, J. Hendrickson, G. Khitrova, H.M. Gibbs, G. Rupper, et al. Vacuum Rabi splitting with a single quantum dot in a photonic crystal nanocavity. *Nature*, 432(7014):200–203, 2004.

25 Y. Sugimoto, H. Ishikawa, and K. Asakawa. Semiconductor-based 2D photonic crystal slab waveguides for ultrafast all-optical switches: PC-SMZ. *Electronics and Communications in Japan (Part II: Electronics)*, 90(6):18–26, 2007.

26 R. Bose, D. Sridharan, H. Kim, G.S. Solomon, and E. Waks. Low-photon-number optical switching with a single quantum dot coupled to a photonic crystal cavity. *Physical Review Letters*, 108(22): 227402, 2012.

27 T. Volz, A. Reinhard, M. Winger, A. Badolato, K.J. Hennessy, E.L. Hu, et al. Ultrafast all-optical switching by single photons. *Nature Photonics*, 6(9):605–609, 2012.

28 Y. Takata, N. Ozaki, S. Ohkouchi, Y. Sugimoto, N. Ikeda, Y. Watanabe, et al. Monolithic growth of InAs-QDs with different absorption wavelengths in different areas for integrated optical devices. In: *LEOS 2007 – IEEE Lasers and Electro-Optics Society Annual Meeting Conference Proceedings*, 30–31, 2007.

29 L. Pang, K. Tetz, Y. Shen, C.-H. Chen, and Y. Fainman. Photosensitive quantum dot composites and their applications in optical structures. *Journal of Vacuum Science & Technology B: Microelectronics and Nanometer Structures Processing, Measurement, and Phenomena*, 23(6): 2413–2418, 2005.

30 Z.-B. Sun, X.-Z. Dong, W.-Q. Chen, S. Nakanishi, X.-M. Duan, and S. Kawata. Multicolor polymer nanocomposites: in situ synthesis and fabrication of 3D microstructures. *Advanced Materials*, 20(5):914–919, 2008.

31 R. Krini, C.W. Ha, P. Prabhakaran, H. El Mard, D.-Y. Yang, R. Zentel, et al. Photosensitive functionalized surface-modified quantum dots for polymeric structures via two-photon-initiated polymerization technique. *Macromolecular Rapid Communications*, 36(11):1108–1114, 2015.

32 B.R. Bennett, B.V. Shanabrook, P.M. Thibado, L.J. Whitman, and R. Magno. Stranski-Krastanov growth of InSb, GaSb, and AlSb on GaAs: structure of the wetting layers. *Journal of Crystal Growth*, 175–176:888–893, 1997.

33 P.M. Petroff, A. Lorke, and A. Imamoglu. Epitaxially self-assembled quantum dots. *Physics Today*, 54(5):46–52, 2001.

34 J.M. García, G. Medeiros-Ribeiro, K. Schmidt, T. Ngo, J.L. Feng, A. Lorke, et al. Intermixing and shape changes during the formation of InAs self-assembled quantum dots. *Applied Physics Letters*, 71(14):2014–2016, 1997.

35 H. Lee, J.A. Johnson, M.Y. He, J.S. Speck, and P.M. Petroff. Strain-engineered self-assembled semiconductor quantum dot lattices. *Applied Physics Letters*, 78(1):105–107, 2001.

36 K. Yamaguchi, K. Yujobo, and T. Kaizu. Stranski-Krastanov growth of InAs quantum dots with narrow size distribution. *Japanese Journal of Applied Physics*, 39(2, 12A):L1245–L1248, 2000.

37 B. Daudin, F. Widmann, G. Feuillet, Y. Samson, M. Arlery, and J.L. Rouvière. Stranski-Krastanov growth mode during the molecular beam epitaxy of highly strained GaN. *Physical Review B*, 56(12):R7069–R7072, 1997.

38 J.X. Chen, A. Markus, A. Fiore, U. Oesterle, R.P. Stanley, J.F. Carlin, et al. Tuning InAs/GaAs quantum dot properties under Stranski-Krastanov growth mode for 1.3 μm applications. *Journal of Applied Physics*, 91(10):6710–6716, 2002.

39 O.G. Schmidt, O. Kienzle, Y. Hao, K. Eberl, and F. Ernst. Modified Stranski–Krastanov growth in stacked layers of self-assembled islands. *Applied Physics Letters*, 74(9):1272–1274, 1999.

40 C. Bayram and M. Razeghi M. Stranski–Krastanov growth of InGaN quantum dots emitting in green spectra. *Applied Physics A*, 96(2):403–408, 2009.

41 N. Ozaki, Y. Takata, S. Ohkouchi, Y. Sugimoto, N. Ikeda, Y. Watanabe, et al. Selective-area growth of self-assembled InAs-QDs by metal mask method for optical integrated circuit applications. *MRS Online Proceedings Library*, 959(1):1703, 2011.

42 N. Ozaki, S. Ohkouchi, Y. Takata, N. Ikeda, Y. Watanabe, Y. Sugimoto, et al. Monolithic fabrication of two-color InAs quantum dots for integrated optical devices by using a rotational metal mask. *Japanese Journal of Applied Physics*, 48(6):65502, 2009.

43 H. Ueyama, T. Takahashi, Y. Nakagawa, K. Morita, T. Kitada, and T. Isu. GaAs/AlAs Multilayer cavity with Er-doped InAs quantum dots embedded in strain-relaxed InGaAs barriers for ultrafast all-optical switches. *Japanese Journal of Applied Physics*, 51:04DG06, 2012.

44 T. Asano, Y. Ochi, Y. Takahashi, K. Kishimoto, and S. Noda. Photonic crystal nanocavity with a Q factor exceeding eleven million. *Optics Express*, 25(3):1769–1777, 2017.

45 K. Kuruma, Y. Ota, M. Kakuda, S. Iwamoto, and Y. Arakawa. Surface-passivated high-Q GaAs photonic crystal nanocavity with quantum dots. *APL Photonics*, 5(4):46106, 2020.

46 W. Bogaerts, V. Wiaux, D. Taillaert, S. Beckx, B. Luyssaert, P. Bienstman, et al. Fabrication of photonic crystals in silicon-on-insulator using 248-nm deep UV lithography. *IEEE Journal on Selected Topics in Quantum Electronics*, 8(4):928–934, 2002.

47 A.V. Gopal, T. Simoyama, H. Yoshida, and J. Kasai. Intersubband absorption saturation in InGaAs – AlAs – AlAsSb coupled quantum wells. *IEEE Journal of Quantum Electronics*, 39(11):1356–1361, 2003.

48 T. Yamamoto, E. Yoshida, and M. Nakazawa. Ultrafast nonlinear optical loop mirror for demultiplexing 640Gbit/s TDM signals. *Electronics Letters*, 34(10):1013–1014, 1998.

49 S. Nakamura, Y. Ueno, and K. Tajima. Femtosecond switching with semiconductor-optical-amplifier-based Symmetric Mach-Zehnder-type all-optical switch. *Applied Physics Letters*, 78(25):3929–3931, 2001.

50 C. Lixue, D. Xiaoxu, D. Weiqiang, C. Liangcai, and L. Shutian. Finite-difference time-domain analysis of optical bistability with low threshold in one-dimensional nonlinear photonic crystal with Kerr medium. *Optics Communications*, 209(4–6):491–500, 2002.

51 Hache A. Ultrafast all-optical switching in a silicon-based photonic crystal. *Applied Physics Letters*, 77:4089, 2000.

52 P.M. Johnson, A.F. Koenderink, and W.L. Vos. Ultrafast switching of photonic density of states in photonic crystals. *Physical Review B – Condensed Matter and Materials Physics*, 66(8):1–4, 2002.

53 T. Tanabe, K. Nishiguchi, A. Shinya, E. Kuramochi, H. Inokawa, M. Notomi, et al. Fast all-optical switching using ion-implanted silicon photonic crystal nanocavities. *Applied Physics Letters*, 90(3):88–91, 2007.

54 T. Tanabe, H. Taniyama, and M. Notomi. Carrier diffusion and recombination in photonic crystal nanocavity optical switches. *Journal of Lightwave Technology*, 26(11):1396–1403, 2008.

55 I. Fushman, E. Waks, D. Englund, N. Stoltz, P. Petroff, and J. Vučković. Ultrafast nonlinear optical tuning of photonic crystal cavities. *Applied Physics Letters*, 90(9):64–66, 2007.

56 K. Nozaki, T. Tanabe, A. Shinya, S. Matsuo, T. Sato, H. Taniyama, et al. Sub-femtojoule all-optical switching using a photonic-crystal nanocavity. *Nature Photonics*, 4(7):477–483, 2010.

57 A. Sharkawy, S. Shi, D.W. Prather, and R.A. Soref. Electro-optical switching using coupled photonic crystal waveguides. *Optics Express*, 10(20):1048, 2002.

58 H. Nakamura, Y. Sugimoto, and K. Asakawa. Ultra-fast photonic crystal/quantum dot all-optical switch for future photonic networks. *Conference on Lasers and Electro-Optics and 2006 Quantum Electronics and Laser Science Conference*, CLEO/QELS 2006. 12(26):6606–6614, 2006.

59 R. Soref. Tutorial: Integrated-photonic switching structures. *APL Photonics* [Internet], 3(2), 2018 doi: 10.1063/1.5017968.

60 Y. Yu, W. Xue, H. Hu, L.K. Oxenløwe, K. Yvind, and J. Mork. All-optical switching improvement using photonic-crystal Fano structures. *IEEE Photonics Journal*, 8(2):1–8, 2016.

61 J. Mork, Y. Yu, A. Sakanas, D.A. Bekele, E. Semenova, L. Ottaviano, et al. Photonic crystal Fano resonances for realizing optical switches, lasers, and non-reciprocal elements. In *Proceedings of SPIE (Vol. 10345)*, 65, 2017. doi: 10.1117/12.2273801.

62 G. Dong, Y. Wang, and X. Zhang. High-contrast and low-power all-optical switch using Fano resonance based on a silicon nanobeam cavity. *Optics Letters*, 43(24):5977, 2018.

63 G. Dong, W. Deng, J. Hou, L. Chen, and X. Zhang. Switches with improved switching dynamic characteristics. *Optics Express*, 26(20):25630–25644, 2018.

64 A. Granpayeh, H. Habibiyan, and P. Parvin. Photonic crystal directional coupler for all-optical switching, tunable multi/demultiplexing and beam splitting applications. *Journal of Modern Optics*, 66(4):359–366, 2019.

65 J. Zhan, Z. Jafari, S. Veilleux, M. Dagenais, and I. De Leon. High-Q nanobeam cavities on a silicon nitride platform enabled by slow light. *APL Photonics* [Internet], 5(6), 2020. doi: 10.1063/5.0007279.

66 D.B.S. Soh, R. Hamerly, and H. Mabuchi. Comprehensive analysis of the optical Kerr coefficient of graphene. *Physical Review A*, 94(2):1–11, 2016.

67 H. Liu, P. Liu, L. Bian, C. Liu, Q. Zhou, and Y. Dong. Electrically tunable switching based on photonic-crystal waveguide loaded graphene stacks. *Optics Communications [Internet]*, 410:565–570, 2017. doi: 10.1016/j.optcom.2017.10.074.

68 M.R.J. Azizpour, M. Soroosh, N. Dalvand, and Y. Seifi-Kavian. All-optical ultra-fast graphene-photonic crystal switch. *Crystals*, 9(9):1–10, 2019.

10

Optical-Electrical-Optical (O-E-O) Switches

Piyali Mukherjee

Department of Electronics and Communication Engineering, University of Engineering & Management, Kolkata, West Bengal, India

10.1   Introduction

The modern-day internet is very much analogous to the road highways that we encounter in our everyday lives. Consequently, the switching technology may be considered similar to the intersections located at those highways. At these major road intersections, each individual has to wait for their turn unless the traffic light signals the vehicles of a particular lane to pass through. The traffic lights are not naturally timed evenly as major lanes are often given more priority than side lanes. Occasionally, bypass lanes are designed to avoid traffic congestion at the busiest intersections.

Network switches for data transmission perform in a very similar fashion. "Switching" is comprehensively used to illustrate information sorting mechanism which is supposed to ingress into a node of the network followed by distribution through the most appropriate egress route. The preliminary designed information switches documented in literature, are the switches that were used in the rotary phones, being just electrical contacts these would spin on the motor shafts. These switches were later modified to be designed as a row of chassis. Additionally, the medium of data transport gradually embellished from copper cables to optical fibers. Optical switching technology has progressed generation after generations with enormous hypes and hubs associated with each progression. Needless to say, each of these transitions even remains today in some form or another through our existing technologies.

The last decade has witnessed the inception and formulation of modular data centers [1, 2], which is nothing but an individualistic system which contains within itself the servers, cooling systems, and the overall network. Several renowned big-shot organizations have been developing larger data centers of their own comprising smaller networks, while several server vendors have also lined up and are offering similar products [3–6]. Pods, often referred to as smaller networks, have recently been the centre of attraction for most systems [7] and networks [8]. Pods typically have the capacity to hold between 250 and 1000 servers, as documented in the literature so far. The only setback of such systems, even today, lies in the fact that interconnection of these pods to design bigger data centers is a significant challenge.

Optical Switching: Device Technology and Applications in Networks, First Edition. Edited by Dalia Nandi, Sandip Nandi, Angsuman Sarkar, and Chandan Kumar Sarkar.
© 2022 John Wiley & Sons, Inc. Published 2022 by John Wiley & Sons, Inc.

The three leading switching categories that are widely employed are the optical-electrical-optical (OEO), optical data unit (ODU), and reconfigurable optical add/drop multiplexers (ROADMs)-based switching systems. These different groups of switches offers combination of bandwidths which may serve a variety of purposes. The bandwidth providers aim at utilizing hybrid switching technologies as the best alternative reach to leverage multiple technology categories.

WDM drives innovation as it has brought about a huge revolution to optical switching. Legacy electrical switching technologies that have been used so far were unable to switch WDM signals on optical fibers.

WDM systems has the ability to merge differing wavelengths onto a single fiber using a combiner, also referred to as a multiplexer, and then separate out the wavelengths again using a splitter, sometimes called a demultiplexer. Since the early optical switches were wavelength-agnostic, signals needed to be split out, switched, and then recombined again. The problem hid behind was the potential for wavelength blocking that arose when a signal was separated and switched to a new wavelength, as it could clash with the wavelength of the combiner channel and had the possibility of being blocked from successful delivery. Finding ways to enhance these WDM networks has been an active area of interest for most researchers even today.

Wavelength Division Multiplexing (WDM) and optical switching technologies have proved to be very favorable technologies as these techniques are responsible for allocating bandwidths very flexibly throughout the data centre. Optical ports uniquely have the capability to carry several multiples of tens of Gb/s with the assumption that the entire data traffic is heading towards the same receiving end. The major setback to such systems is the switching time, which results in it making no sense to apply optical-based switching technologies to transmit data in the form of bursty communication from the transmitting host to a range of receiving hosts [6].

On the contrary, the data centers employing pod-based designs that are being presently used, provides opportunity to uphold the benefits of optical switching technology in terms of its high stability in such a design scenario. On occasions where the traffic demands are bursty, the pod-based data centre design seems prohibitive as there is a huge demand for complex circuitry. Electrical packet switching technology would be better suited under such conditions as compared to its optical counterpart.

10.2 Optical Switching Technologies: Working Principle

With the evolution of the communication systems, the type of information being transported and switched has also changed drastically. The dominant form of WDM traffic carried over fiber-optics became data packets of variable size aggregated and encapsulated into a session, as opposed to fixed-sized voice calls tightly interwoven into a synchronous stream [9–16]. Confronting these challenges triggered a series of optical-switching innovations and provided the impetus for the technological shifts that have resulted in the three major categories of optical switches today: OEOs, ODUs, and ROADMs [17].

10.2.1 Optical-Electrical-Optical Switching

Optical-electrical-optical switching is a broad category of switching technology encompassing a wide range of product sets, depending upon the type of information that has to be switched, and includes synchronous ADMs, cell-based ATM switches, and core IP packet routers, to just name a few. IP over DWDM (IPoDWDM) is merely a subset of OEO switching technology in which the

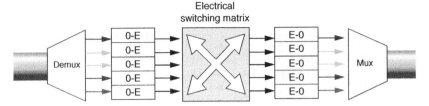

Figure 10.1 Optical-Electrical-Optical (OEO) switching mechanism.

WDM transponder is relocated and placed within the routing equipment to reduce the total number of cable connections that would have been required otherwise.

Irrespective of the type of information being switched, all OEO switches make use of simple, brute force in dealing with WDM traffic. The WDM optical aggregate is split out into individual channels, each of which is then converted to an equivalent electrical signal. Once all the signals are in the electrical domain, they are passed through traditional electronic switch fabric networks as shown in Figure 10.1.

The major problem with electrical switching fabrics is that they are generally massive, increasing as the square of the number of channels that has to be switched. So, considering an example, for an 80-channel, non-blocking square system, the electrical switching matrix would numerically consist of 6,400 switches. The capital and operational investments to deploy, manage and maintain switch matrices of this type are tremendous and often uneconomical.

The drive to make OEO switches less cumbersome still continues and is a thirst for most researchers. With the significant advancement in the planar lightwave circuit (PLC) technology, it has become possible to combine several passive optical elements on a single plate of glass. Photonic integration circuits (PICs), meanwhile, combine many of these active optical elements onto a common indium phosphide (InP) substrate integrated circuit. Despite these efficiencies, OEO switches still remain large, complex, expensive, and power-hungry, which are topics that still remain worth addressing.

10.2.2 Optical Data Unit Switching

Optical data unit switches are basically employed to streamline the size of the switch planes. When researchers were first envisaging forward error correction (FEC) – a technology borrowed from other fields of communications – to improve optical signal-to-noise ratio (OSNR), network architects working at standards bodies recognized it as a valuable possibility. Any extra bits in the overhead could result in traffic encapsulation, thereby enabling node bypass and resulting in switching at a higher level of granularity. In the early days of this digital wrapping, of what is now known more popularly as "ODU switching", network architects imagined that every data ranging from voice calls to packet streams could have been capsulated and switched in many layers of granularity. Today, this vision has become reality and is presently studied by several researchers across the globe.

Optical data units encapsulate information streams in encoded channels which are standardized under ITU-T G.709. Multiple levels of encoding and switching of data are possible. In a similar fashion, traffic streams still must be converted to the electrical domain, as with OEO switches, ODUs enable switching at a comparatively higher level immediately after the conversion from the optical to electrical domain as shown in Figure 10.2. This enables the requirement of a much smaller, simpler switch matrices by using fewer higher-granularity switches for bypass.

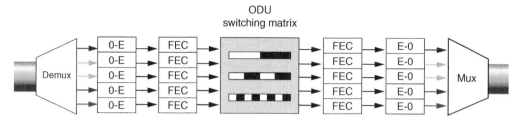

Figure 10.2 Optical Data Unit (ODU) switching mechanism.

However, there is a certain hindrance. If even a single packet of a given encapsulated stream is required at any time at a drop node, then that entire stream has to be sent to the main switch and broken down into its constituents. Now the ambiguity lies in determining how does the switch predict which traffic bypasses a network node, and what traffic gets sent to the main switch. To maximize the effectiveness of node bypass with ODUs, all traffic has to be carefully packaged and layered before transmission. For example, all traffic destined for one location should optimally be encapsulated within the same container.

10.2.3   Reconfigurable Optical Add-Drop Multiplexer (ROADM)-Based Switching

ROADMs are multiplexers or typically a different type of switch that seeks to leverage the advantages of node bypass while affording bandwidth providers an unmatched level of flexibility in network design.

Employing ROADMs does not result in conversion from optical to electrical domain, which can be seen from Figure 10.3, thus minimizing the overall electronics requirement and power consumption. Information streams are switched at the wavelength level of granularity in such switching system. This technique, of course, depends on prior knowledge of what is contained in the incident input wavelength, resulting in a much steeper requirement as compared to ODU switching, given that no electrical conversion has occurred. Network architects have come up with several novel approaches to solve this conundrum, ranging from manually pre-assigning each wavelength, to relaying intelligence from node to node on an optical supervisory channel, to hiding the needed information on an optical sideband to each channel.

Early ROADMs that involved hundreds of discrete optical switches, dielectric filters, and occasionally variable optical attenuators were rats' nests of fiber that rivalled the complexity of OEO switches. In fact, many early ROADMs comprised of solely a specialized subset of OEO switches. Today's ROADMs, however, have benefited greatly from the evolution of technology and typically leverage either PLCs or wavelength-selective switches (WSS) to more tightly integrate necessary functionality, thus helping to reduce ROADM complexity, and boost stability.

Wavelength-selective switch

Figure 10.3   ROADM employing wavelength-selective switching.

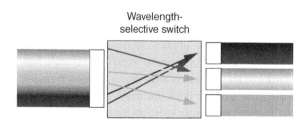

Today, several classes of ROADMs are in use especially by bandwidth network providers, thus enabling wavelengths to be switched as necessary with fluctuations in traffic requirements. Integrated optical attenuators has the ability to automatically rebalance channels as and when networks are reconfigured or new wavelengths are turned up, thus enabling add/drop wavelengths to be assigned to a fixed physical port on network equipment. This helps in minimizing bandwidth and enabling wavelengths to be switched not only between bypass and add/drop, but also in any network direction. State-of-the-art directionless ROADMs convey myriad benefits such as the ability to build mesh networks, to supplement fiber protection with path protection, and to adopt agile optical networking.

10.2.4   A hybrid approach

In the face of relentless growth in bandwidth consumption that threatens to outpace the growth of their own core Generalized Multi-Protocol Label Switching (GMPLS) IP networks, several bandwidth providers are turning to switching approaches which are optically hybrid as the solution to their predicament. Hybrid network architectures combine any and all optical switching technologies in an attempt to find the best balance of features for the intended application.

This section gives an overview of how multiple techniques may be combined to create a more efficient switched network. The core OEO IP routers are scaled as rapidly as possible and are used for packet-level switching granularity. OEOs are surrounded by ODUs that serve two functions. First, all encapsulated IP traffic that is not needed at the network node can bypass the router. Second, legacy SONET/SDH traffic can be switched around the core routers or offloaded to ADMs if needed.

ROADMs are used for switching at the wavelength level of granularity, again serving multiple purposes. First, the IP core routers may be interconnected as a mesh, allowing packets to travel the most direct path rather than around a ring. Second, path protection is provided in addition to traditional ring protection techniques. Third, an entire wavelength may be optically amplified and/or regenerated and passed around the ODU and IP packet switches.

Although at first glance using multiple optical-switching technologies might appear to be overly complex, it actually ends up greatly simplifying the overall network, as each type of information carried is properly routed around the network and only processed when absolutely needed. Coexistence of multiple optical switching technologies is likely to continue for some time now, given that bandwidth providers' capabilities and cost requirements in optical switching are likely to vary across their networks.

10.3   Optical Transponders

10.3.1   WDM Transponders: An Introduction

Owing to the mammoth development of wavelength division multiplexing (WDM) technology for networking, there is an increase in the volume of network traffic and thereby the demand for greater bandwidth is continuously increasing. For conversion of operating wavelength of the incoming bitstream to a wavelength that is compliant to the networking system, WDM transponders proves to be the most vital component in such systems. Being a vital technology in the optical fiber networks, WDM technology is setting itself as the baseline for the future all-optical networking systems. Most of the researchers are presently working on ways and means to optimize the WDM network.

The most important component of the WDM network, for this purpose, is the transponder, which is basically a device which is used to optimize the performance of the WDM network.

Today's optical network mainly relies on the WDM technology. Wavelength Division Multiplexing underlines the principle of assigning an independent dedicated wavelength to a single service of the network which is then multiplexed to a single optical fiber, thus eliminating the requirement of employing several such fibers. Besides, there is also an increase in the capacity of the fibers, which proves the enormous benefits that WDM system provides to both the service providers as well as to the receiving end users. The integral part of these WDM systems is the optical transponders as they serve as the most crucial medium for signal transmission of the entire system.

10.3.2 Basic Working of Optical Transponders

An optical transponder, better termed an OEO (optical-electrical-optical) transponder, is a device unit which basically converts optical to electrical and back to optical wavelength. Optical transponders have been utilized in diverse networks and are a potential candidate for several applications.

An optical transponder consists of a transmitter section and a receiver section, which is herein referred to as a responder, but they are functionally similar. It has the ability to cover wide ranges of transmission owing to wavelength conversion and signal amplification or regeneration. A transponder performs the function of receiving a signal, then amplifying it and finally re-transmitting the signal with a different wavelength, thereby retaining the original data or signal content. The signal that is received by a transponder is optical in nature, which is transduced to its electrical counterpart, and finally the processing of these electrical data takes place. Before transmitting, the transponder is responsible for converting this electrical signal to optical signals mainly in the form of CWDM (coarse WDM) or DWDM (dense WDM) wavelengths. Thus the process of OEO (optical to electrical to optical) conversion is well established by the transducer. In contrast to ordinary electrical transmission, the regeneration in present day WDM transponders is facilitated by the concept of reshaping, retiming and reamplifying the signal to a close proximity of accuracy.

Figure 10.4 depicts the working of a bidirectional transponder where the transponder is placed exactly midway between a transmitting device and an optical DWDM system [18]. It may be well

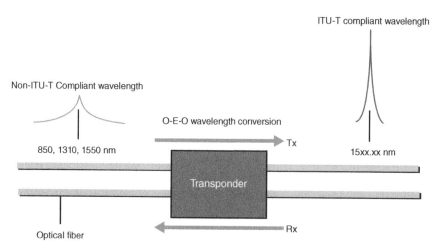

Figure 10.4 Working of an OEO transponder.

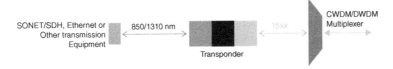

Figure 10.5 Schematic block of optical transponder. Source: [19].

observed that the transponder receives an optical input from the left, which operates at fixed wavelengths of 1310 nm or 1550 nm. The transponder then converts this input wavelength to an ITU-compliant wavelength, as seen on the right, and finally results in transmitting it to the output DWDM system. This entire process gets reversed when looked back from the receiving end. In such case, the ITU-compliant data serves as the input to the transponder, which is finally converted back to the optical signal that is obtained by the client on the receiving end.

An optical transponder basically works as a wavelength regenerator that is responsible for converting an optical signal to its corresponding electrical signal, thereafter generating an exact replica of the input optical signal so as to use this signal to finally generate signals at different optical wavelengths, thus resulting in optical-electrical-optical conversion. The significant characteristics of optical transponders are that it has the ability to receive, amplify and re-transmit a particular signal to an altered wavelength without distorting the content of the input optical signal. The client may be electrical or optical (operating at frequencies 1310 nm or 1550 nm), which may be either co-located or are positioned at a certain distance. The interface may be lined by fibers, Coarse Wavelength Division Multiplexers (CWDM), or Dense Wavelength Division Multiplexers (DWDM). Figure 10.5 depicts a basic block diagram of an optical transponder [19].

10.3.3 Necessity of Optical Transponder (OEO) in WDM System

In recent times, the optical transponder has become very much a necessity in WDM system network for several reasons, as listed below.

Firstly, if a network comprises of different equipment operating at varying wavelengths urge to communicate amongst themselves within the network, a problem of incompatibility may arise. The optical transponder has the capability to address and overwrite such issues.

Additionally, there may be situations where several service providers provide optical fibers that are designed to meet different standards. Under such scenario, a WDM transponder enables traversing from one optical network to another within the system.

10.3.4 Applications of Optical Transponders

Optical transponders are widely used in WDM networking in addition to several other applications. The most commonly employed applications of optical transponders are stated below.

- **Multimode to single-mode conversion:** Optical transponders have the ability to convert multimode fibers to single-mode ones, with designs for short-distance to long-distance lasers, and/or 850 or 1310 nm to 1550 nm wavelengths conversions documented so far. The optical transponder module obeys all networking protocols and independently operates irrespective of the adjoining wavelength channels.
- **Dual fiber to single fiber conversion:** Inter-conversions between dual fiber and single fiber is very much required for design of any networks. In case of dual fibers, a single wavelength is

transmitted over two distinct fiber strands, while in case of single fibers both the wavelengths are transmitted over a single fiber strand, thereby enabling a bidirectional transmission mechanism. Two optical transponders are required for the purpose of dual fiber switching, while a single fiber has the capability to transmit two different wavelengths over a single fiber channel, as they can transmit and receive from either end.

- **Introduction of additional fiber path:** An optical transponder module may also independently include an additional (often treated as redundant) fiber path option for supplementary protection. The additional fiber path allows transmission of the input optical signal to two separate optical channels which are directed towards the two redundant receivers at the receiving terminal. Under any condition, the loss of the primary path would result in switching on the backup receiver. The said phenomena is electronically controlled, which results in much faster and more reliable operation.

- **Optical repeaters:** For the purpose of long-haul communication, WDM transponders have been reported to function as repeaters in order to expand the distance of network by the process of wavelength conversion and optical power amplification. For employment as optical repeaters, the optical transponders regenerate the incoming signal effectively so as to traverse the optimum distance of desired transmission. With this option of signal regeneration, the degraded signal, as an outcome of the transmission process, may be debittered and retransmitted to achieve high quality of signal.

- **Mode conversions:** Optical transponders may be employed for the purpose of mode conversion. Converting modes result in quicker and simpler means of transmitting multimode optical signals to larger network distances employing a single-mode fiber. It may be observed that most of the present day receivers are designed to have the capability of receiving both multimode and single-mode optical signals over the same fiber.

- **Wavelength conversions:** It has been documented in literature that conversion of wavelengths employed in commercial networks even today is only carried out with the aid of optical transponders. It is known that the existing fiber-based optical networks and interfaces naturally operate at the conventional wavelengths of 850 nm, 1310 nm and 1550 nm, thus implying that conversion to its corresponding CWDM or DWDM wavelengths is required to operate within the system.

10.3.5 Network Structure with Optical Transponder

The overall network system immensely benefits with the aid of optical transponders. In this section, two possible WDM ring network configurations which utilizes optical transponder are documented [20, 21].

10.3.5.1 WDM Ring Employing Line Network

A line network is a network formed by two point-to-point links between the nodes A-B and B-C respectively as shown in Figure 10.6. Each of these links are supported with optical transponders at the end points. The system works in such a fashion that if there is a failure in node B, nodes A and C would continue to communicate between each other as there is a bypass path established between the optical transponders which is adjacent to node B, thereby switching the transponders to the protection mode. Figure 10.6 shows the block diagram of a WDM ring line network.

10.3.5.2 WDM Ring Employing Star Network

A star network is so named because of the star-like connection between the nodes, where all the other nodes A, C and D are connected to a star node B. The star node is backed up with a similar

Figure 10.6 Block diagram of line network over WDM ring.

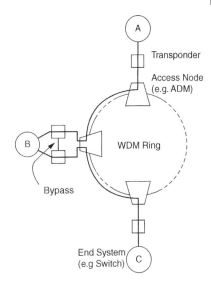

Figure 10.7 Block diagram of star network over WDM ring.

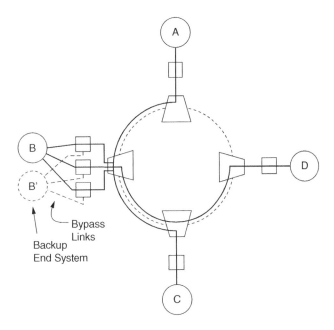

node which basically serves as a redundant node, as may be seen in Figure 10.7. In this type of configuration, the protection mode is switched from the star node to the redundant node once the star node fails to work.

10.3.6 Differences Between Transponder, Muxponder, and Transceiver

In optical fiber networks, the transceiver and the muxponder perform similar functionalities to that of an optical transponder. They are mainly responsible for transmitting and receiving optical signals within the network [22]. Though being functionally similar, there are crucial difference between these three devices when considered from their design perspective and applications.

The transceivers employed to be used in optical fibers have limited functionality in terms of electrical-optical serial transmission only. Optical transponders, on the other hand, may be used in parallel interfaces as well and performs wavelength conversion of optical signals without causing any signal distortion. WDM transponders can be thought of as two back-to-back placed transceivers. The fiber-optic muxponders behave functionally similar to WDM transponders with the additional feature of multiplexing. The muxponder has the ability to multiplex several sub-rate interfaces into an interface of much higher rate.

10.3.7 Summary

In WDM networking systems, optical transponders hold an extremely high position of interest which can never be underestimated. They form the underlying functionalities such as reception of optical signals, followed by amplification or regeneration of such signals and finally retransmission of this signal onto a completely different wavelength, thereby making it the most promising device in networking systems.

The optical transponders have also shown its versatility in several applications such as interconversion of signals including transmission shift from multimode fibers to single-mode ones.

10.4 Performance Analysis Study of All-Optical Switches, Electrical Switches, and Hybrid Switches in Networks

A comparative study of all-optical switches, electrical switches, and hybrid switches in relation to data center networks as well as telecommunication networks is presented in this section. The simulation studies of the response time and the optimum output achieved so far documented reveal the advantages which the hybrid switches have over their electrical and all-optical counterparts [23].

10.4.1 Introduction

The exponentially growing traffic mainly contributes to the demands of the present architecture of the Internet and to support our demands of the service in terms of its quality, class, and its type on an unified podium. The internet service utilized by us today is cored by over-provisioned connection of utilities (technically called nodes), employing the most powerful present-day technology of Wavelength Division Multiplexing (WDM). With the exponential surge of data traffic, there is a continuous and dynamic change of the patterns of data traffic with an enormous growth in data capacity that finally presses the requirement for a more versatile, agile, and scalable networking system which may have the ability to handle this data traffic efficiently.

Additionally, the increasing demand for data and managing data traffic has led to the birth of huge-dimensioned data centers so as to ease the extraction of data easily. The present-day data centers have expanded manyfold, thus leading to consumption of enormous power in the order of tens of megawatts. With the restrictions in power consumption, the data centers aim to result in higher productivities with a paradigm shift to optical switching technology which may be provisioned to be exploited. Today's datacenters mainly rely on the trending components of electronic switches employing Ethernet or Infiniband [24, 25], which results in lower output and delayed system response.

In recent times, it has been reported that both in the telecoms sector as well in data centers, hybrid switch technologies besides all-optical switching have been employed to take care of the huge data traffic effectively. This section aims to collate all similarities and differences of both the stated switching technologies in the light of state-of-the-art electronic switches.

10.4.2 Optical vs. Electrical vs. Hybrid Telecom Switches

Literature reports that the present-day telecom network both at the local and at the global level is formed by the interlocking of nodes, which have connection of the order of two to five distinct associated connections. Though the degree of the links may be thought of to be extremely small, the connection amongst the nodes has immensely high capacity, which ranges from hundreds of gigabytes per second to the order of terabytes per second. Dense WDM technology was the only hope for realizing such high-capacity links within the network, where several varying wavelengths may be transmitted parallelly through the optical fiber link.

The connecting path from the transmitting to the receiving end is often connected through several nodes with multiple hops being present across it. The greater the number of hops within the network, the higher the latency is, thus increasing the overall response time. Typical telecom switches within the network are generally surrounded by the edge routers, which are mainly responsible for realizing functions such as aggregation of data packets and data traffic orientation such that the link is utilized within the network in the most efficient way possible. So if the switching system is designed with the aid of solely all-optical switches, they turn out to be equally beneficial for electronic data processing as well.

An electrical switch, on the other hand, employs principles of Optical-to-Electrical and Electrical-to-Optical conversions at the input and output terminals respectively. The optical signals fed at the input of the switch are first transformed to their electrical counterpart and finally stored. The responsibility for scheduling the process and transmitting the input signals from the input buffer to the output buffer is done by the control plane. This results in extremely high switching capacity once all the incoming packets are made to cross the network traffic.

In order to realize such increased capacity electrical switches, smaller electrical switches are assembled either in serial to parallel configuration mode or in parallel to serial configuration mode to achieve high capacity. Each of the incoming data packets have to traverse several stages of processing, which in turn results in higher power consumption and high response time. In spite of this, the packet loss may be reduced to a greater extent if the buffer capacities can be significantly improved.

On the contrary, in case of all-optical switches [26] as depicted in Figure 10.8, the data is potentially optical in nature while the control plane is devised to remain in the electrical domain. In order to increase the capacity for optical switches, exploiting the property of parallelism of wavelengths within the system is harnessed. To achieve higher speed and to ensure non-blocking switching, an Arrayed Waveguide Grating Router (AWGR) may be employed for different wavelengths which would occupy parallel paths.

At the input of the switching network, tunable wavelength converters (TWCs) are positioned so as to enable proper routing of wavelengths from an input terminal to an output terminal. In order to maintain a synchronization in wavelengths between the input link and the output link, fixed wavelength converters (FWCs) are generally placed at the output end of such networks.

Although no mature optical buffer design has been so far documented in the literature, yet these switches have the ability to effectively use all the wave dimensions such as wavelength, space, and time domains [26]. Additionally, in the case of all-optical switches, there is no concept of

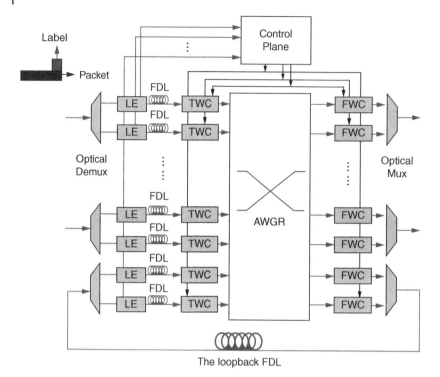

Figure 10.8 All-optical switch design for telecom network. Source: Based on [26].

store-and-transmit processing or bit-by-bit processing, thus resulting in conservation of power as compared to its electrical counterpart. In order to route wavelengths within the network, the AWGR switching technology serves to be more passive and promising as it consumes no power.

In spite of the several similarities that the hybrid optical switch holds with respect to the all-optical switch, there are certain differences. The lookback FLDs in the case of all-optical switches are replaced by a loopback shared buffer system in the case of hybrid switches, which is depicted in Figure 10.9.

An all-optical switch has the capability to handle (w*m) inputs and outputs, where w represents the total number of incoming wavelengths to the switch. In contrast to this system, hybrid switches employ loopback shared buffer system which is generally designed to use only one input and one output line within the network.

Analogous to the all-optical switch, the control plane of the hybrid optical switch upon contention first looks for dissimilar wavelengths which may be made available at the same output link. Once the situation develops such that all the wavelengths get occupied at the output link, the packets in such scenario would be sent back via the loopback shared buffer and would then wait for its turn for retransmission. A negative situation may arise if the shared buffer doesn't have enough space to accommodate the package. Under such scenario, the packet would be dropped.

It is worth mentioning that the loopback shared buffer outperforms loopback FDLs owing to the fact that it introduces arbitrary delay due to the occupancy of fewer ports within the switching network. Once the transmission line is available, the delayed packet are then released. The setback of loopback shared buffer is that it consumes power due to its larger dimension, although researchers are working to reduce the size of this loopback shared system by exploiting multiplexing techniques statistically.

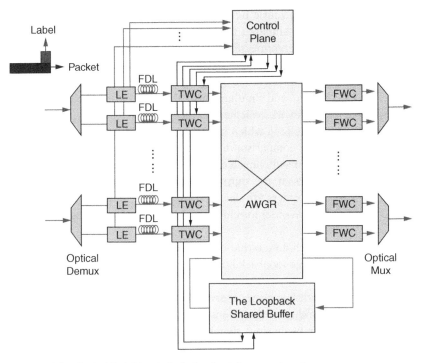

Figure 10.9 Optical hybrid switch design for telecom network.

In contrast to all-optical switches, the optical hybrid switches are documented to produce lower response delay as transmission is initiated as soon as the transmission channel becomes available. In addition to this, the optical hybrid switches offer advantages by lowering the rate of packet dropping since it has the ability to hold the packet before the retransmission is set up.

10.4.3 Optical vs. Electrical vs. Hybrid Data Center Switches

Keeping in pace with the demands of modern technology, data centers have emerged to promote connection among hundreds or thousands of nodes within the network in contrast to the existing traditional telecom network switches. Additionally, data centers have been designed in such a fashion so as to reduce the latency originating between the end nodes. Data centers mainly focus on employing parallel transmission to meet the ever-increasing user demands. In such a situation, latency delay in the order of nanoseconds might cause a drastic degradation in the overall system performance.

Electrical switches face restrictions on the number of ports it can handle. So the existing electrical switching system is based on employing several smaller electrical switches associated with smaller port counts which are assembled in such a fashion so as form larger system by assembling many such smaller counterparts [27], with the final aim of accumulating and interconnecting larger number of nodes. The major drawback of this kind of union structure lies in the fact that though the latency of smaller switch is less, when interconnected the number of hops increases, thus leading to a significant increase in the overall latency of the system.

The performance study documented in the literature makes use of a network called an electrical flatten butterfly network [28], which serves as an alternative to existing electrical switches.

In order to compare, an optical switching fabric network like AWGR is used, which has the ability to provide higher values of ports compared to an electrical switching fabric network, thereby resulting in greater number of interconnections to the end nodes compared to those in an electric switch.

A requirement for a multi-stage switching system may arise, especially when connection of hundreds or even thousands of nodes is desired, and simultaneously the number of switches required would also reduce, thereby making the network design even much simpler.

An optical hybrid switching architecture [29], which has been mainly designed for data centre networks, is depicted in Figure 10.10. In comparison to the optical hybrid switches which are employed for telecom networks, the similar switching technology which is designed to be used for the data centers does not require FWCs at the output terminal of the switching network. Additionally, the switches used in data centers are designed to meet asymmetric link capacity between the incoming and outgoing links in order to achieve the benefits that wavelength parallelism promotes.

Figure 10.11 depicts an all-optical switch design that is typically used in data centre networks. The switch is designed to replace the shared loopback buffer used in telecom networks with multiple fiber links (FDLs).

The electrical flatten butterfly network has been observed to have an extremely poor performance when compared to the other two other combinations of optical switches. The response time in this switching mechanism is comparatively slow, thereby decreasing the effective bandwidth even at lower incoming packet range. On the other hand, both all-optical switches and optical hybrid switches are seen to have very poor latency for most network configurations. The hybrid optical switch is expected to provide its best performance in terms of latency and hence serves to be the best possible candidate for data center switches.

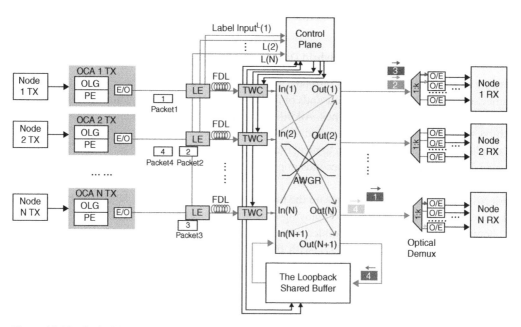

Figure 10.10 Optical hybrid switch (DOS) design for data center network. Source: [29].

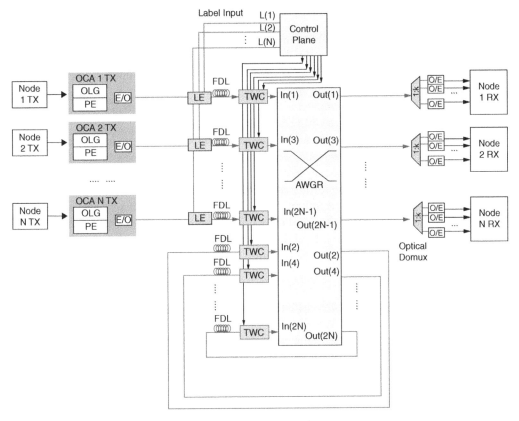

Figure 10.11 All-optical switch design for data center network.

10.4.4 Summary

This section consists of a detailed comparison of all-optical switches, electrical switches, and optical hybrid switches which were designed to work in both telecommunication networks and data center networks. Owing to the significant measurements, such as latency, throughput, and power consumption, optical switches has been observed to have outperformed other electrical switches. The lowest latency is achieved by the optical hybrid switch, but this suffers drawbacks in terms of greater power consumption compared to the all-optical switch.

10.5 Electrical and Optoelectronic Technology for Promoting Connectivity in Future Systems

Optical interconnects have been reported to surpass the most important figure of merit, i.e., the bandwidth-distance product beyond the historic figure of 100 Gbps/meter, while the advances in CMOS technology and signal processing have made it possible for electrical interconnects to break the barrier of 500 Gbps/meter as documented so far, to approach the level of optical interconnects [30].

10.5.1 CMOS Technology

In the last few decades, there has not been much significant improvement in the performance of transistors, and researchers have been able to achieve only a marginal improvement in terms of lowering the power consumption during operation. This has resulted in limits to incorporating larger number of transistors in a smaller space and hence not much refinement as regards density can be achieved. To overcome this, employing multicore parallel architectures has been considered as an alternative to meet the ever-increasing demand [31–35].

In order to achieve the targeted performance, parallel architectures may be required, and interconnect technology is the key to enable it. There are certain drawbacks to these interconnects such as system throughput, increased latency and system density, which pose a major bottleneck to system performance, thus hindering further advances for future architectures.

The solution of choice to define the future of networking has always been a debatable topic ever since the emergence of optical and electrical interconnect technologies as to which should be the best possible candidate in this regard. Both the stated technologies offer trade-offs depending on applications, which may based on power consumption, reliability, density of integration, and even cost.

Network system engineers had been initially reluctant to replace electrical interconnects, which were basically cumbersome, in favour of better optical solutions, which offered advantages such as lower complexity and lower cost (typically when used for longer distances), until in places the electrical interconnect was deemed to be the only possible solution. In the recent times, optics have been seen to outperform its electrical counterpart as most of the present generation of technologies entirely rely on it due to its extremely high-performance matrix. Again, it too suffers drawbacks owing to the cost advantage that the electrical interconnects possess. So optical interconnects have suited themselves for applications in longer distances, where attenuation seems to be a major hindrance for electrical interconnects.

Future technologies might face some other factors for design depending on the future requirements and may smash the existing paradigms. Optimizing the trade-offs for the existing solutions, and examining the requirements and applications for future-generation systems, would be a potential domain of research.

10.5.2 Considerations for Selection of Interconnects

There are several factors which are responsible for the selection of the best alternative to meet the requirements of any system. The selection of interconnect technology is mainly taken into consideration depending on the targeted application.

An important figure to take into consideration for any interconnect selection is the throughput of the device, which may be defined as the throughput of the transmitter when added to the throughput of the receiver that has the possibility of being integrated on a single chip. The power consumed by the transceiver and its total design area pose limitations to obtaining the maximum throughput for any device. Larger dimensions of transceivers possess fabrication and manufacturing issues.

Throughput-distance product (Gbps/meter) is a metric that is being mostly used to evaluate the performance of an interconnect, by in turn evaluating the throughput-distance or bandwidth-distance products [36]. The value of this metric remains constant for a given interconnect technology.

Optical interconnects have always proved to be a better candidate as they offer superior bandwidth compared to other existing technologies. Research reveals the fact that optical links have advantages over electrical links at the throughput of 100 Gbps/meter [37–39]. They have a bottleneck due to the additional area they require to embed the optical functionalities which limits their effective density.

Another important figure to consider is the cost involved with interconnect manufacturing. Economical interconnects would always be at the forefront. The total cost involved depends on several factors such as the cost of the individual components, manufacturing cost, and deployment cost. There is always an additional cost associated with optical interconnects as these rely on components such as lasers, lenses, optical substrates, optical sources, and photodetectors as well as the components, resulting in their interconnections which are non-existent in case of electrical interconnects. Although the optical technology has the capability to integrate most of the components within the chip, the optical source, being the key element, fails to be integrated owing to its cooling purpose [40–42].

The advantages of electrical interconnects lie in the fact that higher level of interconnections is achievable owing to CMOS technology without the requirement of any associated components as was in the case of optical solutions. Signal conditioning may be required, depending upon which electrical links introduce insertion loss compensation [43], and which requires Reed–Solomon (RS) forward error correction (FEC) [44].

The most important concern is the interconnect reliability, which becomes a prime concern in larger systems employing several nodes, as a single nodal failure may result in the entire network failure if they are not addressed correctly on time. The susceptibility of failure of optical interconnects are higher as they involve the use of several additional components.

With the advancement in integrated circuit technology, there is a paradigm shift to faster and more power-efficient FinFet devices, and researchers claim that electrical interconnects will still remain in data centers as well as in the communications chassis well beyond 2020 [45].

10.6 Conclusion

Over the past decade, there has been an enormous hue and cry for higher data rates, which led to the design of switching technologies so as to guide the network with minimum or no system-generated loss. Standing at this juncture, which optical switching technology is the best continues to be a fiercely debated topic. OEOs, ODUs, and ROADMs have each taken a turn as the proposed optical switching technology to end all optical switching technologies.

ROADMs (especially multi-degree varieties) were for a time regarded as the end game in optical switching. Then PICs and PLCs came along, and there was a surge of predictions that higher-density integration would favor the use of Optical-Electrical-Optical (OEO) switching. Recently, ODU switching has again made a comeback in the field of switching technologies.

In reality, each of these switching categories has its pros and cons, and identifying the optimal approach and determining a trade-off as to whether one of the increasingly prevalent hybrid options is appropriate depends upon a particular bandwidth provider's infrastructure and the applications which it is designed to address.

Bibliography

1 J.R. Hamilton. An architecture for modular data centers. CIDR, 2007. arXiv:cs/0612110.

2 H. Wu, G. Lu, D. Li, C. Guo, and Y. Zhang. A high performance network structure for modular data center interconnection. *ACM CoNEXT'09*, 2009.

3 B. Canney. IBM portable modular data center overview. http://www-05.ibm.com/se/news/events/datacenter/pdf/PMDC, Introducion-Brian Canney.pdf, 2009.

4 HP Performance Optimized Data center, ftp://ftp.hp.com/pub/c-products/servers/pod/north_america _pod_datasheet 041509.pdf.

5 S. McNealy, V. Khosla, A. Bechtolsheim, and B. Joy. Sun Modular Data center. http://www.sun.com/service/sunmd, 2010.

6 N. Farrington, G. Porter, S. Radhakrishnan, H.H. Bazzaz, V. Subramanya, Y. Fainman, G. Papen, and A. Vahdat. Helios: A hybrid electrical/optical switch architecture for modular data centers. *SIGCOMM'10*, 2010.

7 K.V. Vishwanath, A. Greenberg, and D.A. Reed. Modular data centers: How to design them? *Proc. of the 1st ACM Workshop on Large-Scale System and Application Performance (LSAP)*, 2009.

8 C. Guo, G. Lu, D. Li, H. Wu, X. Zhang, Y. Shi, C. Tian, Y. Zhang, and S. Lu. A high performance, server-centric network architecture for modular data centers. *ACM SIGCOMM*, 2009.

9 K.J. Barker, A. Benner, R. Hoare, A. Hoisie, A.K. Jones, D.K. Kerbyson, D. Li, R. Melhem, R. Rajamony, E. Schenfeld, S. Shao, C. Stunkel, and P. Walker. On the feasibility of optical circuit switching for high performance computing systems. *SC'05*, 2005.

10 M.F. Tung. An Introduction to MEMS Optical Switches. https://courses.cit.cornell.edu/engrwords/final_reports/Tung_MF_issue_1.pdf, 2001.

11 D. Vantrease, R. Schreiber, M. Monchiero, M. McLaren, N.P. Jouppi, M. Fiorentino, A. Davis, N. Binkert, R.G. Beausoleil, and J.H. Ahn. Corona: system implications of emerging nanophotonic technology. *ISCA'08*, 2008.

12 G. Wang, D.G. Andersen, M. Kaminsky, M. Kozuch. T.S. Eugene Ng, K. Papagiannaki, M. Glick, and L. Mummert. Your data center is a router: The case for reconfigurable optical circuit switched paths. *ACM HotNets'09*, 2009.

13 A.G. Eantc. Arista 7148SX Switch. http://www.aristanetworks.com/en/7100_Series_SFPSwitches, 2009.

14 Calient Networks. http://www.calient.net, 1999.

15 Cisco Data Center Infrastructure 2.5 Design Guide. www.cisco.com/application/pdf/en/us/guest/netsol/ns107/c649/ccmigration_09186a008073377d.pdf, 2013.

16 P.M. Zeitzoff. 2007 International Technology Roadmap: MOSFET scaling challenges, *Solid State Technology*, 51(2):35, 2008.

17 J. Theodoras. Contemporary categories of optical switching. https://www.lightwaveonline.com/network-design/dwdm-roadm/article/16649750/contemporary-categories-of-optical-switching.

18 Tutorials of fiber optic products. https://www.fiber-optic-tutorial.com/category/network-solutions/wdm-optical-network/optical-transponder, 2017.

19 Fiber optic solutions. https://www.fiber-optic-solutions.com/optical-transponder-o-e-o-wdm-network.htm, 2017.

20 A.V. Krishnamoorthy. The intimate integration of photonics and electronics. *Advances in Information Optics and Photonics. SPIE*, 2008.

21 R. Ramaswami and K. Sivarajan. *Optical Networks: a Practical Perspective*. San Francisco, Morgan Kaufmann, 2002.

22 Irving. The versatile optical transponder (OEO) in WDM system. https://community.fs.com/blog/the-versatile-fiber-optic-transponder-oeo-in-wdm-system.html, 2015.

23 X. Ye, V. Akella, and S.J.B. Yoo. Comparative studies of all-optical vs. electrical vs. hybrid switches in datacom and in telecom networks. *Optical Fiber Communication Conference/National Fiber Optic Engineers Conference*, 2011.

24 InfiniBand Architecture Specification, Volume 1, Release 1.0. http://www.infinibandta.org/specs, 2002.

25 Voltaire Vantage 8500 Switch. http://www.voltaire.com/Products/Ethernet/voltaire_vantage_ 8500, 2009.

26 H. Yang and S.J.B. Yoo. All-optical variable buffering strategies and switch fabric architectures for future all-optical data routers. *Special Issue on Optical Networks, IEEE/OSA Journal of Lightwave Technology*, 23(10): 3321–3330, 2005.

27 Al-M. Fares, A. Loukissas, and A. Vahdat. A scalable, commodity data center network architecture. *SIGCOMM '08*, 2008.

28 J. Kim, W.J. Dally, and D. Abts. Flattened butterfly: a cost-efficient topology for high-radix networks. *34th Annual International Symposium on Computer Architecture (ISCA)*, 126–137, 2007.

29 X. Ye, P. Mejia, Y. Yin, R. Proietti, S.J.B. Yoo, and V. Akella. DOS – A scalable optical switch for data centers. *ACM/IEEE Symposium on Architectures for Networking and Communications Systems (ANCS)*, 2010.

30 R. Farjadrad. What's the difference between optical and electrical technology for 100-Gbit/s connectivity in future systems?. https://www.electronicdesign.com/technologies/communications/ article/21800130/whats-the-difference-between-optical-and-electrical-technology-for-100gbits-connectivity-in-future-systems, 2014.

31 G. Kalogerakis, T. Moran, T. Nguyen, and G. Denoyer. A quad 25Gbps 270mW TIA in 0.13um BiCMOS with, 0.15dB crosstalk penalty. *ISSCC Dig of Tech Papers*, 116–117, 2012.

32 T. Tekemoto, et al. A 4x25Gb/s 4.9mW/Gbps-9.7dBm high sensitivity optical receiver based on 65 nm CMOS for board-to-board interconnects. *ISSCC Dig of Tech Papers*, 118–119, 2012.

33 J.Y. Jiang, et al. 100Gb/s Ethernet chipset in 65nm CMOS technology. *ISSCC Dig of Tech Papers*, 120–121, 2012.

34 S. Parikh, et al. A 32Gb/s wireline receiver with a low-frequency equalizer, CTLE and 2-tap DFE in 28nm CMOS. *ISSCC Dig of Tech Papers*, 28–29, 2012.

35 Y. Doi, et al. 32Gb/s data interpolator receiver with 2-tap DFE in 28nm CMOS. *ISSCC Dig of Tech Papers*, 36–37, 2012.

36 A.V. Krishnamoorthy, K.W. Goossen, W. Jan, et al. Progress in low-power switched optical interconnects. *IEEE Journal of Selected Topics in Quantum Electronics*, 2011.

37 M.A. Taubenblatt. Optical interconnects for high-performance computing. *IEEE Journal of Lightwave Technology*, 2012.

38 B.E. Lemoff, et al. Demonstration of a compact low-power 250Gbps parallel-DWM optical interconnect. *IEEE Photonics Technology Letters*, 220–222, 2005.

39 F.E. Doany, et al. Terabit/s-class 24-channel bidirectional optical transceiver module based on TSV Si carrier for board-level interconnects. *Proc 23rd Annual Meeting IEEE Photon Society*, 564–565, 2010.

40 D. Gockenburger, et al. Advantages of silicon photonics for future transceiver applications. *Proceedings of 36th Euro Conference on Optical Communication*, 1–6, 2010.

41 D.V. Thourout. Si Photonics. *Proceeding of Optical Fiber Communication Conference*, 2012.

42 H. Thacker, et al. Hybrid integration of silicon nano-photonics with 40 nm-CMOS VLSI drivers and receivers. *Proc. IEEE 61st Electronic Components Technology Conference*, 829–835, 2011.

43 S. Ibrahim and B. Razavi. A 20Gbps 40mW equalizer in 90nm CMOS technology. *ISSCC Dig of Tech Papers*, 170–171, 2010.

44 P. Dave and J. Petrilla. Proposal for FEC in40G/100G ethernet. *IEEE 802.3 Ethernet Working Group Proposals*, 2008.

45 E. Wu. A framework for scaling future backplanes. *IEEE Communication Magazine*, 188–194, 2012.

11

Quantum Optical Switches

Surabhi Yadav and Aranya B. Bhattacherjee

Department of Physics, Birla Institute of Technology and Science-Pilani, Hyderabad Campus, Telangana, India

11.1 Introduction

Enormous data transfer within optical fiber networks has created a global communication revolution that has impacted human life and modern economic activity. High bit rate optical connections have been extended to numerous other networks over the last few decades, from long-distance transmission networks to short-distance communications inside buildings [1]. A photonic switch with an ultrashort decay time is one of the main components in establishing a high-speed optical signal processing device [2]. The high power requirement of photonic to electrical and electrical to photonic conversions inspired the concept of using photonic networks as a substitute for modern electronic switching. Terahertz switching speed is possible with an optical switch that controls light signals directly through another optical beam, with possible restoration times in the pico- or femtosecond range. The accessibility of such a system together with the use of low-dissipation photonic waveguides and fibers could pave the way for a high bit rate, low-cost optical network over short distances. For efficient all-optical switching, optical non-linearity with low power consumption (low photon number) is a necessary prerequisite. Optical non-linearity at single photon level have been demonstrated in many studies over the years, indicating the possibility of implementing low-power and high bit rate photonic devices for classical information processing [3]. Solid-state devices based on a sole quantum dot coupled strongly to a nano/micro optical cavity are potential candidates for such applications. These devices have a limited carbon footprint and are compliant with standard nano-fabrication procedures.

High difference switching, switching at a high speed, and multi-wavelength configuration were also achieved using photonic switching based on saturable absorbers fabricated using quantum wells in vertical micro cavities [4]. The use of resonant distributive Bragg reflectors (DBR) structure is one way to reduce the energy loss in optical switches. The use of semiconductor quantum dots within DBR micro cavities for high-speed optical switching was studied in a recent experiment [5] and theoretical work [6].

Optical Switching: Device Technology and Applications in Networks, First Edition. Edited by Dalia Nandi, Sandip Nandi, Angsuman Sarkar, and Chandan Kumar Sarkar.
© 2022 John Wiley & Sons, Inc. Published 2022 by John Wiley & Sons, Inc.

The combination of quantum dots and micro cavities allows for an excellent method for studying cavity quantum electrodynamics. The potential of a quantum dot in a mesoscopic cavity to produce non-linear optical effects motivates us to investigate the opportunity of controlling the photon statistics in these systems for practical applications [7]. Single InAs quantum dot in a GaAs photonic crystal cavity was used to explain electro-optical switching at the quantum level [8]. In cavity quantum electrodynamics, the device works in the robust/strong coupling regime. The quantum-confined Stark shift is used to tune the quantum dot by changing the quantum dot frequency [9]. Within nm-sized clusters, quantum dots have three-dimensional confined states [10]. In ultrafast photonics, quantum dots have a wide range of applications. Mode-locked quantum dot lasers and a quantum dot semiconductor saturable absorber mirror have used the ultra-broad optical range of self-assembled quantum dots as an absorption saturator [11].

The rapid growth and versatility of opto-mechanical systems and optomechanics have sparked a lot of interest and study over the last few decades [12]. Cavity optomechanics is a branch of microphysics and quantum optics that studies how electromagnetic field interacts with mechanical systems. Optomechanically induced transparency (OMIT) [13], quantum entanglement [14], quantum synchronization [15], quantum state transfer [16], and optomechanically induced non-reciprocity [17] are only a few of the recent developments and breakthroughs in quantum optomechanics. As a result, it establishes a strong base for quantum optical communication and quantum information processing.

A high level of nonlinearity occurs between the optical and mechanical modes in optomechanical systems, resulting in optical bistability and multistability. This optical phenomenon has applications in optical switching and memory storage in general [18]. Because of the optomechanical nonlinearity, the bistable behavior exhibits switching actions controlled by adjusting the laser strength, rocking parameter, optomechanical coupling, and QD cavity coupling [19]. Through micro-mechanical oscillators, a quantum interface can pass the quantum state between different degrees of freedom. Another promising system for implementing quantum switches is a electro-opto-mechanical system in which a mechanically compliant DBR/membrane is interacting simultaneously with a microwave as well as an optical cavity [20]. Such a hybrid system has been shown to exhibit tunable optical switching behavior. Also, reversible quantum state transfer between microwave and optical photons is possible with hybrid electro-optomechanical systems [21].

11.2 Quantum Dot as an Optical Switch

The speed of electronic components in various electronic communication devices limits the capacity of photonic communication systems that can exceed 10 Tb/s due to the high frequency of the optical carrier. Electronic components can be substituted with high-speed all-optical signal processing components to overcome these limitations [22]. As switches require an extremely fast response, low laser power consumption, and the ability to accommodate multiple channels, photonic switches for future optical networks are looking to be a suitable candidate for 2DPC (two-dimensional photonic crystal) applications [23].

Ultrafast semiconductor devices based on quantum dots (QDs) can operate at speeds of 10 to 100 GHz or even higher. The following are some of the significant benefits of QD-based devices:

- Low power consumption, as demonstrated by low-threshold QD laser and QD-based optical switches.
- Enhanced thermal stability, as shown by temperature-insensitive lasers and QD semiconductor optical amplifiers (SOAs) with ultralow noise and no pattern effect.

• Wide bandwidth, as shown by SOA and QD-based solid-state saturable absorber mirrors, and also used as incoherent broadband sources.

These three benefits provide a strong foundation for QD-based devices to be obviously appealing for use in future less power-consuming optical networks, also known as the "Green internet" on a macro scale, which involve low-power, low-noise, and broadband operation. An optical switch with ultra-short delay time is one of the main systems for establishing a high-speed optical network. The excitation power needed to enter the nonlinear operation regime for most optical materials is relatively high, limiting system efficiency (the well-known "power/speed trade-off"). The necessary energy per pulse for the fabrication of chip-based photonic networks is estimated to be less than one PJ/bit. Owing to the small volume and density of states similar to that of atoms, semiconductor nanostructures like QDs should be able to meet such a system requirement. As a result, quantum dot-based structures can produce high optical non-linearity while consuming very little energy. For the characterization of all-optical switches, two other parameters, differential reflectivity (or transmission) and switching time, are especially important. These parameters refer to two major QD limitations, namely the degree of the optical nonlinearity and the speed of the carrier dynamics, in the case of QD switches. Due to their dispersed distribution in both actual and frequency space, the interaction length and effective cross-section of self-assembled QDs are extremely limited. In order to generate high optical nonlinearity and, as a result, high differential reflectivity in practical devices, the light-QD interaction must be improved. Integrating QDs with either two-dimensional waveguides made out of photonic crystal or vertical cavities will partially solve the problem of light's poor interaction with QDs. The photonic crystal waveguide of μm scale with hundreds of photonic air holes results in a system with a wide surface area and potentially high cost. Optical switches fabricated using QD embedded in vertical cavities will be a good option in terms of power efficiency. One-dimensional photonic crystals, which are also known as vertical cavity, improve the optical mode-QD interaction while preserving the low power consumption advantage. In comparison to Quantum Well (QW) devices, QD interband transitions are not inherently easy. The existence of such slow recombination rates can limit the device's switching efficiency. Non-radiative channels may be introduced into QW products using methods such as impurity doping combined with low-temperature processing. These techniques, on the other hand, minimize absorption intensity and hence optical nonlinearity. Such techniques could be used in QDs, but defect-related methods must be controlled with extreme caution due to the low absorption values. The existence of discrete energy levels in QDs, fortunately, provides another diverting procedure for manipulating switching dynamics.

11.2.1 Vertical Cavities

The function of a QD optical switch using a Febry–Perot cavity is illustrated schematically in Figure 11.1. Two distributed Bragg reflector (DBR) mirrors, each with several pairs of alternating high and low refractive index layers, make up the semiconductor-based optical cavity. According to the output direction, the two DBR mirrors are called back and front mirrors. Each independent layer of the DBR mirror has a thickness of $\dfrac{k}{4n}$. Here n is the refractive index of the DBR layer and k is the operation wavelength. The cavity area between two layers of DBR has a thickness $L = \dfrac{mk}{2n}$ (m is an integer), referred as the k cavity [11].

$$R_{CM} = R_F \left(\frac{1-\left(R_B / R_F\right)^{1/2} e^{-\Lambda}}{1-\left(R_B R_F\right)^{1/2} e^{-\Lambda}} \right)^2 ,$$

$$(11.1)$$

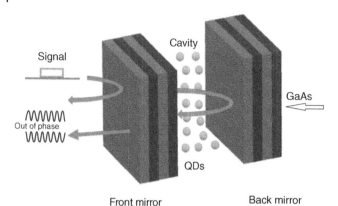

where the reflectivity of the back and front mirrors are R_B and R_F respectively and Λ is the collective absorption in QDs. As laser light is pumped into the optical cavity, the light reflected by one mirror can be ultimately canceled by the reflection from the second mirror in the cavity mode, as long as the two reflected beam are out of phase. This mechanism is used in both QW/QD and bulk materials in vertical-geometry optical switches. The relationship between front and back reflectivity explains the so-called "zero reflectivity state".

$$R_F = R_B e^{-2\Lambda} \tag{11.2}$$

A photonic bandgap area with high reflectivity close to 100% is seen. When the zero reflectivity requirement is met, the light beam will completely penetrate the cavity without reflection in the resonant cavity mode. The absorption of quantum dots saturates when heavy pumping of laser power occurs at the resonant cavity mode. This disrupts the device's zero reflectivity state and causes it to switch modes. A large differential reflectivity between the operation states with and in the absence of pumping is needed for efficient switching.

For several years, the limitations imposed by physical dimensions of optical switches have been a source of contention. Keyes and Armstrong [24] presented the first significant discussion of this subject in the late 1960s, concluding that optical power utilization will be a major issue preventing the use of optical devices. This contention persisted until the 1980s when technology based on QW and optical bistability were discovered. Using the quantum-confined stark effect (QCSE), a high degree of optical nonlinearity was demonstrated in semiconductors, leading to a series of active realizations of photonic logic device.

There are certain important properties of optical microcavities that is taken into account while designing and fabricating these structures. The quality factor (Q-FACTOR) defined as the ratio of a resonant cavity frequency ω_c to the Full Width at Half Maxima (FWHM) of cavity mode $\Delta\omega_c^H$,

$$Q = \frac{\omega_c}{\Delta\omega_c^H} \tag{11.3}$$

The Q factor is a parameter which determines the rate at which optical energy decays from the cavity.

The finesse F_c of the cavity is

$$F_c = \frac{\Delta\omega_c^F}{\Delta\omega_c^H} = \frac{\pi\sqrt{R}}{1-R}, \tag{11.4}$$

where $\Delta\omega_c^F$ is free spectral range or, in other words, the frequency seperation between the successive longitudinal optical modes of the microcavity. R is the total power reflectivity. The on-resonance optical intensity enhancement factor is defined as:

$$I_E = \frac{I_{intracavity}}{I_{incident}} = \frac{1}{1-R} = \frac{F}{\pi\sqrt{R}} \tag{11.5}$$

In a microcavity, this enhancement is non-uniform and spatially localized. One can couple a two-level system such as a QD directly to this localized and enhanced optical field to harness strong coupling and optical nonlinear interaction.

11.2.2 Power Density

The coupled oscillator model widely used in cavity quantum electrodynamics (CQED) research can theoretically reflect the QD cavity structure. Figure 11.2 shows a two-level system (QD) in a photonic crystal cavity with an interaction strength g, cavity mode decay rate κ and carrier relaxation time, T_1. The mechanism is assumed to operate in the weak coupling regime, i.e. $g << \kappa$. The semi-classical method using Maxwell–Bloch equations yields the absorption and phase nonlinearity of QDs.

$$\frac{d\sigma}{dt} = \left(\iota\Delta - \frac{1}{T_2}\right)\sigma - \iota g\delta \tag{11.6}$$

Here, σ is the off-diagonal component of the density matrix, δ is the population inversion term, Δ is the detuning of the optical field from resonance, g is the QD-field coupling strength and T_2 is the dipole phasing time.

We present a brief quantum optical analysis of a QD cavity system coupled strongly to show optical switching effect. The study is valid for any single-mode cavity and excludes any photonic crystal-specific charateristics. The Hamiltonian H describing the quantum dynamics of the system of a QD embedded in a photonic crystal cavity is [3]

$$H = \hbar w_q \sigma^\dagger \sigma + \hbar w_o a^\dagger a + \hbar g\left(a + a^\dagger\right)\left(\sigma + \sigma^\dagger\right) \tag{11.7}$$

where w_o and w_q are the resonance frequencies of the cavity and the QD respectively, $a(\sigma)$ is the lowering operator for the cavity mode (Quantum dot). σ is equal to $|g><e|$ with $|g>$ and $<e|$ being the ground and excited state of the QD respectively. The strength of interaction between the QD and the optical cavity mode is $g = \Omega/2$, where Ω is the vacuum Rabi splitting. This Hamiltonian can be diagonalized to find the system's eigen-energies. The bare cavity resonance splits into two resonances due to the QD-cavity interaction, corresponding to the coupled system's eigen state, also known as polaritons, in a strongly coupled QD-cavity system when $g > (\kappa - \gamma)/2$, where κ is the field decay rate and γ is decay rate of the QD. As a result, when the cavity contains a single strongly coupled QD, the transmission of a laser whose frequency is resonant with the frequency of the empty cavity is significantly reduced.

As a function of laser power, normalized cavity output transmission is calculated. The coupled

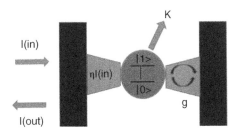

Figure 11.2 Photonic cavity coupled with a QD.

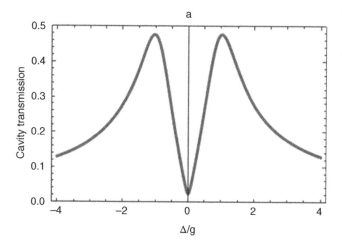

a

Figure 11.3 The DC transmission characteristics of an all-optical switch constructed from a strongly coupled QD-cavity system.

QD-cavity system's nonlinear transmission characteristic is depicted in Figure 11.3. This is the switch's direct current behavior. The split resonance is a result of the QD-cavity coupling. On changing the detuning around the resonance condition of Δ/g, the output transmission changes between zero transmission (OFF state) to maximum transition (ON state).

The switching speed of any switch is a critical parameter. One can numerically measure the optical response of the coupled dot-cavity method to estimate the switching speed. Two pulses drive the coupled device, and complete optical transmission through the cavity is detected. When the two pulses are separated by a significant amount of time, they are transmitted separately via the QD-cavity device. We assume a steady transmission as a function of time. However, as the pulses get closer in time, they begin to overlap, resulting in a more excellent transmission due to the system's non-linearity.

Figure 11.4 depicts a diagram of the planned tunable optical transmission. Two DBR mirrors made of quarter-wavelength tick alternating layers of different dielectric materials make up the unit. A cavity spacer area, forming a DBR microcavity, separates the two mirrors. The cavity spacer contains QDs which act as saturable absorbers to generate a light intensity-dependent optical cavity.

Two independent laser signals, an input laser signal and a control laser, are used to activate the switch, as shown in Figure 11.4. The carrier frequency of the two signals is chosen to be the same. The input signal comes in from the top surface of the DBR and is supposed to be of low-intensity

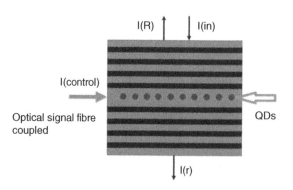

Figure 11.4 DBR-QD switch.

such that the QDs are not saturated. The control signal is strong enough to saturate the QDs, changing the cavity's absorption. Except for random emission noise, the weak signal and strong control pulses move in orthogonal directions and are entirely distinct and separable. The DBR structure gives rise to low reflectivity since all the light is transmitted at low QD density. The DBR makes uninterrupted transitions from transmitting to highly reflecting as the QD density increases. The increased absorption inside the cavity causes this transition. The reflectivity is affected in two ways by increasing the number of DBR layers. First, the density of QDs needed to change the cavity reflectivity diminishes as the number of QDs increases and second, the maximum reflectivity obtained in the high-density regime also grows.

11.3 Quantum Well as an Optical Switch

For Quantum Wells (QW), conduction band electrons and valence band holes behave as particles with a certain effective mass (m_e and m_h corresponding to that of electron and hole) which is not the same as the free electron/hole mass. According to the "k.p" band theory, m_e and m_h are approximately equal and proportional to the band gap energy.

Quantum wells are an example of a hetero-structure, which is a structure formed by fusing dissimilar materials, typically in layers, and with the materials being fused right at the level of atoms. Heterostructures in general have a wide range of applications. State-of-the-art electronic devices (e.g. resonant tunneling devices, modulation-doped field-effect transistors, hetero junction bipolar transistors), optical materials (e.g., wave guides, mirrors, micro resonators), and optoelectronic devices and structures (e.g., laser diodes, photo detectors, quantum wells) all benefit from hetero-structures. Although hetero-structures are helpful in electronics, they also play an essential role in many optoelectronic devices (e.g., lasers). Perhaps their most valuable technical feature is that they can be used for many of these electronic, optical, and optoelectronic applications, allowing them to be integrated.

11.3.1 Optical Properties

To comprehend the interband linear optical absorption in quantum wells, we ignore the "excitonic" effects. This is an excellent conceptual first model that describes some of the critical properties. Unlike in bulk semiconductors, excitonic effects in quantum wells are very visible at room temperature and significantly impact device performance. By absorbing a photon, an electron can rise from the valence band to a state in the conduction band having the same momentum (a "vertical" transition). The state in the conduction should have similar momentum because the photon has no momentum on the scale, usually of interest in the semiconductor. In this simple model, we also assume that all such transitions have the same strength, even though they will have dissimilar energies. We have a selection rule corresponding to the direction perpendicular to the layers in quantum wells rather than momentum conservation, as shown in Figure 11.5. The transition is allowed between states with the same quantum number in the conduction and valence bands (to lowest order). The strength of optical absorption is directly proportional to the overlap of the valence and conduction wave functions, resulting in this rule. In quantum wells, holes and electrons are still free to move in a plane parallel to the layers. There are no discrete energy states for holes and electrons; instead, there are "subbands" that begin at the calculated confined state energies. For a particular confined state, the electron can aquire any kinetic energy corresponding to its 2D motion in the QW. Also, it can have any energy which is larger than or equal to the energy of a confined state for that subband.

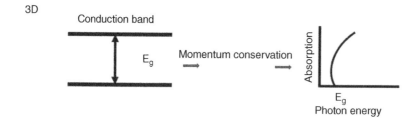

Figure 11.5 Optical absorption in bulk and in QW.

Absorption in the quantum well occurs in step, and particle-in-a-box calculations give the position of these steps correctly. However, there are peaks in the spectra that are not estimated or described by this simple "non-excitonic" model. These spectral peaks have a significant influence and will be especially noticeable near the band-gap energy.

We must now introduce the concept of excitons in understanding these peaks (electron-hole pair). The correct approach is to consider creating an electron-hole pair rather than raising an electron from the valence band to the conduction band. First, we must analyze and understand the states corresponding to an electron-hole pair in a crystal. The creation of a particle, the exciton, is used to explain optical absorption. It is critical to understand that we are not lifting an existing hydrogen-like particle to an excited state, as is done with usual atomic absorption; rather, we are creating the particle. The absorption of electron-positron pairs in the vacuum is an analogy that could help explain this difference. The quantum well differs from bulk material in two important ways, both of which arise from the quantum well's confinement as shown in Figure 11.6. When we create an exciton in a 100 Å-thick quantum well, the exciton shrinks in all three directions.

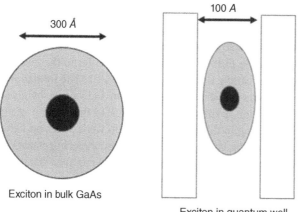

Figure 11.6 Comparison of bulk and QW exciton sizes and shapes.

Quantum wells can exhibit a wide range of nonlinear optical effects. Only one type of effect, saturation of optical absorption near the band-gap energy, will be discussed in this section. This set of effects has been extensively studied for various systems and is suitable for applications in laser mode-locking [25]. In the most basic instance, we direct a laser on the material, causing the associated optical absorption to generate a large population of holes and electrons, resulting in "excitons" or "free carriers". Absorbing at the peaks corresponding to the exciton generates excitons, whereas absorbing at greater photon energies produces free carriers.

We cannot make two identical excitons in the same spatial region. As we start to create space excitons, we run out of space and as a result, the likelihood of creating more excitons must decrease, as must the optical absorption associated with it. As a result, the exciton absorption line will reach saturation. Screening effects are the second type of mechanism that can alter absorption. When we increase the density of free carriers, the dielectric constant changes, and thus the size of the exciton changes, usually increases. When the exciton size increases, the possibility of finding the hole and electron in the same place is diminished; as a result, the strength of optical absorption also diminishes, resulting in an effect similar to saturation. In quantum wells, the saturation effects corresponding to the exciton peak are very sensitive. A variety of nonlinear optical switching devices have been investigated using them.

11.3.2 Self-Electro-Optic-Effect Devices

The theory behind the self-electro-optic effect system (SEED) is to combine a quantum well modulator (or group of photodectors) to create an optically operated system with an optical output (or outputs). Although the conversion from optics to electronics and back is often inefficient and costly, this does not have to be if the devices are well integrated. The quantum well devices can be effectively integrated, allowing innovative optoelectronic units to be created. Such devices open up new avenues for data processing and switching architectures. There are two types of SEEDs: those that only use diodes and those that also use transistors. Figure 11.7 shows the most basic SEED configuration. It exhibits optical bistability and is a "resistor-biased" SEED (R-SEED).

A positive feedback system in SEED is responsible for the bistability. Assume that there is not much light shining on the diode at first. The result is a low photocurrent and as a consequence, a low voltage drop around the resistor. Hence, the maximum supply voltage appear across the diode, and the diode absorption is strong. We get more photocurrent as we shine more light on the diode. We now have less voltage across the diode due to the voltage decrease across the resistor, and hence more absorption, and thus more photocurrent. We can switch into a highly absorbing state as a result of this mechanism. Decreasing to a lower power level eventually makes the diode revert to its high-voltage, low-transmission state as shown in Figure 11.8.

11.4 Optomechanical Systems as Optical Switch

11.4.1 Optical Nonlinearity

Within the framework of classical optics, optical bistability belongs to a category of non-linear optical phenomena in

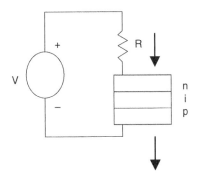

Figure 11.7 Basic SEED configuration.

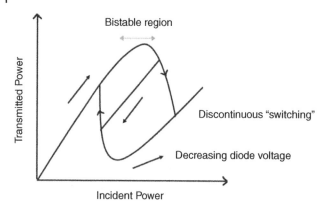

Figure 11.8 Bistable curve for SEED operating as an optical switch.

which the refractive index of the material through which an intense beam of light is allowed to pass depends on the intensity of the incident light,

$$n = n_0 + n_2 I, \tag{11.8}$$

where n_0 is the linear refractive index, n_2 is the nonlinear index, and I is the intensity of light. Optical bistability leads to the possibility of designing "all-optical switches" (also called photonic switching). The nonlinear coefficient n_2 depends on the third-order susceptibility $\chi^{(3)}$ as [26]:

$$n_2 = \frac{\chi^{(3)}}{2c\epsilon_0}, \tag{11.9}$$

where c is the speed of light in vacuum. In optical/photonic switching, the output state of the device depends on the current state of the device. Coherently controlling the system parameters, the device can be made to switch between "high" and "low" state. Optical nonlinearity can also lead to multistability, thus leading to the possibility of a new type of logic beyond binary logic.

Photonic switching in a nonlinear Fabry–Pérot etalon has been described earlier [27]. The transmitted intensity as a function of incident intensity in a nonlinear Fabry–Pérot etalon exhibits a hysteresis loop (optical switching) as shown in Figure 11.9.

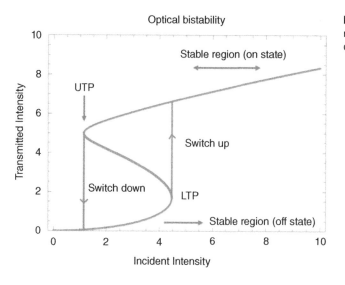

Figure 11.9 Schematic representation of a system showing optical bistability.

Figure 11.9 shows the hysteresis loop that the transmitted intensity follows as the incident intensity is gradually increased or decreased. Initially if transmitted intensity (I_0) is on the lower stable region (off state) and as incident intensity (I_i) increases past its lower turning point (LTP), I_0 switches upto the upper stable region (on state). As long as I_0 remains larger than its value at LTP, I_0 is given by the upper stable region solution. Now as we gradually decrease I_i below the upper turning point (UTP), I_0 switches down to the initial lower stable region (off state.)

Photonic switching devices have applications in optical computing and experimental efforts are targeted towards realizing novel materials which can lead to optical devices based on optical bistability. Experimental progress in nano-materials and photonic crystals has led to a revival of photonic devices [28].

11.4.2 Hybrid Optomechanics

An optical field is used in optomechanical systems to measure and control the dynamics of the mechanical resonator. The optical field is typically confined within a cavity, which allows for resonant enhancement of field strength and mechanical displacement sensitivity. **Radiation pressure** is a scattering force caused by light reflection, which has momentum associated with it. Optical gradient forces result from the spatial variation of optical intensity. The bistability of an optomechanical cavity is the result of the static effect of radiation pressure. A shift in frequency and modified dissipation of the mechanical resonator are the dynamic effects of radiation pressure combined with a finite cavity lifetime.

A Fabry–Pérot optical cavity in which one of the mirrors is mechanically compliant is the most basic optomechanical system in which optical pressure provides the optomechanical coupling as shown in Figure 11.10. The mechanical resonator's motion changes the length of the optical cavity and thus the resonance frequency of the cavity. The coupling is calculated by determining the relationship between the cavity resonance frequency and the mechanical component's displacement.

Let us consider the most basic optomechanical system as depicted in Figure 11.10.

The Hamiltonian of the system is given as

$$H = \hbar \Omega_c \left(q\right) a^\dagger a + \hbar \Omega_m b^\dagger b \tag{11.10}$$

where $\hat{a}$ and $\hat{b}$ are the annihilation operators for the cavity mode and the mechanical mode, respectively. Ω_c and Ω_m are the optical and mechanical frequency respectively. The length of the cavity L, is shifted by the mechanical oscillator's motion, with $L = L - q$, where q is the mechanical oscillator's displacement away from its equilibrium location. Since the cavity length is not fixed due to one of the mirrors being movable, the cavity frequency is dependent on the length of the cavity $\Omega_c(q)$. The mode frequency of the cavity is

$$\Omega_c = \frac{2\pi c}{\lambda} = \frac{\pi c j}{L-q} \approx \Omega_c \left(1 + \frac{q}{L}\right), \tag{11.11}$$

where $\lambda = 2\dfrac{(L-q)}{j}$ is the longitudinal optical mode wavelength with mode number j. Therefore, the Hamiltonian of the system is rewritten as

$$H = \hbar \Omega_c a^\dagger a + \hbar \Omega_m b^\dagger b - \hbar g_0 a^\dagger a \left(b + b^\dagger\right) \tag{11.12}$$

where $g_0 = G x_{zp}$ is the single photon optomechanical coupling, G is the cavity mode frequency shift caused by the mechanical resonator's zero-point motion x_{zp} and the last term is defined as the

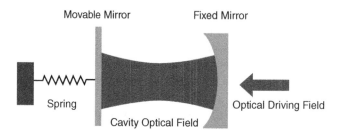

Figure 11.10 Cavity optomechanical system consists of an optical cavity formed of two high-reflectivity mirrors. One of the mirrors is fixed while other is movable.

interaction part of the Hamiltonian (remember $b^\dagger + b = \hat{q}/x_{zp}$; note that now $\hat{q}$ is a position operator). The Hamiltonian shows that a movable mirror's interaction with the radiation field is essentially a nonlinear mechanism involving three operators.

We rotate the Hamiltonian of the above equation in the frame of reference of the pump frequency Ω_L, as a result of which we get the following Hamiltonian:

$$H = \hbar \Delta a^\dagger a + \hbar \Omega_m b^\dagger b - \hbar g_0 a^\dagger a \left(b^\dagger + b \right). \tag{11.13}$$

Here, Δ is the detuning between the optical cavity and the incident laser frequency as

$$\Delta = \Omega_c - \Omega_L. \tag{11.14}$$

The open system dynamics of a cavity optomechanical system are described by the Langevin equation of motion.

$$\dot{a} = -\left[\frac{k}{2} + i \left(\Delta a + \sqrt{2} g_0 \hat{Q} \right) \right] a + \sqrt{\kappa} \alpha_{in}, \tag{11.15}$$

$$\dot{\hat{Q}} = \Omega_m \hat{P}$$

$$\dot{\hat{P}} = -\Omega_m \hat{Q} + \sqrt{2\Gamma} P_{in} - \Gamma \hat{P} - \sqrt{2} g_0 a^\dagger a,$$

where $\hat{Q} = \frac{1}{\sqrt{2}} \frac{\hat{q}}{x_{zp}}$ is the position and $\hat{P}$ is the momentum quadrature of the mechanical oscillator while κ and Γ are the decay rates of the optical cavity and mechanical oscillator respectively.

Consider that the mechanical oscillator is guided incoherently by its bath ($< P_{in} >= 0$) and describes the cavity's coherent amplitude $\alpha = <\hat{a}>$, proportional to the mean optical cavity occupancy N introduced by coherent driving by $N = |\alpha|^2$. Hence,

$$\dot{\alpha} = -\left[\frac{k}{2} + i \left(\Delta a + \sqrt{2} g_0 \hat{Q} \right) \right] \alpha + \sqrt{\kappa} \alpha_{in} \tag{11.16}$$

These equations show a nonlinear coupling between the mechanical oscillator and the optical field, resulting in a wide variety of classical behaviour, including parametric instability regimes marked by exponential growth in mechanical oscillation amplitude. The steady states of α and Q are represented as α_s and Q_s respectively and are derived as

$$\alpha_s = \sqrt{\kappa} \alpha_{in} / \left[\frac{k}{2} + i \left(\Delta a + \sqrt{2} g_0 Q_s \right) \right], \tag{11.17}$$

$$Q_s = -\sqrt{2} g_0 |\alpha_s|^2 / \Omega_m \tag{11.18}$$

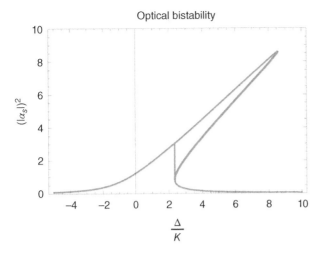

Figure 11.11 Intracavity photon number as a function of cavity detuning. The figure demonstrates optical bistability due to the inherent optomechanical nonlinearity present in the system.

Equations (17) and (18) represent two coupled equations giving rise to nonlinear effect observed as optical bistability which is shown in Figure 11.11.

Bistability is a common occurrence in many nonlinear systems. The nonlinearity of the equations of motion indicates that such effects can be observed through optomechanical coupling. Experiments have shown that optical bistability can exist in semiconductor microcavities. In these systems, several mechanisms could lead to bistable behavior. One possible mechanism is to introduce nonlinearity into the system by increasing exciton density, which results in exciton-exciton scattering. Bleaching the Rabi splitting is another method for achieving bistable behavior [29]. Optomechanical bistability can be understood better due to a competition between the mechanical restoring force and the radiation pressure force, which is highest at the cavity resonance. The optomechanical coupling can squeeze the cavity mode, and this effect is maximal near the bistable regime. It has also been observed that quantum entanglement is maximum at the bistability threshold under certain conditions [30].

A hybrid system is made up of an optomechanical cavity and a second optical cavity containing an ensemble of ultra-cold atoms that serves as feedback to the first cavity. The bistable response of the intracavity field in the optomechanical cavity can be coherently tuned by varying the frequency of the single control laser, which drives the cavity-field detuning in the optomechanical cavity and the atom-field detuning in the atomic cavity. Compared to a single-cavity optomechanical system, this allows for more flexibility in tuning bistability [31]. The optical spring effect generated due to optomechanical coupling between the cavity field and the oscillator is indicated by the multistability of the displacement of a mechanical oscillator coupled to a cavity field. In non-zero detuning, multistability is a sign of dynamical back-action, which causes the mechanical oscillator to heat or cool. This is because multistable behavior indicates a change in the oscillator's decay rate and resonance frequency. Multistability is also an essential factor in the design of all-optical switches, logic gates, and memory devices. The photon tunneling rate between two coupled cavities controls optomechanical bistability and multistability, which occurs at low input power levels [32].

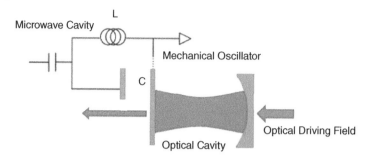

Figure 11.12 A schematic figure of an Electro-opto mechanical system.

11.4.3 Electro-opto Mechanics

Figure 11.12 shows a hybrid electro-optomechanical system (EMOS). It consists of a mechanical resonator (MR) capacitively coupled to the microwave field of a superconducting microwave cavity (MC) on one side and an optical cavity on the other (OC). By tuning the various interactions, the system can exhibit optical switching behaviour at low input power levels [21]. The system's controllable bistable behaviour demonstrates that it can be used as an all-optical switch, logic gates, and memory device for quantum information processing with low energy power input.

The fact that silicon nitride membranes have remarkable optical and mechanical properties was a key development discovery. The oscillations of the silicon nitride membrane interact with the electromagnetic field confined in the cavity when placed between the two mirrors that form a high-quality factor Fabry–Pérot cavity. State-of-the-art experiments use metallic membranes to create low-loss capacitors that naturally couple membrane vibrations to electrical signals at microwave frequencies. Cryogenic temperatures are required for these experiments in order to minimize Johnson noise at microwave frequencies and to create low energy-loss superconducting circuits. The integration of the two electromagnetic resonators is the most difficult aspect of building an electro-optomechanical system.

The optical cavity must function at low temperatures required by the microwave circuit. The microwave circuit must operate near the Fabry–Pérot cavity, where light from the cavity may couple with the microwave circuit. High-frequency light absorbed by a superconducting microwave circuit, in particular, can harm its performance. Photons with energies larger than the superconducting gap can separate Cooper pairs into their electrons. The frequency converter is built with a physically large silicon nitride membrane. The superconducting microwave circuit and optical cavity can couple to spatially distant portions of the silicon nitride membrane. Standard photolithographic techniques are used to fabricate a silicon nitride membrane fixed by a silicon frame. A portion of the membrane is also covered in niobium, used in the superconducting circuit. Figure 11.13 shows the schematic of the electro-opto mechanical system fabricated and used for microwave to optical frequency conversion [33].

11.5 Conclusion and Future Outlook

We have discussed the current technological progress of optical switching systems based on semiconductor nanostructures. Quantum photonic devices can be designed by focusing and guiding light into semiconductor nanostructures, which play an essential role in future generation

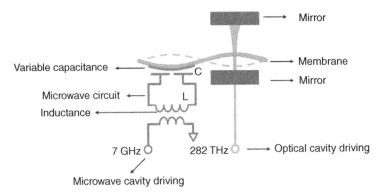

Figure 11.13 Schematic of a optical-to-microwave frequency converter. Source: Modified from Andrews [33].

power-efficient optical networks. The present development of optical switches shows excellent potential to meet future optical communication systems requirements between electronic chips or within chips. A hybrid optomechanical system is capable of storing and transferring information, thus forming a part of a quantum information processing unit. All-optical switches have received much interest because they can overcome the speed limitation of electric switches. A solid-state-based optomechanical system in the presence of an additional second-order nonlinearity is found to be highly tunable and can be used to implement low-power switching devices. The mean-field optical bistability of a hybrid electro-optomechanical system in the presence of a qubit displays optical switching characteristics which can be tuned to consume low input power.

Bibliography

1 C.-Y. Jin and O. Wada. Photonic switching devices based on semiconductor nano-structures. *Journal of Physics D: Applied Physics*, 47 (13):133001, 2014. doi: 10.1088/0022-3727/47/13/133001.

2 O. Wada. Femtosecond all-optical devices for ultrafast communication and signal processing. *New Journal of Physics*, 6:183–183, 2004.

3 A. Majumdar, M. Bajcsy, D. Englund, and J. Vuckovic. All optical switching with a single quantum dot strongly coupled to a photonic crystal cavity. *IEEE Journal of Selected Topics in Quantum Electronics*, 18:1812–1817, 2012. doi: 10.1109/JSTQE.2012.2202093.

4 D. Sridharan and E. Waks. All-optical switch using quantum-dot saturable absorbers in a dbr microcavity. *IEEE Journal of Quantum Electronics*, 47(1):31–39, 2011. doi: 10.1109/JQE.2010.2070487.

5 C.-Y. Jin, O. Kojima, T. Kita, O. Wada, M. Hopkinson, and K. Akahane. Vertical-geometry all-optical switches based on inas/gaas quantum dots in a cavity. *Applied Physics Letters*, 95:021109, 2009. doi: 10.1063/1.3180704.

6 T. Kitada, T. Kanbara, K. Morita, and T. Isu. A GaAs/AlAs multilayer cavity with self-assembled InAs quantum dots embedded in strain-relaxed barriers for ultrafast all-optical switching applications. *Applied Physics Express*, 1:092302, 2008. doi: 10.1143/ apex.1.092302.

7 S. Mahajan and A.B. Bhattacherjee. Controllable nonlinear effects in a hybrid optomechanical semiconductor microcavity containing a quantum dot and Kerr medium. *Journal of Modern Optics*, 66(6):652664, 2019. doi: 10.1080/09500340.2018.1560510.

8 A. Majumdar, N. Manquest, A. Faraon, and J. Vuckovic. Theory of electro-optic modulation via a quantum dot coupled to a nano-resonator. *Optics Express*, 18(5):3974, 2010. doi: 10.1364/ oe.18.003974.

9 A. Faraon, A. Majumdar, H. Kim, P. Petroff, and J. Vukovi. Fast electrical control of a quantum dot strongly coupled to a photonic-crystal cavity. *Physical Review Letters*, 104(4), 2010. doi: 10.1103/ physrevlett.104.047402.

10 R.C. Ashoori. Electrons in artificial atoms. *Nature*, 379(6564):413–419, 1996.

11 C.-Y. Jin, M. Hopkinson, O. Kojima, T. Kita, K. Akahane, and O. Wada. *Quantum Dot Switches: Towards Nanoscale Power-Efficient All-Optical Signal Processing*, pages 197–221. 2012. doi: 10.1007/978-1-4614-3570-9_10.

12 L. Du, Y.-M. Liu, B. Jiang, and Y. Zhang. All-optical photon switching, router and amplifier using a passive-active optomechanical system. *Europhysics Letters*, 122(2):24001, 2018. doi: 10.1209/0295-5075/122/24001.

13 S. Weis, R. Rivire, S. Delglise, E. Gavartin, O. Arcizet, A. Schliesser, and T.J. Kippenberg. Optomechanically induced transparency. *Science*, 330(6010):15201523, 2010. doi: 10.1126/ science.1195596.

14 D. Vitali, S. Gigan, A. Ferreira, H.R. Böhm, P. Tombesi, A. Guerreiro, V. Vedral, A. Zeilinger, and M. Aspelmeyer. Optomechanical entanglement between a movable mirror and a cavity field. *Physical Review Letters*, 98:030405, 2007. doi: 10.1103/PhysRevLett.98.030405.

15 A. Mari, A. Farace, N. Didier, V. Giovannetti, and R. Fazio. Measures of quantum synchronization in continuous variable systems. *Physical Review Letters*, 111:103605, Sep 2013. doi: 10.1103/ PhysRevLett.111.103605.

16 Y.-D. Wang and A.A. Clerk. Using interference for high fidelity quantum state transfer in optomechanics. *Physical Review Letters*, 108(15):153603, 2012.

17 X.-W. Xu and Y. Li. Optical nonreciprocity and optomechanical circulator in three-mode optomechanical systems. *Physical Review A*, 91:053854, 2015. doi: 10.1103/PhysRevA.91.053854.

18 A.B. Bhattacherjee and M.S. Hasan. Controllable optical bistability and fano line shape in a hybrid optomechanical system assisted by kerr medium: possibility of all optical switching. *Journal of Modern Optics*, 65(14):16881697, 2018. doi: 10.1080/09500340.2018.1455917.

19 V. Bhatt, S.A. Barbhuiya, P.K. Jha, and A.B. Bhattacherjee. Controllable normal mode splitting and switching performance in hybrid optomechanical semiconductor microcavity containing single quantum dot, 2019.

20 Sh. Barzanjeh, D. Vitali, P. Tombesi, and G.J. Milburn. Entangling optical and microwave cavity modes by means of a nanomechanical resonator. *Physical Review A*, 84:042342, 2011. doi: 10.1103/ PhysRevA.84. 042342.

21 T. Kumar, S. Yadav, and A.B. Bhattacherjee. Optical response properties of hybrid electro-opto-mechanical system interacting with a qubit, 2021.

22 Y. Ben Ezra, B.I. Lembrikov, and M. Haridim. Ultrafast all-optical processor based on quantum-dot semiconductor optical amplifiers. *IEEE Journal of Quantum Electronics*, 45(1):34–41, 2009. doi: 10.1109/JQE. 2008.2003497.

23 H. Nakamura, Y. Sugimoto, K. Kanamoto, N. Ikeda, Y. Tanaka, Y. Nakamura, S. Ohkouchi, Y. Watanabe, K. Inoue, H. Ishikawa, and K. Asakawa. Ultra-fast photonic crystal/quantum dot all-optical switch for future photonic networks. *Optics Express*, 12(26):6606–6614, 2004. doi: 10.1364/OPEX.12.006606.

24 R.W. Keyes and J.A. Armstrong. Thermal limitations in optical logic. *Applied Optics*, 8(12):2549–2552, 1969. doi: 10.1364/AO.8.002549.

25 D. Miller. Quantum well optical switching devices. 340, 1995. doi: 10.1007/978-1-4615-1963-8_22.

26 E. Hecht and A. Zajac. *Optics*, volume 5. San Francisco, Addison Wesley, 2002.

27 H.M. Gibbs. *Optical Bistability: Controlling Light with Light*. 1985.

28 C.-H. Chen, S. Matsuo, K. Nozaki, A. Shinya, T. Sato, Y. Kawaguchi, H. Sumikura, and M. Notomi. All-optical memory based on injection-locking bistability in photonic crystal lasers. *Optics Express*, 19(4):3387–3395, 2011. doi: 10.1364/OE. 19.003387.

29 A. Baas, J.Ph. Karr, H. Eleuch, and E. Giacobino. Optical bistability in semiconductor microcavities. *Physical Review A*, 69:023809, 2004. doi: 10.1103/PhysRevA.69.023809.

30 R. Ghobadi, A.R. Bahrampour, and C. Simon. Quantum optomechanics in the bistable regime. *Physical Review A*, 84:033846, 2011. doi: 10.1103/PhysRevA.84.033846.

31 B. Sarma and A.K. Sarma. Controllable optical bistability in a hybrid optomechanical system. *Journal of the Optical Society of America B*, 33(7):1335–1340, 2016. doi: 10.1364/JOSAB.33.001335.

32 V.N. Prakash and A.B. Bhattacherjee. Negative effective mass, optical multistability and fano line-shape control via mode tunnelling in double cavity optomechanical system. *Journal of Modern Optics*, 66(15):16111621, 2019. doi: 10.1080/09500340.2019.1650208.

33 R.W. Andrews. Quantum signal processing with mechanical oscillators. *Quantum*, 2015.

12

Nonlinear All-Optical Switch

Rajarshi Dhar[1], Arpan Deyasi[2], and Angsuman Sarkar[3]

[1] Department of Electronics and Telecommunication Engineering, IIEST Shibpur, Howrah, West Bengal, India
[2] Department of Electronics and Communication Engineering, RCC Institute of Information Technology, Kolkata, West Bengal, India
[3] Department of Electronics and Communication Engineering, Kalyani Government Engineering College, Kalyani, West Bengal, India

12.1 Introduction

The limitations of electrical communication systems have led people to move to optical communication systems. The optical communication systems have proven to be much more advantageous over the electrical systems as they overcome the problems or limitations like bandwidth, speed, security, reduced system noise, and several other factors which are undesirable for a faithful and sustainable communication system [1, 2]. But one of the problems in optical systems is that of optical processing. The optical processors that have been developed until now can only process low-speed signals or signals with low bit rates. The processors need electrical signals which are converted from optical signals are processed and then converted back to optical signals. This is done using the Optical-Electrical-Optical Switches or OEO Switches [3, 4]. Now this conversion and back-conversion takes up a lot of power and time which are not desirable for high-speed systems. Thus this leads to the need for all-optical processing systems, which include the all-optical switches.

12.2 Classification of All-Optical Switches

There are several types of all-optical switches, namely thermo-optical switches, acousto-optical switches, liquid crystal optical switches, and nonlinear optical switches.

12.2.1 Thermo-Optical Switch

There are several types of all-optical switches, including thermo-optical switches [5–7]. The operation of thermo-optical switches is based upon the principle of thermal effects on optical properties [8]. The thermo-optic effect is a basic property which is available in all optical materials. The thermo-optic coefficient or TOC given by $d\varepsilon/dT$ is the fundamental parameter which depicts how the refractive index depends on temperature, where ε is the complex dielectric function of the material at temperature T. The TOC is a very important factor for applications of optics and optoelectronics. The several applications include light guidance, coupling of light and radiation modulation. It is very much important and necessary to obtain a theoretically correct result, i.e. its absolute magnitude and its sign, over a wide range of frequencies, based on the little available values at a few frequencies. A number of thermo-optical switches have been developed over the years. They include interferometric devices, such as Mach–Zender IF, directional coupler, and optical digital switches.

For all isotropic materials in the transparent regime, the macroscopic Clausius–Mossotti formula is applicable:

$$\left(\frac{\varepsilon-1}{\varepsilon+2}\right)=\frac{4\pi\alpha_m}{3V}, \tag{12.1}$$

where α_m is the polarizability of a bulk small sphere (measured in macroscopic scale) with a volume V which is much larger when compared with the lattice dimensions. The numbers of effects which are responsible for the temperature dependence of dielectric constant are three in total. They are: direct volume expansion effect, dependence of polarizability on volume expansion, and dependence of polarizability on temperature.

Differentiating Equation 12.1 with respect to temperature one gets

$$\frac{1}{(\varepsilon-1)(\varepsilon+2)}\left(\frac{\partial\varepsilon}{\partial T}\right)_P=\frac{1}{3V}\left(\frac{\partial V}{\partial T}\right)_P+\frac{1}{3\alpha_m}\left(\frac{\partial\alpha_m}{\partial V}\right)\left(\frac{\partial V}{\partial T}\right)_P+\frac{1}{3\alpha_m}\left(\frac{\partial\alpha_m}{\partial T}\right)_V=A+B+C, \tag{12.2}$$

where the terms given by A, B and C represent the following:

A: With increase in temperature, the volume increases and thus the inter-atomic spaces increase too which causes a decrease of dielectric constant. This is direct effect of volume expansion.
B: Increase of polarizability with the volume expansion.
C: Temperature dependence of polarizability at constant volume.

The temperature derivative of volume can be called the linear thermal expansion coefficient, given by

$$\frac{\partial V}{\partial t}=3V\,\alpha. \tag{12.3}$$

Thus Equation 12.2 can be written as

$$\frac{1}{(\varepsilon-1)(\varepsilon+2)}\left(\frac{\partial\varepsilon}{\partial t}\right)_P=-\alpha\left[1-\frac{V}{\alpha_m}\left(\frac{\partial\alpha_m}{\partial V}\right)_P\right]+\frac{1}{3\alpha_m}\left(\frac{\partial\alpha_m}{\partial T}\right)_V. \tag{12.4}$$

Thus this is the equation that governs the working of the thermal-optical switches. The permittivity of the material is related to the RI of the material by $e=n^2$. Thus by putting $e=\sqrt{n}$ in the above equations, one can easily get the temperature dependence of the RI to temperature.

12.2.2 Acousto-Optic Switch

The second type of all-optical switch is the acousto-optic switch. These work on the principle of the elasto-optic effect. This is the occurrence of periodic modulations within the RI of the material when a sound or acoustic wave moves through the transparent medium. This provides a grating which has a moving phase which causes some portion of the incident light to move in different directions [9]. This phenomenon which is known as the acousto-optic diffraction, conceptually utilized to manifest various optical devices that can be used to perform space-domain, time-domain, and frequency-domain modulations of electromagnetic spectrum [10]. These devices find extensive applications in optoelectronic systems starting from controlling of light beam to signal and data processing applications to wavelength routing [11, 12].

The theory for acousto-optic switches is given in [9]. As mentioned above, the elasto-optic effect is the most important mechanism for acousto-optic operation. To describe the effect in crystals Pockels gave the phenomenological theory which helps in the introduction of elasto-optic tensor. Deformation gradient is a symmetric term defined when wave is propagated in elastic medium

$$S_{ij} = \left(\frac{\partial u_i}{\partial x_j} + \frac{\partial u_j}{\partial x_i} \right) / 2i, j = 1 \ to \ 3, \tag{12.5}$$

where u_i is the displacement. Six independent components of symmetric strain sensor are defined in following way:

odd scattering parameters:

$$\begin{aligned} S_1 &= S_{11} \\ S_3 &= S_{33} \\ S_5 &= S_{13} \end{aligned} \tag{12.6.1}$$

even scattering parameters:

$$\begin{aligned} S_2 &= S_{22} \\ S_4 &= S_{23} \\ S_6 &= S_{12} \end{aligned} \tag{12.6.2}$$

The conventional elasto-optic effect that was introduced by Pockel said that the impermeability tensor ΔB_{ij} is directly proportional to strain tensor in linear fashion

$$\Delta B_{ij} = p_{ijkl} S_{kl} \tag{12.7}$$

where p_{ijkl} is the elasto-optic tensor. If it is written in contracted notation one gets

$$\Delta B_m = p_{mn} S_n \tag{12.8}$$

In modern research, however, the conventional theory has been changed and the elasto-optic effect introduced. This is basically the nonlinear polarization which arises owing to the modulation of dielectric tensor $\Delta \varepsilon_{ij}$. Since ε_{ij} and B_{ij} are inversely proportional to each other in a principal axis system, hence

$$\Delta \varepsilon_{ij} = -\varepsilon_{ii} \Delta B_{ij} \varepsilon_{jj} = -n_i^2 n_j^2 \Delta B_{ij}, \tag{12.9}$$

where n_i is the refractive index. Replacing Equation 12.7 into Equation 12.9, we can write

$$\Delta \varepsilon_{ij} = \chi_{ijkl} S_{kl}, \tag{12.10}$$

where the elasto-optic susceptibility tensor i

$$\chi_{ijkl} = -n_i^2 n_j^2 p_{ijkl}. \tag{12.11}$$

There are two other modifications of the elasto-optic effect, namely the roto-optic effect and the indirect elasto-optic effect.

The roto-effect was introduced by Nelson and Lax [13], where they found that the elasto-optic effect is not originated from the classical concept of working principle of birefringent crystals and described that due to the anti-symmetric part of the deformation gradient, there exists an additional roto-optic susceptibility given by

$$\Delta B'_{ij} = p'_{ijkl} R_{kl},$$ (12.12)

where $R_{ij} = (S_{ij} - S_{ji})/2$.

The indirect-optic effect arises in piezoelectric crystals as a result of the piezoelectric effect succeeded by the elasto-optic effect. Thus the original elasto-optic tensor changes and the effective elasto-optic tensor is given by [14]

$$p^*_{ij} = p_{ij} - \frac{r_{im} S_m e_{jn} S_n}{\varepsilon_{mn} S_m S_n},$$ (12.13)

where p_{ij} is the direct elasto-optic tensor, r_{im} is the electro-optic tensor, e_{jn} is the piezoelectric tensor, ε_{mn} is dielectric tensor, and S_m is unit acoustic wave vector.

12.2.3   Liquid Crystal Optical Switch

It is the third type of all-optical switch. Refractive index and absorption are the two most important and practical parameters of a liquid crystal mixture or compound. All the light modulation mechanisms are dependent on the change in refractive indices. The absorption coefficient has a very important crucial effect on the photostability. The electronic structure of the crystals determines the refractive indices and absorption property of the crystal in the visible spectral region. Now the physical mechanisms that are used for modulating light are described briefly in the next part:

i) **Dynamic Scattering:** When a nematic liquid crystal is subjected to a DC or low-frequency AC field [15], electro-hydrodynamic flow is induced by the conductivity anisotropy ionic motion. Through viscous friction, electro-hydrodynamic flow is connected with alignment at molecular level. This makes liquid crystal in turbulent mode results in strong scattering of light. Large electric field of 10^4 V/cm is required to produce such effect. The contrast ratio of minimum-to-maximum intensities is around 1:20 and the response time is 200 ms.

ii) **Guest-Host Effect:** These systems are produced by dissolving 1–5% of dichroic dye in a liquid crystal [16]. The substrate material should be unclouded enough in the spectrum of the desired area. It is important for the dichroic molecules to have weak absorption for one polarization and strong absorption for the other in order to obtain higher contrast ratio. To obtain high absorption in the field-off state, molecules of the dye are closely aligned to the polarization of the incident wave. Absorption can be reduced when directors of liquid crystal [17] are re-oriented by the field, and consequent re-orientation of dye molecules. The guest-host effects in ferroelectric liquid crystals are published in various literatures.

iii) **Field-induced Nematic-cholesteric Phase Change:** This mechanism has been observed experimentally [18, 19]. The liquid crystal at the initial stage is at the cholesteric phase where it attains the same structure as a helical one and has its axis aligned to the substrate (made by glass). As a result, incident light suffers from scattering and therefore a whitish layer appears on the glass substrate. The nematic phase is observed in the helix when applied bias exceeds 10^5 V/cm. In the nematic condition, alignment is seen in the helix. The result is that that cell becomes transparent.

iv) **Field-induced Director Axis Re-orientation:** This is a common example of electro-optic effect on aligned nematic and ferroelectric liquid crystal for light modulation. Different methods have already been applied for perfect alignment of nematic crystals, and some are reported as very successful [20, 21]. Those are later developed [22] for several applications involving nematic liquid crystals [21, 22].

v) **Laser Addressed Thermal Effect:** This effect has been observed in liquid crystals of type cholesteric [23] and sematic-A [24]. For sematic-A crystal, a small dot on the cell is identified and thereafter focused by IR laser. As a result of absorption, the area which is under the influence of the radiation is heated over the sematic-nematic phase transition temperature. As the particular spot cools down, it is transferred into a perfectly scattered state or may be a well-aligned non-scattered state also where degree of cooling level and applied bias plays a pivotal role in determining the final state. If the cooling process is slower, then the disordered molecules reorient themselves into the preliminary homogeneous state of configuration which turns it into the non-scattering state. If a comparatively higher field is applied during the cooling process, it also helps in the alignment process.

vi) **Light Scattering by Micron-sized Droplets:** Scattering is produced in significant amount when micron-sized droplets are dispersed in polymer matrix [25, 26]. Refractive index mismatch between the droplets of the liquid crystal along with the substrate polymer is the major reason behind the scattering in the voltage-off state. In the opposite (voltage-on) state, droplets align themselves according to the direction of applied field, so that the desired matching takes place. Owing to this, magnitude of scattering decreases and therefore the transmission of light becomes maximum. For this mechanism the advantage is that non-polarized light can be used, hence the optical efficiency for this mechanism is greatly enhanced.

12.2.4 Nonlinear Optical Switch

The concept of "light controlling with light" is a difficult thing to achieve as the photons are electrically neutral. The way of controlling them is an indirect way and that is by using the concepts of nonlinear optics [27–31]. The nonlinear property can be used by applying a pump light which modulates the transmission properties of the signal and that helps in the switching of the signal light. One of the examples is the RAMAN Scattering which is used for optical amplification. The hole mechanism is depended on the suitable utilization of refractive index property of the medium (nonlinear). Owing to Kerr effect, as light with very high intensity passes through the nonlinear medium, it induces a change in the refractive index Δn. Now this modification in RI causes a tuning in the signal phase that is a phase shift occurs between the original signal and the one that passed through the nonlinear medium. The contrast ratio or the intensity ratio is dependent upon the amount of phase shift; obviously, the greater the phase shift, the greater the ratio.

12.3 Classification of Nonlinear All-Optical Switches

The nonlinear optical switches are based mainly on two nonlinear mechanisms – nonlinear mechanism and nonlinear materials. The various nonlinear mechanisms or nonlinear optical effect include the nonlinear refraction, reflection, absorption, polarization, frequency modulation, and phase transition. Other than these, a lot devices based on optical properties are also used for these purposes like the nonlinear interferometer, coupler, grating, attenuator, amplifier etc.

Table 12.1 Fundamental properties of widely used nonlinear optical materials.

Type of material	Material type	Nonlinear mechanism	$n_2(cm^2/W)$	$a_0(cm^{-1})$	$\tau(s)$	$Q(cm^3/\tau W)$
Nematic LC	Liquid crystal	Molecular orientation	10^{-3}	10^3	1	10^{-6}
GaAs	Direct bandgap semiconductor	Free exciton nonlinearity	10^{-6}	10^4	10^{-8}	10^{-2}
CS$_2$	Organic material	Molecular orientation	10^{-13}	10^{-1}	10^{-12}	10^1
Si	Organic material	Electronic polarization	10^{-14}	10^{-5}	10^{-14}	10^5

The other optical property for realization of nonlinear AOS is the optical material itself. The Kerr effect based AOS required the modulation of nonlinear refraction coefficient n_2, the nonlinear response time τ and the absorption α_0. These are the influencing optical parameters. The quality factor is defined as

$$Q = \frac{n_2}{\alpha_0 \pi}.$$

Table 12.1 lists the different types of nonlinear materials along with their nonlinear properties. The four typical nonlinear materials are

i) nonlinear molecular orientation based nematic liquid crystal
ii) free exciton nonlinear mechanism as available in direct bandgap semiconductor
iii) nonlinear molecular orientation based organic material
iv) nonlinear electronic polarization as available in indirect bandgap semiconductor.

From Table 12.1 it can be easily seen that the different optical materials exhibit different properties and depending upon those properties the material is chosen for being used for optical switches. In the table the values of the different properties n_2, a_0, and τ are all decreasing from top to bottom. Fast switching times are not ideal for fast operating optical switches hence liquid crystals are not ideal materials for optical switches although they possess the highest nonlinearity. The direct bandgap semiconductor has strong nonlinearity but due to its high absorption it is not ideal for cascaded operations where devices are connected end to end. The organic material too, in spite of having high nonlinearity and low absorption, is not suited as a good material for optical switches because their chemical stability is less. Thus the best material for realization of optical switches is Silicon (Si), which has small response time, low absorption, and good quality factor.

12.3.1 Optical Coupler AOS

One of the basic and important elements of the optical interferometer is the optical coupler. The optical coupler is used to couple light in and out of the interferometer. Even the optical coupler can be itself considered as an interferometer based on the concept of interference principle between two beams. An optical coupler is another version of the directional coupler but based on light technology and thus is known as the Optical Directional Coupler [32–34]. It comprises of two linear waveguides parallel to each other with single-mode propagation of e.m waves, generally named as waveguide I and waveguide II. Two waveguides having same length l in the coupling regime are

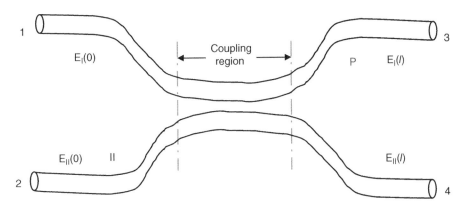

Figure 12.1 Basic structure of an optical coupler.

in very close range of each other, and their dimensions are in the micron range. Figure 12.1 shows the basic construction of an optical coupler. In the coupled region, wave can propagate from one waveguide to another in evanescent mode.

The waveguide pairs are mainly made of silica by fused taper technology. Other than that, various semiconductors, organic materials, lithium niobate etc. may be used to construct the strip-type plane waveguides with the help of linear integration techniques.

The incident light is assumed to be a monochromatic plane wave whose electric field amplitude is defined as $E(z,t)e^{-(\omega t - \beta z)}$, where $E(z,t)$ is slowly varying electrical amplitude with complex magnitude propagates along z-axis. The rest of the variables are self-explanatory. As the incident wave falls at Port I of the coupler it is partitioned into two unequal waves. They respectively propagate along the waveguides 1I and II respectively, and both of them naturally become parallel to the z direction. The amplitudes in both the beams are given by $E_I(z,t)$ and $E_{II}(z,t)$ respectively. Under the condition of weak coupling, light absorption becomes insignificant, and therefore the variation of E_I and E_{II} can be respectively given by the following equations, which are similar to the Nonlinear Schrodinger Equations:

$$\frac{\partial E_I}{\partial z} + \beta_I \frac{\partial E_I}{\partial t} + \frac{i\beta_{II}}{2}\frac{\partial^2 E_I}{\partial t^2} = i\kappa_{12}E_{II} + i\delta E_I + i\left(\gamma_I |E_I|^2 + \psi_{12}|E_{II}|^2\right)E_I \tag{12.14}$$

$$\frac{\partial E_{II}}{\partial z} + \beta_I \frac{\partial E_{II}}{\partial t} + \frac{i\beta_{II}}{2}\frac{\partial^2 E_{II}}{\partial t^2} = i\kappa_{21}E_I - i\delta E_{II} + i\left(\gamma_{II}|E_{II}|^2 + \psi_{21}|E_I|^2\right)E_{II}, \tag{12.15}$$

where $\beta_I \equiv \frac{1}{v_g}$ and v_g is usually termed the group velocity of the wave. β_{II} is GVD (group velocity dispersion) parameter; κ_{12} and κ_{21} are the coupling coefficients respectively; γ_I and γ_{II} are the SPM (self-phase modulation) parameters; C_{12} and C_{21} are the parameters describing XPM (cross-phase modulation). The degree of asymmetry δ is defined as

$$\delta = \frac{1}{2}\left(\beta_{0I} - \beta_{0II}\right), \tag{12.16}$$

where β_{0I} and β_{0II} are termed as constants of propagation with respect to wave flowing inside waveguide I and II respectively.

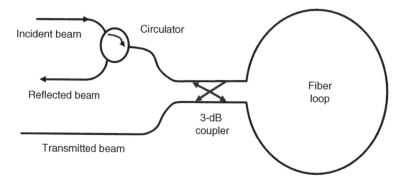

Figure 12.2 Basic structure of a Sagnac interferometer.

12.3.2 Sagnac Interferometer AOS

Sagnac interferometers (SIs) are all-optical interferometers which work on the principle of two-beam interference principle [35, 36]. Sagnac interferometers are composed of a optical waveguide coupler which is physically coupled with a loop of waveguide. The basic construction of the SI is shown in Figure 12.2.

From the figure it can be seen that the SI consists of two output ports: Port 1 and Port 2. Port 1 is the reflection port and Port 2 is the transmission port. The incident light beam also comes through Port 1. The circulator is placed so that the reflected light does not disturb the incident light beam. The ports 3 and 4 are connected to a fiber loop. If the loop is made of nonlinear material, then it is known as nonlinear Sagnac Interferometer. If the 3-dB coupler is symmetric, then the counter-propagating waves in the loop are equal and it is known as the symmetric Sagnac Interferometer. If the 3-dB coupler is asymmetric then it is known as asymmetric Sagnac Interferometer.

Since in a SI, both beams travel the same path in directions opposite to each other, hence SIs are extremely stable and aligns easily even with an extended broadband light source. SIs are generally used for measurement of rotation. Phase shift is obtained by rotating the interferometer with angular velocity ω at an angle θ with respect to a predefined axis, and the plane of rotation should be normal with the plane of interferometer as

$$\varphi = \frac{8\pi\omega A\cos\theta}{\lambda c}, \tag{12.17}$$

where A is the area which the light path encloses, λ is the wavelength, and c is the speed of light.

12.3.3 M–Z Interferometer AOS

Mach–Zender interferometers or M–Z interferometers are also based upon the two-beam interference principle. Their basic construction includes two 3-dB couplers DC_1 and DC_2 which are connected to two straight waveguides I and II. Thus the device consists of total of 4 ports as shown in Figure 12.3.

In a self-pump M–Z interferometer, the two arms L_1 and L_2 are made of different materials with varying refractive indices n_1 and n_2 respectively. Along with that, the arm lengths are also of two different sizes. When light is incident in Port 1, the beam gets divided by the DC_1 and two light beams with same intensity flow into arms L_1 and L_2. Since the two arms are different, two different phase shifts occur in them, with notations ϕ_1 and ϕ_2. These two-phase shifted light signals come at DC_2 and interfere and are output from Ports 3 and 4. The distribution of power among these two

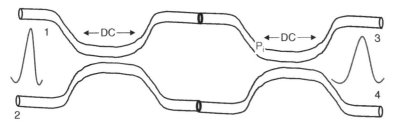

Figure 12.3   M–Z interferometer AOS.

ports depends on the phase difference of the individual phase shifts, $\phi = \phi_1 - \phi_2$. It can be established that for low input power, an output wave can be obtained from Port 4, but for high input powers, the phase shift nearly equals π and light is output from Port 3 and this phenomenon represents the all-optical switching. If both the arms become identical, i.e. symmetric in nature, the phase shift equals 0 and then the interferometer cannot work as an all-optical switch.

Because the beams travel the measurement path only once and the beam separation can be changed according to requirement, this interferometer is thus well suited for applications like to study of flow of gases, transfer of heat, and the distribution of temperature in plasmas and flames.

12.3.4   Ring Resonator AOS

A ring resonator or RR is a type of interferometer which is based on the principle of multiple-beam interference, similar to the Fabry–Pérot interferometer. The main working differences between the two is that the RR works on travelling-wave interference while the FP works on standing wave interference; in the case of FP, the wave travels back and forth multiple times and in case of RR, the wave travels multiple times inside the ring. The basic structures of both interferometers are given in Figure 12.4.

Figure 12.4   (a) Multiple-bean interference inside FP resonant cavity. (b) Multiple-bean interference inside ring resonator.

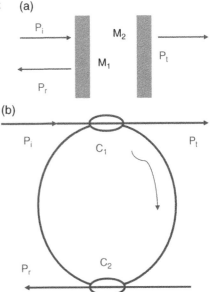

The function of couplers C_1 and C_2 in the RR are same as the mirrors M_1 and M_2 in case of the FP interferometer. The couplers fed back the wave along the input and output paths, which mirrors are unable to do. The incident light with power P_i enters through C_1 or M_1 and outputs through C_2 or M_2 with power P_t, which are the transmitted output, and outputs from C_1 or M_1 with power P_r are called the reflected output.

There are two types of RRs as available from literature:

i) single coupler Ring Resonator (SCRR): one common example is the all-optical switch, which is a 1×1 optical switch (one straight waveguide coupled with ring through single coupler)
ii) double coupler Ring Resonator (DCRR): such an example is a 1×2 all-optical switch.

12.3.5   Fiber Grating AOS

All macro-scale all-optical switches cannot fully serve the very fine practical requirements like fine tunings of wavelength and other fine jobs. For those purposes the fiber grating all-optical switches have proved to be fruitful. The fiber grating all-optical switches have a lot of advantages like (i) lower insertion loss, (ii) smaller dimension, (iii) comparatively less complicated structure from a fabrication point of view, and (iv) mapped with fiber system. There are a few major types of fiber gratings like Fiber Bragg grating (FBG), Chirp Fiber Grating (CFG), Phase Shift Fiber Grating (PSFG), Long-period fiber grating (LPFG) etc.

Based on the magnitude of grating period, fiber grating can be broadly be classified into following two categories:

i) **Fiber Bragg Grating (FBG):** Grating constant (Λ_B) is less than 0.1 μm. The method of mode coupling in FBG is the interference of forward propagating and counter-propagating waves. They together constitute reflected spectrum with a sharper bandwidth with the grating wavelength at the center. Thus these are known as optical grating of reflection type.
ii) **Long-Period Fiber Grating (LPFG):** Grating constant (Λ_L) is greater than 100 μm. Waves propagated inside core and cladding regions independently cause interference when they are superposed, provided directions of propagations are identical. It forms a transmitted profile with 20 nm approximated bandwidth with the grating wavelength at the center. Thus these are known as transmission type optical grating.

12.4   Working Methodology of Different Types of Nonlinear All-Optical Switches

The previous section gave an overall idea of all the types of nonlinear all-optical switches. This section provides the working principle of these switches, describing their conceptual working, their types and their various fields of applications.

12.4.1   Optical Coupler AOS

Optical coupler AOSs are classified into three types: symmetric couplers working in low incident power, symmetric couplers working in high-power incident light with self-phase modulation, and asymmetric couplers working with high-power incident light with cross-phase modulation. These are individually explained below

12.4.1.1 Symmetric Coupler Working at Low Incident Power

Symmetric couplers are those where the two wavelengths are the same as each other in all properties, so that the geometrical configurations of two waveguides are the same i.e. $\kappa_{12} = \kappa_{21} = \kappa$, and they have the same refractive indices i.e. $\eta_1 = \eta_2 = \eta$ and $\beta_{0I} = \beta_{0II}$ or $\delta = 0$ (from Equation 12.16). When the incident light has low power and is continuous in nature, then the nonlinear effects do not come into play, that is, the light is unable to change the refractive index of the materials. Hence the parameters γ_1, γ_2, C_1, and C_2 from Equation 12.14 and Equation 12.15 are all zero and the equations reduce to

$$\frac{dE_I(z)}{dz} = i\kappa E_{II}(z) \tag{12.18}$$

$$\frac{dE_{II}(z)}{dz} = i\kappa E_I(z), \tag{12.19}$$

whose solutions are given by

$$E_I(z) = \cos(\kappa z)E_I(0) + i\sin(\kappa z)E_{II}(0) \tag{12.20}$$

$$E_{II}(z) = i\sin(\kappa z)E_I(0) + \cos(\kappa z)E_{II}(0). \tag{12.21}$$

Now $\cos(\kappa z)$ is defined as the coupler reflectivity denoted by r and $\sin(\kappa z)$ is defined as the coupler transmissivity denoted by t. Hence

$$r = \cos(\kappa z) \tag{12.22}$$

$$t = \sin(\kappa z), \tag{12.23}$$

where r and t are reflectance and transmittance for wave amplitude, r^2 and t^2 are power magnitudes of those parameters respectively. From the above notations, it can be said that they follow the relation (if absorption is neglected)

$$r^2 + t^2 = 1. \tag{12.24}$$

Also

$$t^2 = C_t \text{ and } r^2 = C_r \tag{12.25}$$

respectively define the transmittance split factor and reflectance split factor for output powers.

If light enters only from Port 1 such that $E_1 \neq 0$ and $E_2 = 0$ as shown in Figure 12.5, the coupling region length is taken to be l, and the amplitudes of the output waves at port 3 from waveguide I and at port 4 from waveguide II respectively are:

$$E_I(l) = rE_I(0) \tag{12.26}$$

$$E_{II}(l) = itE_I(0). \tag{12.27}$$

From Figure 12.5 it can be seen that the transmitted path is called t and the reflected part is called r. From Equations 12.26 and 12.27 it can be seen that the transmitted light suffers a 90° phase shift with respect to the reflected light in the direct arm. To find the respective power outputs

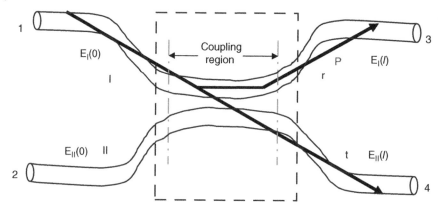

Figure 12.5 Transmitted path (t) (cross path) and reflected path (r) (bar path) inside the optical coupler.

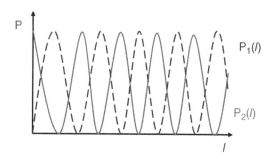

Figure 12.6 Output light travels from I to II as a function of coupling length.

from P_3 to P_4, we use the basic power-energy relation $P = |E|^2$. Since we know the individual energy outputs from P_3 and P_4, thus the respective powers can be easily calculated as

$$P_1\left(l\right) = P_{in}\cos^2\left(\kappa l\right)$$

(12.28)

$$P_2\left(l\right) = P_{in}\sin^2\left(\kappa l\right).$$

(12.29)

Thus it can be seen that the power distribution between the arms completely depends upon the product of the coupling coefficient κ and the length of the arm l. For a certain κ, the power output versus length variation is shown in Figure 12.6.

Here L_C refers to the coupling length which is expressed by $L_c = \frac{\pi}{2\kappa}$.

In the above case we can observe that the power outputs from the respective arms do not depend on the incident light power, but rather on the geometry of the system, hence this cannot be used to realize an all-optical switch. The realization of all-optical switches requires high powers which are discussed in the succeeding sections where these couplers are used with the concept of self-phase and cross-phase modulation.

12.4.1.2 Symmetric Coupler Working in High-Power Incident Light with SPM

In this case a quasi-continuous high-power incident light enters the coupler from port 1. The input power is denoted by P_0 and there exists a critical power P_c such that when the input power is lower than the critical power i.e. $P_0 << P_c$, then the total signal emits from waveguide II. On the other hand if $P_0 > P_c$, that is the input power is greater than the critical power, then the optical Kerr effect

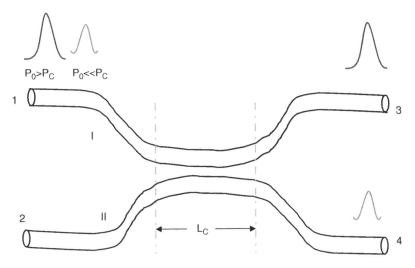

Figure 12.7 AOS with SPM technique based on symmetric coupler.

occurs in both the waveguides and power distribution occurs on both the arms. Since the powers are different in the two arms, hence the change in refractive indices are also different, hence different phase shifts occur within the two arms. If the phase shift crosses the threshold, then total output can only be obtained from waveguide I, as shown in Figure 12.7.

Since the coupler is symmetric, hence $\kappa_{12}=\kappa_{21}=\kappa$, and also the Kerr effect in both the arms is the same, hence, $\gamma_1=\gamma_2=\gamma$ and $C_{12}=C_{21}=\gamma\sigma$, thus the Equations 12.14 and 12.15 reduce down to

$$\frac{\partial E_I}{\partial z}=i\kappa E_{II}+i\gamma\left(\left|E_I\right|^2+\sigma\left|E_{II}\right|^2\right)E_I \tag{12.30}$$

$$\frac{\partial E_{II}}{\partial z}=i\kappa E_I+i\gamma\left(\left|E_{II}\right|^2+\sigma\left|E_I\right|^2\right)E_{II}. \tag{12.31}$$

where the nonlinear parameter γ is defined by

$$\gamma=\frac{k_0 n_2}{S}, \tag{12.32}$$

where n_2 is the nonlinear refraction coefficient, S is the effective area of light field in waveguides and k_0 is the wave vector at free space. Denoting the phase and power in the individual arms by φ_i and P_i ($I=$ I, II,. . .), then the amplitude of the light wave may be expressed as

$$E_i=\sqrt{P_i}e^{i\phi}. \tag{12.33}$$

Substituting Equation 12.33 into Equation 12.30 and Equation 12.31 and setting the phase shift difference to $\phi=\phi_1-\phi_2$, one obtains the following equations:

$$\frac{dP_I}{dz}=2\kappa\sqrt{P_I P_{II}}\sin\phi \tag{12.34}$$

$$\frac{dP_{II}}{dz}=-2\kappa\sqrt{P_I P_{II}}\sin\phi \tag{12.35}$$

$$\frac{d\phi}{dz} = \frac{P_{II} - P_I}{\sqrt{P_{II}P_I}} \kappa \cos\phi + \frac{4\kappa}{P_C}\left(P_I - P_{II}\right) \tag{12.36}$$

and the critical power P_C is given by

$$P_C = \frac{4\kappa}{\gamma(1-\sigma)}. \tag{12.37}$$

If cross-phase conduction is ignored between the arms, that is $\sigma = 0$, and using $k_0 = \frac{2\pi}{\lambda_0}$, and $L_C = \frac{\pi}{2\kappa}$, one obtains the final expression for critical power:

$$P_C = \frac{4\kappa}{\gamma(1-\sigma)}. \tag{12.38}$$

Equation 12.38 depicts that for specific values of coupling length, wavelength off incoming beam and input light, critical power reduces with enhancement of nonlinear refraction coefficient and also with the decrease of waveguide working area.

Equations 12.34–12.36 can be solved analytically using elliptical functions. When light is incident on waveguide I of the coupler with a power P_0, then the powers are divided into two ports, and termed as P_I and P_{II}. Then at any arbitrary point z, powers can be written as

$$P_I(z) = \left|E_I(z)\right|^2 = \frac{1}{2}P_0\left[1 + cn\langle 2\kappa | \tau\rangle\right] \tag{12.39}$$

$$P_{II}(z) = P_0 - P_I(z), \tag{12.40}$$

where P_0 is input power, and $cn\langle x | \tau\rangle$ is Jacobi elliptic function where $\tau = \left(\frac{P_0}{P_C}\right)^2$.

The relative power output from the two ports versus the input power curve is shown in Figure 12.8.

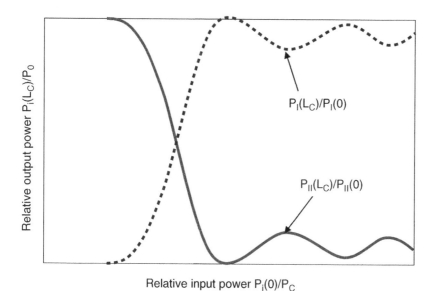

Figure 12.8   Relative output powers with relative input power.

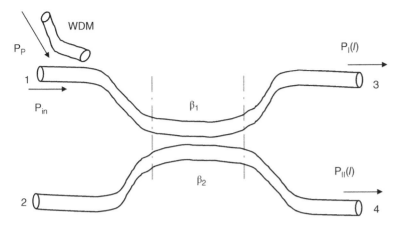

Figure 12.9 AOS in asymmetric coupler under XPM mode.

Figure 12.10 Reflectivity and transmittance profiles with phase for low input signal.

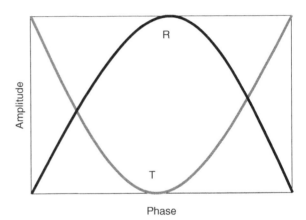

From the plot, it may be observed that threshold switching power of the device can be given by

$$P_{0C} = 1.25 P_C = 1.25 \frac{\lambda S}{n_2 L_C} = 2.5 \frac{\kappa \lambda S}{\pi n_2}. \tag{12.41}$$

12.4.1.3 Asymmetric Coupler Working in High-Power Pump Light with Cross-phase Modulation

The setup for cross-phase modulation using optical coupler is shown in Figure 12.9. The light as denoted by P_{in} is at low power and hence the nonlinearities do not occur in the waveguide for the incident signal wave. The other signal that is the pump signal denoted by P_p, which is at a different wavelength from the signal one, is a high-power signal and thus nonlinearities start to occur in the waveguides. The corresponding reflectivity and transmittivity profiles are exhibited in Figure 12.10.

After introduction of the pump beam into the coupler, the refractive index changes in the waveguides become unequal. Since low power and quasi-continuous signal light is used it cannot generate nonlinearities in the system and hence the terms related to nonlinearity can be neglected. Since the geometries of the guides are same, then $\kappa_{12} = \kappa_{21} = \kappa$, but since the pump is of different frequency, the coupler suffers nonlinearities, hence the power in the two different arms are different and due to the Kerr effect the refractive indices of the arms are also different, thus the propagation constant for the two are also different, i.e. $\beta_{01} \neq \beta_{02}$ i.e. $\delta \neq 0$.

Putting these into Equation 12.14 and Equation 12.15 we get

$$\frac{dE_I}{dz} = i\kappa E_{II} + i\delta E_I \tag{12.42}$$

$$\frac{dE_{II}}{dz} = i\kappa E_I - i\delta E_{II}, \tag{12.43}$$

where E_I and E_{II} are the signal amplitudes. Taking derivatives of Equation 12.42 and Equation 12.43, we get

$$\frac{d^2 E_I}{dz^2} = i\kappa \frac{dE_{II}}{dz} + i\delta \frac{dE_I}{dz} \tag{12.44}$$

$$\frac{d^2 E_{II}}{dz^2} = i\kappa \frac{dE_I}{dz} - i\delta \frac{dE_{II}}{dz}. \tag{12.45}$$

With a few mathematical computations

$$\frac{d^2 E_I}{dz^2} + \kappa_\varepsilon^2 E_I = 0 \tag{12.46}$$

$$\frac{d^2 E_{II}}{dz^2} + \kappa_\varepsilon^2 E_{II} = 0, \tag{12.47}$$

where

$$\kappa_\varepsilon = \sqrt{\kappa^2 + \delta^2}. \tag{12.48}$$

Solving the above equations using the initial values, we can get the transmission and reflection variables as

$$t' = \left(\frac{\kappa}{\kappa_\varepsilon} \right) \sin\left(\kappa_\varepsilon z\right) \tag{12.49}$$

$$r' = \cos\left(\kappa_\varepsilon z\right) + i\left(\frac{\delta}{\kappa_\varepsilon} \right) \sin\left(\kappa_\varepsilon z\right). \tag{12.50}$$

One can, therefore, easily find the expression for the amplitudes E_I (l) and E_{II} (l) as

$$E_I\left(l\right) = \left[\cos\left(\kappa_\varepsilon l\right) + i\left(\frac{\delta}{\kappa_\varepsilon} \right) \sin\left(\kappa_\varepsilon l\right) \right] E_{in} \tag{12.51}$$

$$E_{II}\left(l\right) = \left(\frac{\kappa}{\kappa_\varepsilon} \right) \sin\left(\kappa_\varepsilon l\right) E_{in}. \tag{12.52}$$

Using the power-energy relation, the respective powers can also be found out and is given by

$$P_I\left(l\right) = \left[\cos^2\left(\kappa_\varepsilon l\right) + i\left(\frac{\delta}{\kappa_\varepsilon} \right)^2 \sin^2\left(\kappa_\varepsilon l\right) \right] P_{in} \tag{12.53}$$

$$P_2\left(l\right) = \left(\frac{\kappa}{\kappa_e} \right)_2 \sin\left(\kappa_e l\right)^2 P_{in}. \tag{12.54}$$

If $L_C = \frac{\pi}{2\kappa}$, then threshold switching power of pump wave can be written as

$$P_{pc} = \frac{\sqrt{3}\kappa\lambda S}{2\pi n_2} \frac{1}{|1 - 2C_{rp}|}\left(C_{rp} \neq \frac{1}{2}\right) \tag{12.55}$$

12.4.2 Sagnac Interferometer AOS

The Sagnac Interferometer AOS are classified into four types as given below.

12.4.2.1 Sagnac Interferometer (SI) Under Low Incident Power

As in the case for coupler switches, SIs also cannot function as all-optical switches. Figure 12.2 shows the basic structure of a SI and the same structure are used as reference for present analysis. When the light is incident on Port 1 of the coupler, it gets divided equally into two light beams with powers Pin/2 which travel into the bar and cross paths of the and enter the fiber loop from Port 3 or Port 4. After one trip through the loop, no phase shift occurs. At the ports they are again separated into two light beams, which then again pass through the bar and cross paths and output from Port 1 and Port 2. At each output port, the power of the two beams depends on the difference of phases of the two light beams. If the phase difference between the beams is an integral multiple of π or $(2m\pi)$, then it is called constructive interference and the power output is the added power of the individual light beams. If the phase difference is, however, an odd multiple of π or $(2m+1)\,\pi$, then destructive interference occurs and the output power is the difference of the individual light beam powers. Symmetric Sagnac Interferometer cannot be used to make the all-optical switch. The mathematical explanation is given as follows:

SI is made of symmetric coupler having coupling length l and coefficient κ. The nonlinear loop has a length L. Using coupled mode theory, we can obtain

$$E_I(l) = \cos(\kappa l) E_I(0) + i\sin(\kappa l) E_{II}(0) \tag{12.56}$$

$$E_{II}(l) = \cos(\kappa l) E_{II}(0) + i\sin(\kappa l) E_I(0). \tag{12.57}$$

If wavelength of input wave from Port 1 is λ, then we can formulate boundary conditions as

$$E_1(0) = \sqrt{P_{in}} \tag{12.58}$$

$$E_2(0) = 0. \tag{12.59}$$

Wave amplitudes of port 3 and port 4 are respectively written in the form

$$E_I(l) = \sqrt{P_{in}} \cos(\kappa l) \tag{12.60}$$

$$E_{II}(l) = i\sqrt{P_{in}} \sin(\kappa l). \tag{12.61}$$

The two light beams propagate back and forth within the loop, that is, in the clockwise and anti-clockwise directions. Let the respective refractive indices of the loop medium for the light beams be n_{cl} and n_{ccl}. After complete rotation, phase shifts are

$$\phi_{cl} = \left(\frac{2\pi}{\lambda}\right) n_{cl} L$$

$$\phi_{ccl} = \left(\frac{2\pi}{\lambda}\right) n_{ccl} L$$

Therefore, field amplitudes of two beams passing through bar path and cross paths respectively to reach Port 1 are

$$E_{rl} = i\sqrt{P_{in}} \sin(\kappa l)\cos(\kappa l)e^{i\left(2\pi/\lambda\right)n_{cl}L}$$

(12.62)

$$E_{tl} = i\sqrt{P_{in}} \sin(kl)\cos(\kappa l)e^{i\left(2\pi/\lambda\right)n_{ccl}L}.$$

(12.63)

Therefore, net field at Port 1 is

$$E_{rnet} = E_{rl} + E_{tl}.$$

(12.64)

Consequently the power at Port 1 is given by

$$P_r = E_{rnet} \cdot E_{rnet}*.$$

(12.65)

The reflectivity and transmissivity of the system are defined by

$$R = \sin^2\left(2\kappa l\right)\cos^2\left(\frac{\pi\Delta nL}{\lambda}\right) = \sin^2\left(2\kappa l\right)\cos^2\left(\frac{\Delta\phi}{2}\right)$$

(12.66)

$$T = 1 - R = 1 - \sin^2\left(2\kappa l\right)\cos^2\left(\frac{\pi\Delta nL}{\lambda}\right) = 1 - \sin^2\left(2\kappa l\right)\cos^2\left(\frac{\Delta\phi}{2}\right),$$

(12.67)

where Δn indicates difference of refractive indices, $\Delta\phi$ is the phase difference.

For low power signal, $\Delta n = 0$, thus $\Delta\phi = 0$. Then

$$R = \sin^2\left(2\kappa l\right)$$

(12.68)

$$T = 1 - \sin^2\left(2\kappa l\right).$$

(12.69)

Thus it can be seen that for low input incident light power, the output depends on the κl product and not on the input power and hence cannot be realized as all-optical switch.

12.4.2.2 Sagnac Interferometer AOS with Non-3dB Coupler

A non-3dB coupler is used with the SI to induce asymmetry to the system. The setup is shown in Figure 12.11. A light incident on the input port of the SI reaches the output ports 3 and 4 with

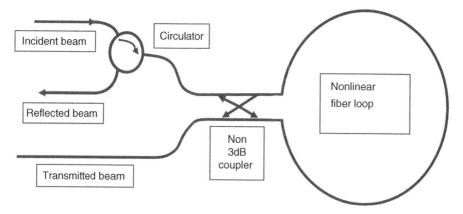

Figure 12.11 AOS in nonlinear SI with non-3dB coupler working at SPM mode.

different powers which enter the loop in counter-propagating directions. After one complete cycle, the lights produce a phase shift which is equal to π for high signal powers. When the beams reach the input ports of the system, they interfere with each other, and all the light transfers from the reflected port to the transmitted port. This is how an AOS is realized. However, under high nonlinearity and large power splitting conditions, the required signal power can be a lot less.

For splitting ratio C_r, clockwise beam and the anticlockwise beam powers can be respectively expressed as

$$P_{cl} = C_r P_{in}$$ (12.70)

$$P_{ccl} = (1 - C_r) P_{in},$$ (12.71)

where $C_r = r^2 = \cos^2(\kappa l)$, $1 - C_r = t^2 = 1 - r^2 = 1 - \cos^2(\kappa l) = \sin^2(\kappa l)$.

For high incident power, both the cross-phase and self-phase modulations have to be considered, therefore the RI of the beams are given by

$$n_{ccl} = n_0 + n_2 \frac{P_{ccl} + 2P_{cl}}{S}$$ (12.72)

$$n_{cl} = n_0 + n_2 \frac{P_{cl} + 2P_{ccl}}{S}.$$ (12.73)

Therefore change in RI can be given by

$$\Delta n = n_{ccl} - n_{cl} = (2C_r - 1) \frac{n_2 P_{in}}{S}.$$ (12.74)

Thus if $C_r = 1/2$, R=1 and T =0 which indicates total power output from reflected port only. For other values of C_r, the power in the two beams are different and hence the difference in RI as well as the phase difference are also non-zero i.e. $\Delta n = 0$ and $\Delta\phi = 0$. Thus light will output in parts from both the reflected and transmitted ports. To achieve complete switching, the condition for is that the difference in RI should be $\Delta n = \lambda/2L$. From that the expression for threshold input power can be given as

$$P_{inc} = \frac{\lambda S}{n_2 L} \cdot \frac{1}{2|2C_r - 1|} \quad \left(C_r \neq \frac{1}{2}\right)$$ (12.75)

12.4.2.3 Sagnac Interferometer AOS in Cross-Phase Modulation

The signal and pump waves are input of the system through a WDM coupler. The wavelengths of the signal and the pump are not same. Moreover, the power of the signal is lower than that of the pump. Hence the 3-dB coupler is symmetric to the signal light but asymmetric to the pump light since it is of high power. Thus, as described above, the pump wave splits into two different beams with different powers and hence with different RIs.

When the input wave is very strong, output power is directed from Port 1 to Port 2. Assuming the pump power to be P_p and the splitting ratio to be C_{pr}, clockwise light is indicated by P_{pcl} and anticlockwise by P_{pccl}.

$$P_{pcl} = C_{pr} P_p$$ (12.76)

$$P_{pccl} = (1 - C_{pr}) P_p$$ (12.77)

For high-power pump light, both xPM (cross-phase modulation) Kerr effect and SPM (self-phase modulation) Kerr effect are taken into consideration. Moreover there is an additional XPM Kerr effect induced from the addition of the signal and pump power, hence the RIs for the clockwise and anticlockwise directions are given by Equations 12.72 and 12.73.

$$n_{ccl} = n_0 + n_2 \frac{P_{pccl} + 2P_{pcl}}{S} \tag{12.78}$$

$$n_{cl} = n_0 + n_2 \frac{P_{pcl} + 2P_{pccl}}{S} \tag{12.79}$$

From the above equations, the difference in RI can be obtained as

$$\Delta n = n_{ccl} - n_{cl} = \left(2C_{pr} - 1\right)\frac{n_2\left(2P_p\right)}{S}. \tag{12.80}$$

Following the same argument for C_{pr} as given in the preceding section, the threshold input power is given by

$$P_{pcl} = \frac{\lambda S}{n_2 L} \cdot \frac{1}{4|2C_r - 1|} \quad \left(C_{pr} \neq \frac{1}{2}\right). \tag{12.81}$$

12.4.2.4 Sagnac Interferometer AOS with Optical Amplifier

The setup is made by simply attaching a bidirectional EDFA in the fiber loop near the end of the coupler [37] as shown in Figure 12.12. All the operations remain same as before, just the amount of the two separate beam powers change, which are given by

$$P_{cl} = GC_r P_{in} \tag{12.82}$$

$$P_{ccl} \approx \left(1 - C_r\right)P_{in}, \tag{12.83}$$

where P_{in} is input power, C_r is the splitting ratio for bar path of coupler and G is amplifier gain.

The difference in RIs is given by

$$\Delta n \left(C_r \left(1 + G\right) - 1\right)\frac{n_2 P_{in}}{S}. \tag{12.84}$$

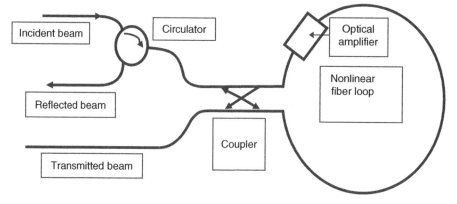

Figure 12.12 AOS in SI with optical amplifier placed asymmetrically. Source: Richardson et al. [37] / Institution of Electrical Engineers.

Considering the condition for switching, $\Delta n = \lambda/2L$, the threshold switching power is given by

$$P_{inc} = \frac{\lambda S}{n_2 L} \cdot \frac{1}{2|2C_r(1+G)-1|}.$$

(12.85)

When $C_r = 1/2$, the expression reduces to

$$P_{inc} = \frac{\lambda S}{n_2 L} \cdot \frac{1}{|G-1|}.$$

(12.86)

12.4.3 M–Z Interferometer AOS

Keeping the basic structure same the M–Z Interferometer is mainly classified into types which are

i) M–Z interferometer AOS with different arm materials
ii) M–Z interferometer AOS with different arm lengths.

Both of these are explained in detail [38–41]. The basic structure of an M–Z interferometer is shown in Figure 12.3.

12.4.3.1 M–Z Interferometer AOS with Different Arm Materials

This setup is same as the one shown in Figure 12.3 but with both the arms made of different non-linear optical materials. Hence the lengths of the arms $L_1 = L_2 = L$, but due to different materials, the nonlinear RIs are not same to each other $n_{12} \neq n_{22}$. The incident light has a power P_{in} which induces unequal changes in the RIs and induces different phase shifts $\phi_1 \neq \phi_2$ where $\phi_1 = k_0 \Delta n_1 L$ and $\phi_2 = k_0 \Delta n_2 L$. Thus the phase difference $\phi = \phi_2 - \phi_1 \neq 0$.

A 3-dB coupler is used, hence $\kappa z = \pi/4$ and $r = t = 2\sqrt{2}$. Hence the respective amplitude and power outputs (using the power-energy relation) from Ports 3 and 4 can be written as

$$E_{III} = \frac{1}{2}\left(e^{-i\phi_1} - e^{-i\phi_2}\right)E_I$$

(12.87)

$$E_{IV} = i\frac{1}{2}\left(e^{-i\phi_1} + e^{-i\phi_2}\right)E_I$$

(12.88)

$$P_{III} = P_{in}\sin^2\left(\frac{\phi}{2}\right)$$

(12.89)

$$P_{IV} = P_{in}\cos^2\left(\frac{\phi}{2}\right).$$

(12.90)

From the above equations it can be seen that for $\phi = 0$, the output $P_{III} = 0$ and all the output is obtained from Port 4 that is $P_{IV} = P_{in}$. For $\phi = \pi$, $P_{III} = P_{in}$ and $P_{IV} = 0$, thus the condition for all-optical switching is that the difference of phase should be equals to π.

Assuming that the linear RIs of the two arms are nearly equal to each other, and then the change in nonlinear RIs can be written as

$$\Delta n_1 = \frac{n_{21} P_{in}}{2S}$$

(12.91)

$$\Delta n_2 = \frac{n_{22} P_{in}}{2S},$$

(12.92)

where S is the cross-sectional area. The individual beam phase shifts are given by

$$\phi_1 = k_0 \Delta n_1 L = \frac{k_0 n_{21} P_{in} L}{2S}$$

(12.93)

$$\phi_2 = k_0 \Delta n_2 L = \frac{k_0 n_{22} P_{in} L}{2S}$$

(12.94)

phase difference is given by

$$\phi = \phi_2 - \phi_1 = k_0 L \left(n_{22} - n_{21} \right) \frac{P_{in}}{2S}.$$

(12.95)

If $n_{21} \gg n_{22}$ and $k_0 = \frac{2\pi}{\lambda_0}$, then the phase difference can be written as

$$\phi \approx \frac{\pi n_{21} L}{\lambda_0 S} P_{in}.$$

(12.96)

Since the condition for all-optical switching is, hence the threshold input power for all-optical switching is

$$P_{inc} = \frac{\lambda_0 S}{n_{21} L}$$

(12.97)

12.4.3.2 M–Z Interferometer All-Optical Switch with Different Arm Lengths

The setup for this system is shown in Figure 12.13. Both the arms have same nonlinear RIs, hence $n_1 = n_2 = n$, but the arms have different lengths and the difference in length is given by $\Delta L = L_1 - L_2 \approx L_1$ (assuming $L_1 \gg L_2$).

Since a 3-dB coupler is used, the light splits equally in both the arms and the individual phase shifts and the phase difference are given by

$$\phi_1 = \frac{k_0 n_1 P_{in} L_1}{2S}$$

(12.98)

$$\phi_2 = \frac{k_0 n_2 P_{in} L_2}{2S}$$

(12.99)

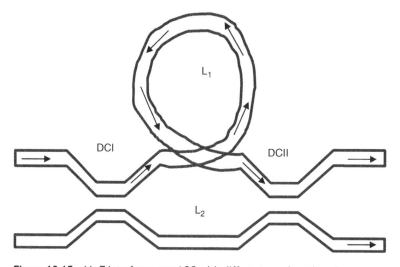

Figure 12.13 M–Z interferometer AOS with different arm length and same arm material.

$$\phi = \phi_2 - \phi_1 = \frac{2\pi}{\lambda_0} n_2 \Delta L \frac{P_0}{2S} \approx \frac{\pi n_2 L_1 P_{in}}{\lambda_0 S}. \tag{12.100}$$

The condition for all-optical switching is given by $\phi = \pi$, therefore threshold input power can be given by

$$P_{inc} = \frac{\lambda_0 S}{n_2 L_1}. \tag{12.101}$$

Henceforth, as the difference of arm length widens, switching power starts to decrease.

12.4.4 Ring Resonator AOS

Ring Resonators [42–46] are mainly divided into two types:

i) AOS in M–Z Interferometer coupled with SCRR
ii) AOS in DCRR.

12.4.4.1 AOS in M–Z Interferometer Coupled with SCRR

This setup is a modification to the M–Z interferometer to reduce the required input power. The setup is shown in Figure 12.14. It consists of a single coupler ring resonator (SCRR). The light inputs from Port 1 through the straight waveguide with amplitude E_I. E_{IV} is field amplitude at Port 4, where input light comes into cavity of the ring from the coupler. E_{II} and E_{III} are the amplitudes at Ports 3 and 4. The output E_{III} is the transformed output E_1 with phase shift $\phi = \phi_3 - \phi_1$.

Therefore field amplitude equation may be formulated as

$$E_{III} = rE_I + itE_{II} \tag{12.102}$$

$$E_{IV} = itE_I + rE_{II} \tag{12.103}$$

$$E_{III} = e^{i\phi} E_I \tag{12.104}$$

$$E_{II} = ae^{i\varphi} E_{IV} = e^{-\alpha/2} e^{i\varphi} E_{IV}, \tag{12.105}$$

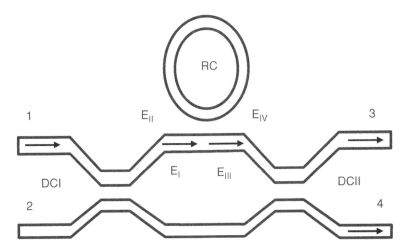

Figure 12.14 Structure of AOS in M–Z interferometer coupled with ring cavity (RC).

where r and t have usual significances mentioned earlier, φ is phase shift for unit circle propagation, α is the absorption coefficient, 'a' is the loss rate for unit propagation, and l is ring perimeter.

From the above equations, the field amplitude ratios are obtained as

$$\frac{E_{II}}{E_I} = \frac{itae^{i\varphi}}{1 - rae^{i\varphi}} \tag{12.106}$$

$$\frac{E_{III}}{E_I} = \frac{r - ae^{i\varphi}}{1 - rae^{i\varphi}}. \tag{12.107}$$

The amplification coefficient and finesse of the ring cavity are given by the expression

$$M = \frac{P_{II}}{P_I} = \frac{|E_{II}|^2}{|E_I|^2} = \frac{(1 - r^2)a^2}{1 - 2ra\cos\varphi + r^2a^2} \tag{12.108}$$

$$F = \frac{2\pi}{\delta\varphi} = \pi \left[2\sin^{-1}\left(\frac{1 - ar}{2\sqrt{ar}}\right) \right]^{-1}. \tag{12.109}$$

Now from Equation (104) and from Equation (105), ϕ can be defined as

$$\phi = \arg\left[\frac{r - a\exp(i\varphi)}{1 - ra\exp(i\varphi)} \right]. \tag{12.110}$$

Thus from here the transmittance can be defined as

$$T = \frac{|E_{III}|^2}{|E_I|^2} = \frac{r^2 - 2ar\cos\varphi + a^2}{1 - 2ar\cos\varphi + a^2r^2}. \tag{12.111}$$

The T versus φ curve is shown in Figure 12.15.

Now to calculate the threshold power, let the length of the loop be l and the Kerr effect induced nonlinear phase shift φ is proportional to the input power P_2, thus the change of φ with respect to P_{II} is given by

$$\frac{d\phi}{dP_{II}} = \frac{d}{dP_{II}}\left(\varphi_0 + \frac{2\pi}{\lambda_0}n_2l\frac{P_{II}}{S} \right) = \frac{2\pi n_2l}{\lambda_0 S}. \tag{12.112}$$

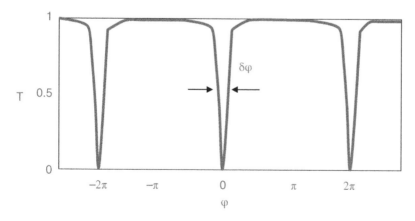

Figure 12.15 Transmittance (T) profile with phase shift (φ) for SCRR.

Since $\dfrac{dP_{II}}{dP_I} = \dfrac{P_{II}}{P_I} = M_{max}$ hence

$$\frac{d\phi}{dP_I} = \frac{d\phi}{d\varphi}\frac{d\varphi}{dP_{II}}\frac{dP_{II}}{dP_I} = \frac{2\pi n_2 l}{\lambda_0 S}M_{max}^2. \tag{12.113}$$

Integrating with the limits we get

$$\int_0^\pi d\phi = \frac{2\pi n_2 l}{\lambda_0 S}M_{max}^2 \int_0^{P_{lc}} P_I. \tag{12.114}$$

Thus the threshold power can be given as

$$P_{inc} = 2P_{lc}. \tag{12.115}$$

Thus

$$P_{inc} = \frac{\lambda_0 S}{n_2 l}\frac{1}{M_{max}^2}. \tag{12.116}$$

In terms of maximum finesse

$$P_{inc} = \frac{\lambda_0 S}{n_2 l}\left(\frac{\pi}{2}\right)^2 \frac{1}{F_{max}^2}. \tag{12.117}$$

12.4.4.2 AOS in DCRR

Basic structure of a DCRR is shown in Figure 12.16. For these structures the reflectivities r_1 and r_2 are kept high ($\rightarrow 1$) so as to keep maximum energies within the ring cavity. If the light input power is low then the phase shift is zero and thus the light outputs from Output Port 2, however if the light power is high, then the phase shift is π and the output is obtained from Output Port 1.

Assuming the input field amplitude to be E_{in}, the individual port field amplitudes are given by

$$E_r = r_1 E_{in} + it_1 E_{IV} \tag{12.118}$$

$$E_I = it_1 E_{in} + r_1 E_{IV} \tag{12.119}$$

$$E_{II} = \exp\left(-\alpha_1 l_1 / 2\right)\exp\left(i\varphi_1\right)E_I \tag{12.120}$$

Figure 12.16 Double-coupler-ring-resonator AOS with one IP and two OPs.

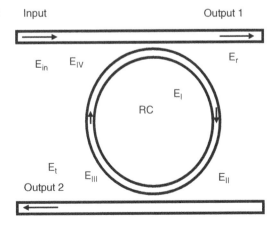

$$E_{III} = r_2 E_{II} \tag{12.121}$$

$$E_t = it_2 E_{II} \tag{12.122}$$

$$E_{IV} = \exp\left(-\alpha_2 l_2 / 2\right) \exp\left(i\varphi_2\right) E_{III}, \tag{12.123}$$

where r_i and t_i (i=1, 2) are the reflectivity and transmissivity of the couplers 1 and 2 respectively. Similarly $a_1 = \exp\left(-\alpha_1 l_1/2\right)$ and $a_2 = \exp\left(-\alpha_2 l_2/2\right)$ are the loss coefficients of the respective couplers where α_1, α_2 and l_1, l_2 are the absorption and lengths of the left half ring and right half ring respectively. Thus for the whole ring, $\alpha = \alpha_1 + \alpha_2$, $l = l_1 + l_2$ and $\varphi = \varphi_1 + \varphi_2$.

Assuming $a_2 \approx a \approx 1$, $r_1 = r_2 = r$, and neglecting absorption loss, we get

$$R = \frac{2r^2\left(1 - \cos\varphi\right)}{1 - 2r^2 \cos\varphi + r^4} \tag{12.124}$$

$$T = \frac{\left(1 - r^2\right)^2}{1 - 2r^2 \cos\varphi + r^4}. \tag{12.125}$$

As before the amplification factor and finesse are defined by

$$M = \left|\frac{E_{IV}}{E_{in}}\right|^2 = \frac{r_2^2(1 - a^2 r_1^2)}{1 - 2a r_1 r_2 \cos\varphi + a^2 r_1^2 r_2^2} \tag{12.126}$$

$$F = \frac{2\pi}{\delta\varphi} = \pi\left[2\sin^{-1}\left(\frac{1 - a r_1 r_2}{2\sqrt{a r_1 r_2}}\right)\right]^{-1}. \tag{12.127}$$

With $r_1 = r_2 = r \to 1$ and $a \to 1$ and $\varphi = 2m\pi(m = 1, 2, 3. . .)$, the maximum amplification factor and finesse can be approximated to

$$M_{max} \approx \frac{2}{1 - r^2} \tag{12.128}$$

$$F_{max} \approx \frac{\pi}{1 - r^2} = \pi M_{max}. \tag{12.129}$$

Taking into consideration of optical Kerr effect, and length of half-ring as $l_1 = l_2 = l/2$, phase shift for unit circle propagation is

$$\varphi = \varphi_0 + \frac{\pi n_2 l}{\lambda S}\left(\left|E_{II}\right|^2 + \left|E_{IV}\right|^2\right), \tag{12.130}$$

where S is the cross-sectional area.

For all-optical switching condition that is $\phi = \pi$, the threshold input power can be given as

$$P_{inc} = \frac{\lambda_0 S}{n_2 l} \frac{\pi}{2} \frac{1}{F_{max}}. \tag{12.131}$$

Apart from switching power, ring resonator AOS also has switching time property index (*s*). It depends on two factors: the photon lifetime of ring cavity (τ_c) and the nonlinear response time of material (τ_f):

$$\tau = \tau_c + \tau_f. \tag{12.132}$$

12.4.5 Fiber Grating AOS

Fiber Grating all-optical switches [47–51] are mainly divided into two groups, namely:

i) single nonlinear FBG AOS
ii) single nonlinear LPFG AOS.

Other than these there are two special types of grating optical switches, which are described below.

12.4.5.1 Single Nonlinear FBG AOS

This is the simplest FBG structure to be used as a nonlinear switch. Every Bragg Grating has a particular central wavelength which is known as the Bragg Wavelength (λ_B). If the incident signal contains any wavelength which is equal to the Bragg Wavelength, then according to the Bragg Condition, that particular wavelength is reflected back while the others are transmitted. Thus if a narrowband signal with a wavelength equal to Bragg wavelength is incident in the input port of the FBG (λ_i), then by the principle of energy conservation, whole incident light (λ_f) is reflected back and the transmission is zero.

The relationship between the propagation constants according to the conservation of momentum principle is given by

$$\beta_i - \beta_f = \beta, \tag{12.133}$$

where β_i and β_f are propagation constants of incident and reflected waves respectively. Under Bragg condition and principle of energy conservation

$$\lambda_i = \lambda_f = \lambda$$

which implies

$$\beta_i = \beta_f = \frac{2\pi n_{eff}}{\lambda_B}. \tag{12.134}$$

Using the Bragg grating constant notation (Λ_B) it can be written as

$$\beta = \frac{2\pi}{\Lambda_B}. \tag{12.135}$$

Since the incident and reflected signals are opposite to each other, hence

$$\beta_i = -\beta_f. \tag{12.136}$$

Grating wavelength can be written as

$$\lambda_B = 2n_{eff}\Lambda_B. \tag{12.137}$$

Refractive index of uniform fiber Bragg grating can be expressed as

$$n(z) = n_{eff} + \overline{\delta n_{eff}} \cos\left(\frac{2\pi}{\Lambda} z\right), \tag{12.138}$$

where z is displacement along fiber axis direction, $\overline{\delta n_{eff}}$ is the modulation amplitude of average effective grating refractive index.

The field variations are given by

$$A^+(z) = A(z)\exp\left(i\delta z - \phi/2\right) \tag{12.139}$$

$$B^+(z) = B(z)\exp\left(-i\delta z + \phi/2\right),$$

(12.140)

where $\delta = 2\pi n_{eff}\left(\dfrac{1}{\lambda} - \dfrac{1}{\lambda_B}\right)$ is the detuning parameter.

In steady state, FBG couple-mode equations are given by

$$\frac{\partial A^+}{\partial z} = i\delta A^+ + i\kappa B^+$$

(12.141)

$$\frac{\partial B^+}{\partial z} = -i\delta B^+ - i\kappa A^+,$$

(12.142)

where κ is the coupling coefficient given by

$$\kappa = \kappa^* = \frac{\pi}{\delta} s\overline{\delta n}_{eff},$$

(12.143)

where s is the contrast ratio of refractive index modulation.

Assuming the grating length to be L, the amplitude reflectivity of grating is given by

$$|r| = \frac{\kappa \sinh\left(\sqrt{\kappa^2 - \delta^2}\, L\right)}{\sqrt{\left(\kappa^2 - \delta^2\right)\cosh^2\left(\sqrt{\kappa^2 - \delta^2}\, L\right) + \delta^2 \sinh^2\left(\sqrt{\kappa^2 - \delta^2}\, L\right)}}.$$

(12.144)

Energy reflectivity of grating given by, $R = |r|^2$ is given by

$$R = \frac{\kappa^2 \sinh^2\left(\sqrt{\kappa^2 - \delta^2}\, L\right)}{\kappa^2 \cosh^2\left(\sqrt{\kappa^2 - \delta^2}\, L\right) - \delta^2}.$$

(12.145)

For $\delta=0$, FBG has maximum reflectivity is obtained at $\lambda = \lambda_B$ which is given by

$$R_{max} = \tanh^2\left(\kappa L\right).$$

(12.146)

The reflected spectrum bandwidth of FBG ($\Delta\lambda_0$) is described as interval between two wavelengths where null reflectivity occurs at both sides of grating wavelength. It is given by

$$\Delta\lambda_0 = \frac{\lambda_B^2}{n_{eff}L}\sqrt{1 + \left(\frac{\kappa L}{\pi}\right)^2}.$$

(12.147)

Assuming $\kappa L << \pi$

$$\Delta\lambda_0 \approx \frac{\lambda_B^2}{n_{eff}L} \approx \frac{2\lambda_B\Lambda_B}{L} = \frac{2\lambda_B}{N},$$

(12.148)

where N is the total number of grating given by

$$N = \frac{L}{\Lambda_B}.$$

(12.149)

Introducing optical Kerr effect under self-phase modulation, change of refractive index of the fiber is given by

$$n_{eff} = n_0 + n_2 \frac{P}{S},$$

(12.150)

where n_0 is linear refractive index, n_2 is the nonlinear refraction coefficient, P is input light power.

Change of grating wavelength is

$$\Delta\lambda_B = 2\Delta n_{eff}\Lambda_B = 2\Lambda_B n_2 \frac{P}{S}. \tag{12.151}$$

The necessary condition for realizing switching process under input power is the shift of grating wavelength, i.e.

$$\Delta\lambda_B = \frac{\Delta\lambda_0}{2} = \frac{\lambda_B\Lambda_B}{L}. \tag{12.152}$$

Thus the threshold switching power can be given as

$$P_c = \frac{1}{2}\left(\frac{\lambda_B S}{n_2 L}\right) \tag{12.153}$$

12.4.5.2 Single Nonlinear LPFG AOS

The name suggests LPFG refers to the fiber gratings with the period of grating longer than the normal gratings. Coupling between core and cladding modes are considered for light propagation inside LPFG. For coupling between fundamental mode LP_{01} and m^{th} order mode LP_{0m}, their propagation constants β_{co} and β_{cl} are related as

$$\beta_{co} - \beta_{cl} = \beta_g = \frac{2\pi}{\Lambda_L}, \tag{12.154}$$

where β_g is the propagation constant of fiber grating, kL is the grating constant of LPFG.

In terms of difference between the RIs of core and cladding, it can be written as

$$\beta_{co} - \beta_{cl} = \beta_g = \frac{2\pi}{\lambda_L}\left(n_{co} - n_{cl}\right) = \frac{2\pi}{\lambda_L}\Delta n_g, \tag{12.155}$$

where $\lambda_L = \Lambda_L \Delta n_g$ is the grating wavelength of LPFG; Δn_g is the effective refractive index difference between fundamental mode and cladding mode of fiber, i.e.,

$$\Delta n_g = n_{co} - n_{cl}. \tag{12.156}$$

The detuning is defined as

$$\delta = \frac{1}{2}\left[\left(\beta_{co} - \beta_{cl}\right) - \frac{2\pi}{\Lambda_L}\right] = \pi\Delta n_g\left(\frac{1}{\lambda} - \frac{1}{\lambda_L}\right), \tag{12.157}$$

where

$$\kappa = \frac{\pi\overline{\delta n_{eff}}}{\lambda}. \tag{12.158}$$

The reflected spectrum bandwidth $\Delta\lambda_0$ is given by

$$\Delta\lambda_0 = \frac{2\lambda_L^2}{\Delta n_g L}\sqrt{1 - \left(\frac{\kappa L}{\pi}\right)^2}. \tag{12.159}$$

With the assumption $\kappa L \ll \pi$, it can be expressed as

$$\Delta\lambda_0 = \frac{2\lambda_L^2}{\Delta n_g L} = \frac{2\lambda_L\Lambda_L}{L} = \frac{2\lambda_L}{N}, \tag{12.160}$$

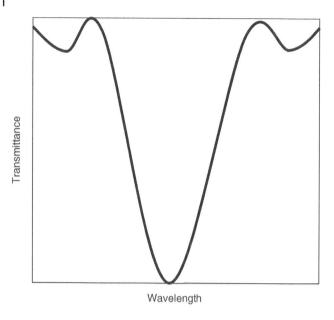

Figure 12.17 Transmission spectrum of LPFG.

where

$$N = \frac{L}{\Lambda_L} \text{ is the period number.} \tag{12.161}$$

The reflection curve of LPFG is given in Figure 12.17.

In SPM mode, when grating wavelength becomes equal to incident wavelength, transmittance of grating structure is minimum. With increasing signal power, the transmission spectrum moves towards longer wavelength by the effect of the optical Kerr effect. If grating wavelength becomes half of bandwidth, then transmittance becomes maximum, and a corresponding similar change is attained by transmitted power.

Due to the induced Kerr effect by the signal power, changes in core and cladding RIs are given by

$$\Delta n_{co} = n_2 \frac{P}{S}. \tag{12.162}$$

The change in cladding can be neglected, hence

$$\Delta n_{cl} = 0. \tag{12.163}$$

Therefore

$$\Delta(\Delta n_g) = \Delta n_{co} - \Delta n_{cl} \approx n_2(P/S). \tag{12.164}$$

Hence the shift in grating wavelength is given by

$$\Delta \lambda_L = \Delta(\Delta n_g)\Lambda_L = n_2 \Lambda_L \frac{P}{S}. \tag{12.165}$$

At $\Delta \lambda_L = \frac{\Delta \lambda_0}{2}$, optical switching is accomplished, hence

$$\Delta \lambda_L = \frac{\Delta \lambda_0}{2} = \frac{\lambda_L \Lambda_L}{L}. \tag{12.166}$$

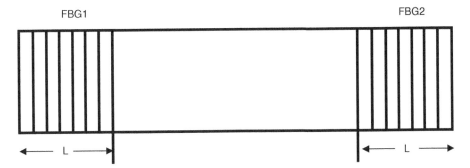

Figure 12.18 Optical bistable switching device consisted of two same FBGs.

Finally the threshold switching power is given by

$$P_c = \frac{\lambda_L S}{n_2 L}.$$ (12.167)

There are two other types of special configurations for FBGs, which are:

i) nonlinear fiber connected LPFG-pair AOS
ii) nonlinear fiber connected FBG-pair optical bistable switch.

The first one uses the same concept as the amplifier connected Sagnac interferometer, whose main purpose is to decrease the input threshold power considerably. The nonlinear fiber here consists of an EDF which acts as an amplifier at the 1550 nm window with a pump source at 980 nm. With the use of an amplifier with the FBG, the same properties of an FBG all-optical switch can be achieved but can be operated with much lower input optical power which thus makes the system more economical.

The second type is the nonlinear FC FBG-pair OBS, which consists of two symmetric FBGs connected by a rare-earth-doped nonlinear fiber. This ultimately constitutes nonlinear F-P interferometer. Multiple-beam interference is formed through back-and-forth movement of signal beam. Change of both refractive index and also of phase take place when the beam interacts with the nonlinear medium. Figure 12.18 shows the optical bistable switch.

For lower switching power, Yb^{3+}-doped fiber is referred by most researchers, owing to its higher speed of switching and smaller absorption.

12.5 Nanoscale AOS

From all the above description of different all-optical switches, we can see that there is a common factor in every formula for the input power, which is

$$P_{inc} = \frac{\lambda S}{n_2 L},$$ (12.168)

where all symbols have known significances. In addition, for RR, there is ring finesse F_{max} in the denominator. From the above expression, certain conclusions can be derived as

i) *The threshold power can be reduced by increasing the RI of the material.* But by increasing the RI, the absorption increases which can be harmful for the signal transmission as greater absorption greatly reduces the signal power throughout the transmission.

ii) *To decrease effective cross-section of waveguide S.* Reduction of switching power from kW to mW can happen if S can be reduced from μm^2 to nm^2. Thus nanoscale waveguides are required for the realization of all-optical switching device.

iii) *Extension of length L is beneficial for switching power reduction.* But overall size of device will increase considerably. Thus micro-ring structures can be used. To offset the shortage of material with small n_2, the ring structure can increase the nonlinearity. High finesse F_{max} ring is expected to be used for this purpose.

Therefore, using nanoscale materials and structures is a fantastic way to reduce switching power of all-optical switches. In addition, small-scale devices can also help in reducing the processing time as light has to travel lesser distances in small-scale devices. The nanoscale all-optical switch is based on nanophotonics [52–59]. Research on photonic crystals leads the path [60–64] towards development of filter-based optical switches.

The main focus of nanophotonics is to study the interaction between optical radiations in the nanoscale with nano-scale optical materials and devices. The most important nano-materials are the materials which have periodicity in the microcosm scale, such as the low-dimensional nanoelectronic materials which have the properties of quantum confinement (owing to their dimensional similarity with de-Broglie wavelength) which include quantum 2D device (well), quantum 1D device (wire) and quantum dots etc. They in general consist of two alternative bandgap materials (preferably compound semiconductor so that tailoring of bandgap can be achieved as per requirement), and also grating optical structures (e.g., photonic crystals) materials consisting of two dielectric materials alternately with diverse refractive indices; in addition the conductive nano-materials with a conductor-insulator/semi-insulator interface in which transfer of information occurs between incident optical wave and periodic e.m. field to form stable/quasi-stable Surface Plasmon Polariton (SPP). One of the widely acceptable nano-structure for all-optical switch is confined transmission line in submicron dimension (waveguide in that regime) with similar cross-section and the interferometers made by it, particularly the micro-ring.

12.6 Future Scope and Conclusion

Optical communication is the future of telecommunication. There are several advantages of optical systems over electrical systems. Moreover, with the help of optical systems, quantum communications can be achieved, which can be even more useful and faithful for long-distance and minimum-loss communications. But the full deployment of optical systems has not not possible till now due for various reasons such as long fiber losses, bending losses, unavailability of good processing systems, and lack of good optical hardware like switches, modulators, amplifiers at all frequencies. Though some of these are available for macro systems, for micro systems they are still being developed and hence fully independent optical systems are yet to be fully deployed.

All-optical switches play a very important role in the hardware processing of optical signals. They are used to route signals through different channels and send them to several measurement devices, make divisions in optical power and much more. After studying and analyzing all the above all-optical switches, it can be seen that all-optical switches have been greatly developed in recent research and hence are paving the way for modern-day optical communications as well as quantum communications.

With the advent of more advanced technologies all other optical devices like nano-amplifiers and nano-processors will be extremely crucial for the operation of an optical circuit. Thus the study of nanophotonics – photonic crystals – which has been mentioned in the preceding section is the main interest of research in the field of photonics. Silicon Integrated Photonics paves the way for low-cost and efficient nanophotonics, and in coming days more such materials with better optical properties can be used for the research and manufacturing of nano-optical circuits and devices.

To conclude, we can say that the present electrical systems are running out of opportunities to be as useful as they used to be with respect to the requirement of bandwidths, higher security, undisturbed connections, and much more. Optical systems help to provide those features and thus prove much more advantageous than electrical systems. Moreover, if free-space optics can be made more efficient, then they would completely replace the present electrical systems and fully independent optical systems can be deployed and used both commercially and industrially.

Bibliography

1 D. Botez and G.J. Herskowitz. Components for optical communications systems: A review. *Proceedings of the IEEE*, 68(6):689–731, 1980.

2 R. Alferness. Guided-wave devices for optical communication. *IEEE Journal of Quantum Electronics*, 17(6):946–959, 1981.

3 N. Tsukada and T. Nakayama. Polarization-insensitive integrated-optical switches: A new approach. *IEEE Journal of Quantum Electronics*, 17(6):959–964, 1981.

4 A. Milton and W. Burns. Mode coupling in tapered optical waveguide structures and electro-optic switches. *IEEE Transactions on Circuits and Systems*, 26(12):1020–1028, 1979.

5 T.H. Chu, S. Yamada Ishida, and Y. Arakawa. Compact 1 × N thermo-optic switches based on silicon photonic wire waveguides. *Optics Express*, 13(25):10109–10114, 2005.

6 H.C. Tapalian, J.-P. Laine, and P.A. Lane. Thermooptical switches using coated microsphere resonators. *IEEE Photonics Technology Letters*, 14(8):1118–1120, 2002.

7 S. Ghafari, M.R. Forouzeshfard, and Z. Vafapour. Thermo optical switching and sensing applications of an infrared metamaterial. *IEEE Sensors Journal*, 20(6):3235–3241, 2019.

8 L. Sirleto, G. Coppola, M. Iodice, M. Casalino, M. Gioffrè, and I. Rendina. Thermo-optical switches. In B. Li and S.J Chua, editors, *Optical Switches*, 61–96. Cambridge, Woodhead, 2010.

9 M. Bass, E.W. Van Stryland, D.R. Williams and W.L. Wolfe. *Handbook of Optics, Volume II – Devices, Measurements, and Properties*. McGraw-Hill.

10 M. Delgado-Pinar, D. Zalvidea, A. Diez, P. Perez-Millan, and M. Andres. Q-switching of an all-fiber laser by acousto-optic modulation of a fiber Bragg grating. *Optics Express*, 14(3):1106–1112, 2006.

11 D.A. Smith, R.S. Chakravarthy, Z. Bao, J.E. Baran, J.L. Jackel, A. d'Alessandro, D.J. Fritz, S.H. Huang, X.Y. Zou, Hwang, S.-M., A.E. Willner, and K.D. Li. Evolution of the acousto-optic wavelength routing switch. *Journal of Lightwave Technology*, 14(6):1005–1019, 1996.

12 S. Antonov, A. Vainer, V. Proklov, and Y. Rezvov. Switch multiplexer of fiber-optic channels based on multibeam acousto-optic diffraction. *Applied Optics*, 48(7):C171–C181, 2009.

13 D.F. Nelson and M. Lax. New symmetry for acousto-optic scattering. *Physical Review Letters*, 24:378–380, 1970.

14 J. Chapelle and L. Tauel. Theorie de la diffusion de la lumiere par les cristeaux fortement piezoelectriques. *Comptes Rendus de l'Académie des Science*, 240:743, 1955.

15 G.H. Heilmeier, L.A. Zanoni, and L.A. Barton. Dynamic scattering: A new electrooptic effect in certain classes of nematic liquid crystals. *Proceedings of IEEE*, 56(7):1162–1171, 1968.

16 G.H. Heilmeier and L.A. Zanoni. Guest-host interactions in nematic liquid crystals. A new electro-optic effect. *Applied Physics Letters*, 13:91–92, 1968.

17 S.-T. Wu and J.D. Margerum. Liquid crystal dyes with high solubility and large dielectric anisotropy. *Applied Physics Letters*, 64:2191–2193, 1994.

18 J.J. Wysocki, J. Adams, and W. Haas. Electric-field-induced phase change in cholesteric liquid crystals. *Physical Review Letters*, 20(19):1024, 1968.

19 G.H. Heilmeier and J.E. Goldmacher. A new electric-field-controlled reflective optical storage effect in mixed-liquid crystal systems. *Applied Physics Letters*, 13:132–133, 1968.

20 M. Schadt and W. Helfrich.Voltage-dependent optical activity of a twisted nematic liquid crystal. *Applied Physics Letters*, 18:127–128, 1971.

21 M.F. Schiekel and K. Fahrenschon.Deformation of nematic liquid crystals with vertical orientation in electrical fields. *Applied Physics Letters*, 19:391–393, 1971.

22 K. Lu and B.E.A. Saleh. Complex amplitude reflectance of the liquid crystal light valve. *Applied Optics*, 30:2354–2362, 1991.

23 A. Sasaki, K. Kurahashi, and T. Takagi. Liquid-crystal thermo-optic effects and two new information display devices. *Journal of Applied Physics*. 45:4356, 1974.

24 F.J. Khan. Orientation of liquid crystals by surface coupling agents. *Applied Physics Letters*, 22:386, 1973.

25 H. Ren and S.-T. Wu. Inhomogeneous nanoscale polymer-dispersed liquid crystals with gradient refractive index. *Applied Physics Letters*, 81:3537–3539, 2002.

26 D.-K. Yang, L.-C. Chien, and J.W. Doane.Cholesteric liquid crystal/polymer dispersion for haze-free light shutters. *Applied Physics Letters*, 60:3102–3104, 1992.

27 C. Li. *All-Optical Switches Based on Nonlinear Optics*. Beijing, Science Press, 2015.

28 H.M. Gibbs.*Optical Bistability: Controlling Light with Light*. New York, Academic Press, 1985.

29 S. Divya, I. Sebastian, V.P.N. Nampoori, P. Radhakrishnan, and A. Mujeeb.Power and composition dependent non-linear optical switching of TiO_2-SiO_2 nano composites. In *International Conference on Optical Engineering*, Belgaum, India, 1–5, 2012.

30 A. Elgamri and B. Rawat.Optical switching using optical Kerr non-linear effect on photonic crystal fiber. In *International Conference on Signal Processing and Communication*, Noida, 4–9, 2016.

31 W. Liu, C. Yang, M. Liu, W. Yu, Y. Zhang, M. Lei, and Z. Wei. Bidirectional all-optical switches based on highly nonlinear optical fibers. *Europhysics Letters*, 118(3):34004, 2017.

32 A. Granpayeh, H. Habibiyan, and P. Parvin. Photonic crystal directional coupler for all-optical switching, tunable multi/demultiplexing and beam splitting applications. *Journal of Modern Optics*, 66(4):359–366, 2018.

33 Y. Akihama and K. Hane. Single and multiple optical switches that use freestanding silicon nanowire waveguide couplers. *Light: Science & Applications*, 1(6), e16, 2012.

34 C. Setterlind and L. Thylen. Directional coupler switches with optical gain. *IEEE Journal of Quantum Electronics*, 22(5):595–602, 1986.

35 S. Slussarenko, V. D'Ambrosio, B. Piccirillo, L. Marrucci, and E. Santamato. The Polarizing Sagnac Interferometer: a tool for light orbital angular momentum sorting and spin-orbit photon processing. *Optics Express*, 18(26):27205–27216, 2010.

36 M. Jinno and T. Matsumoto. Demonstration of laser-diode-pumped ultrafast all-optical switching in a nonlinear Sagnac interferometer. *Electronics Letters*, 27(1):75–76, 1991.

37 D.J. Richardson, R.L. Laming, and D.N. Payne. Very low threshold Sagnac switch incorporating an Erbium doped fiber amplifier. *Electronics Letters*, 26(21):1779–1781, 1990.

38 A. Stanley, G. Singh, J. Eke, and H. Tsuda. Mach-Zehnder interferometer: A review of a perfect all-optical switching structure. In *Proceedings of the International Conference on Recent Cognizance in Wireless Communication & Image Processing*, 415–425, 2016.

39 M. Islam and M. Barsha. Mach Zehnder interferometer (MZI) as a switch for all optical network. *IEEE International Conference on Innovation in Engineering and Technology*, 27–28 Dec, Dhaka, Bangladesh, 2018.

40 Z. Lu, D. Celo, H. Mehrvar, E. Bernier, and L. Chrostowski. High-performance silicon photonic tri-state switch based on balanced nested Mach-Zehnder interferometer. *Scientific Reports*, 7(1), 2017.

41 T. Hirokawa, M. Saeidi, S. Pillai, A. Nguyen-Le, L. Theogarajan, A. Saleh, and C. Schow. A wavelength-selective multiwavelength ring-assisted Mach-Zehnder interferometer switch. *Journal of Lightwave Technology*, 38(22):6292–6298, 2020.

42 J.E. Heebner and R.W. Boyd. Enhanced all-optical switching by use of a nonlinear fiber ring resonator. *Optics Letters*, 24(12):847–849, 1999.

43 M. Ghadrdan and M.A. Mansouri-Birjandi. Implementation of all-optical switch based on nonlinear photonic crystal ring resonator with embedding metallic nanowires in the ring resonators. *Optical and Quantum Electronics*, 48(5), 2016.

44 A.H.J. Yang and D. Erickson. Optofluidic ring resonator switch for optical particle transport. *Lab on a Chip*, 10(6):769–774, 2010.

45 J.K. Rakshit, T. Chattopadhyay, and J.N. Roy. Design of ring resonator based all optical switch for logic and arithmetic operations – A theoretical study. *Optik – International Journal for Light and Electron Optics*, 124(23):6048–6057, 2013.

46 J.E. Heebner and R.W. Boyd. Enhanced all-optical switching by use of a nonlinear fiber ring resonator. *Optics Letters*, 24(12):847–849, 1999.

47 G.R. Broderick, D. Taverner, and D.J. Richardson. Nonlinear switching in fiber Bragg gratings. *Optics Express*, 3(11):447–453, 1998.

48 D. Taverner, G.R. Broderick, and D.J. Richardson. Nonlinear self-switching and multiple gap-soliton formation in a fiber Bragg grating. *Optics Letters*, 23(5):328–330, 1998.

49 R.H. Stolen, W.A. Reed, K.S. Kim, and G.T. Harvey. Measurement of the nonlinear refractive index oflong dispersion-shifted fibers by self-phase modulation at 1.55 μm. *Journal of Lightwave Technology*, 16(6):1006–1012, 1998.

50 M. Janos, J. Arkwright, and Z. Brodzeli. Low power nonlinear response of Yb3+ doped optical fibre Bragg gratings. *Electronics Letters*, 33(25):2150–2151, 1997.

51 B. Guan and S. Liu. Erbium-doped fiber Bragg grating based all-optical switch. In *IEEE CLEO/ Pacific Rim 2003, The 5th Pacific Rim Conference on Lasers and Electro-Optics*, Taipei, Taiwan, 2003.

52 R.V. Almeida, A. Carlos, and M. Lipson. All-optical control of light on a silicon chip. *Nature*, 431(28):1081–1084, 2004.

53 M. Lipson. Overcoming the limitations of microelectronics using Si nanophotonics: solving the coupling, modulation and switching challenges. *Nanotechnology*, 15(10):S622–S627, 2004.

54 C. Koos, P. Vorreau, T. Vallaitis, and P. Dumon. All-optical high-speed signal processing with silicon–organic hybrid slot waveguides. *Nature Photonics*, 3(4):216–219, 2009.

55 A. Martınez, J. Blasco, P. Sanchis, et al. Ultrafast all-optical switching in a silicon-nanocrystalbased silicon slot waveguide at telecom wavelengths. *Nano Letters*, 10(4):1506–1511, 2010.

56 K. Tajima. All-optical switch with switch-off time unrestricted by carrier lifetime. *Japanese Journal of Applied Physics*, 32(12A):Ll746–Ll748, 1993.

57 K. Asakawa. Fabrication and characterization of photonic crystal slab waveguides and application to ultra-fast all-optical switching devices. *IEEE ICTON*, 1:193–197, 2003.

58 H. Nakamura et al. Ultra-fast photonic crystal/quantum dot all optical switch for futurephotonic networks. *Optics Express*, 12(26):6606–6614, 2004.

59 Z. Qiang, W. Zhou, and R.A. Soref. Optical add-drop filters based on photonic crystal ring resonators. *Optics Express*, 15(4):1823–1831, 2007.

60 A. Deyasi, U. Dey, S. Das, S. De, and A. Sarkar. Computing photonic bandgap from dispersion relation for TM mode propagation inside metamaterial-based 1D PhC. *Micro and Nanosystems*, 12:201–208, 2020.

61 A. Deyasi and A. Sarkar. Computing optical bandwidth of bandpass filter using metamaterial-based defected 1D PhC. *AIP Conference Proceedings*, 2072:020003, 2019.

62 A. Deyasi and A. Sarkar. Variation of optical bandwidth in defected ternary photonic crystal under different polarization conditions. *International Journal of Nanoparticles*, 10(1&2):27–34, 2018.

63 A. Deyasi and A. Sarkar. THz bandpass filter design using metamaterial-based defected 1D photonic crystal structure. In A. Biswas, A. Banerjee, A. Acharyya, H. Inokawa, J. Nath Roy, editors, *Emerging Trends in Terahertz Solid-State Physics and Devices*, 1:1–21, 2020.

64 A. Deyasi. Computation of electromagnetic bandgap in two-dimensional photonic crystal. In A. Deyasi, editor, *Foundations in Photonics and Fiber-Optics*, 12:257–278, 2019.

13

Silicon Photonic Switches

Nadir Ali, Mohammad Faraz Abdullah, and Rajesh Kumar

Department of Physics, Indian Institute of Technology Roorkee, Roorkee, India

13.1 Introduction

A switch is a crucial component of optical networks in long-haul fiber-optic communication and data centers. A major function of the switch is routing signals from multiple input ports to multiple output ports. Over the years, the exponential rise in data traffic over optical communication networks has resulted in unsustainable energy usage and complexity in the network. The use of electrical switches in optical networks is energy-inefficient and a major source of heat generation in data centers. Photonic switches offer numerous advantages over electrical switches, including higher energy efficiency, faster switching response time, larger bandwidth, and minimal heat generation. Silicon photonics has emerged as the preferred technology platform for integrated, reconfigurable, and low-cost photonic switching devices due to silicon's excellent material properties and existing mature fabrication process.

In this chapter, we will cover some of the emerging switching technologies based on silicon photonics. The switching technologies covered in this chapter are organized as follows. Section 13.2 lists the crucial performance parameters of switching devices. Section 13.3 presents an introduction and motivation for the silicon photonics platform for switching applications. Section 13.4 describes the associated physical principles for enabling switching functionalities in silicon waveguides. Major photonic switch configurations based on a directional coupler, microring resonator, Mach–Zehnder interferometer, and micro-electro-mechanical-systems are presented in Section 13.5. In Section 13.6, the hybrid switch configurations based on III/V materials, 2D materials, and phase change materials are presented. Section 13.7 presents switch fabrics realized using silicon photonic waveguide components. Finally, the chapter is summarized in Section 13.8.

13.2 Performance Parameters

Switch performance is evaluated using several performance parameters. These parameters depend on the underlying physical principles and material system used for switching operation. A switch with an optimum trade-off between the different parameters can be selected considering the

Optical Switching: Device Technology and Applications in Networks, First Edition. Edited by Dalia Nandi, Sandip Nandi, Angsuman Sarkar, and Chandan Kumar Sarkar.
© 2022 John Wiley & Sons, Inc. Published 2022 by John Wiley & Sons, Inc.

application requirements. Some of the crucial performance parameters of a photonic switch are presented below [1].

- **Insertion Loss (IL):** IL is the fraction of the input power lost due to the placement of switch in the network. The switch IL is calculated as the ratio of the power at its output port (P_o) to the power at its input port (P_i) and is defined as

$$IL\left(dB\right) = -10log\left(\frac{P_o}{P_i}\right) \tag{13.1}$$

The value of IL should be as low as possible.

- **Extinction Ratio (ER):** The switch states can be distinguished either with bar and cross or alternatively with on and off. The ER is the ratio of the output power in the on-state (P_{on}) to the output power in the off-state (P_{off}) and can be written as

$$ER\left(dB\right) = 10log\left(\frac{P_{on}}{P_{off}}\right) \tag{13.2}$$

It defines the contrast between the output states of the switch and should be as high as possible.

- **Cross Talk (CT):** This is the ratio of the power at a specific output from the desired input to the power from all other inputs.
- **Switching Time:** This is the time required to switch the input signal to the desired output port.

13.3 Silicon Photonic Platform

Silicon photonics has been extensively used for various photonic applications since its inception in the 1980s [2]. The main drive for silicon as an integrating platform is its compatibility with the mature complementary-metal-oxide-semiconductor (CMOS) manufacturing process and other desirable features such as low power consumption, low-loss propagation of optical wave, and low cost [3]. Silicon photonic platforms are of various material systems such as silicon-on-insulator (SOI), SiN, Ge-on-Si, silicon-on-sapphire, and native CMOS etc. Out of these materials systems, the SOI platform has been widely used for photonic applications including switching devices.

The material system of SOI used for making waveguides for photonic device applications is shown in Figure 13.1. The SOI waveguides are formed by crystalline silicon layers placed on top of the buried silicon dioxide substrate layer. A standard SOI active and passive device is fabricated in semiconductor foundries using multi-project wafer services, consisting of a 220 nm thick silicon layer on top of a 2 μm buried silicon dioxide layer [4] for devices operating at telecommunication wavelength of 1.55 μm. The SOI offers high refractive index contrast between silicon ($n = 3.47$) and silica ($n = 1.45$), enabling high light confinement in nanoscale waveguides. Such waveguides make it possible to integrate photonic components with high density and enable strong light-matter interaction to realize nonlinear all-optical effects. The switching of light in silicon waveguide structures is usually achieved by inducing refractive index modulation via various physical mechanisms. Therefore, silicon's material properties play an important role in determining the characteristics of the switching devices. Although great effort has been put into realizing efficient switching devices using pure silicon, it is a challenging task since silicon lacks some intrinsic material properties needed for switching applications. Various other materials have been brought in to

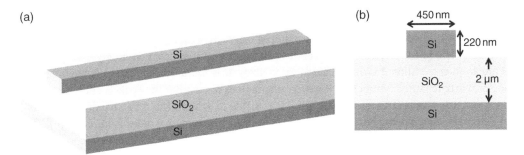

Figure 13.1 SOI material system used for photonic waveguides: (a) slant view and (b) cross-section of a waveguide.

compensate for the drawbacks of silicon and enhance switching performance. Some of these materials include 2D materials, III/V semiconductors, and phase change materials.

Silicon-based photonic switches can be broadly divided into two categories: volatile and nonvolatile. The volatile switches are the ones that require a continuous power source to sustain the swicthed state. Examples include switches based on conventional effects such as thermo-optic, electro-optic, and carrier modulations. Nonvolatility in photonic switches implies that no holding power or static power is needed to sustain the switch state, i.e., the nonvolatile switch state is self-sustained. Recently, these switches have been realized by exploiting the self-holding bi-stability of phase change materials. Such features further accelerate the potential of the hybrid silicon photonics platform by providing new directions for achieving reconfigurable and low-power-consuming photonic devices [5].

13.4 Physical Principles for Operation of Switches

In a switching device, a suitable mechanism is required to induce the refractive index change in the medium to change the path of light. Some of the most common refractive index modulation techniques used in silicon waveguide-based switching devices are shown in Figure 13.2 and are described in this section.

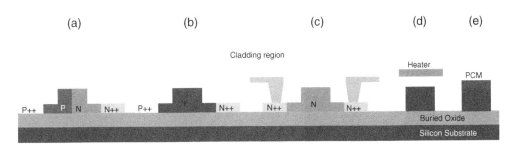

Figure 13.2 Physical mechanisms and techniques implemented for realizing the switching functionalities in SOI: (a) carrier extraction; (b) carrier injection; (c) carrier accumulation; (d) thermo-optic effect; (e) refractive index modulation via hybridization of a silicon waveguide with other materials such as PCM.

13.4.1 Electro-optic Effect

The electro-optic effect is the change in the material's refractive index under the application of an electric field. The linear electro-optic effect or Pockels effect is the most commonly used physical mechanism in conventional switches based on LiNbO$_3$. It is found in the materials having zinc-blende or wurtzite lattice symmetry. However, silicon does not exhibit the Pockels effect due to its centrosymmetric lattice structure. Although it is possible to achieve the refractive index modulation in strained silicon using Pockels effect, the change in the refractive index is minimal [6]. The quadratic effect (Kerr effect) is weak in silicon and is not usually employed in silicon since it induces small refractive index change, consumes high power, and requires a large interaction length to achieve a sizeable effect. The electro-optic effect has the advantage of high speed (GHz range), but it is energy-inefficient.

13.4.2 Carrier Injection/Extraction

The carrier injection/extraction mechanisms have been widely used for the refractive index modulation in silicon. In these effects, the carrier concentration of the semiconductor material is manipulated to alter the effective refractive index of the material. The carrier depletion using the PN diode is the most common mechanism to modulate the refractive index of the silicon waveguides. The PN semiconductor structure used for the carrier depletion is shown in Figure 13.2(a). A reverse-biased PN junction alters the depletion region characteristics, changing the carrier concentration inside the junction leading to the refractive index modulation. The carrier concentration of the semiconductor can be further increased through the carrier injection mechanisms by electron-hole injection by PIN structure, as shown in Figure 13.2(b). In carrier injection, forward-biased PIN diodes are used. Electrons and holes are injected into the intrinsic region of silicon, leading to the change in the refractive index near the active region. The accumulation of electrons and holes at two silicon interfaces separated by a thin SiO$_2$ film in a semiconductor-insulator-semiconductor (SIS) type structure is shown in Figure 13.2(c). In case of carrier accumulation, a silicon-insulator-silicon type structure is used to form the capacitor, which induces the opposite charges on the two sides of the insulator on applying a bias voltage and changes the carrier concentration of material. This kind of structure is difficult to fabricate and introduces large losses due to metal electrodes.

The tuning of the refractive index (Δn) and the absorption coefficient ($\Delta\alpha$) in silicon due to the change in the carrier density (ΔN_e or ΔN_h) can be estimated for the 1550 nm using the equations [7]

$$\Delta n = -\left[8.8\times10^{-22} \times \Delta N_e + 8.5\times10^{-18} \times \Delta N_h^{0.8} \right] \tag{13.3}$$

$$\Delta\alpha = 8.5\times10^{-18} \times \Delta N_e + 6.0\times10^{-18} \times \Delta N_h. \tag{13.4}$$

The free holes are more effective in tuning the refractive index and less effective in absorption modulation as compared to the free electrons. These effects are fast on the scale of ps, but they require large active lengths, which leads to high insertion losses [8]. The devices based on these effects have good performance as stand-alone devices when used for applications as switches.

13.4.3 Thermo-optic Effect

The thermo-optic effect is the variation in the material refractive index (Δn) due to the change in temperature (ΔT) of the material itself. The response of a material is dependent upon the thermo-optic coefficient, dn/dT, where n and T are the refractive index and temperature, respectively. The phase shift ($\Delta\varphi$) of the light traveling through the heated waveguide of length L, induced by a temperature change of ΔT, can be calculated using the equation

$$\Delta\varphi = \frac{2\pi}{\lambda}\left(\frac{dn_{eff}}{dT}\right)\Delta TL \tag{13.5}$$

where dn_{eff}/dT is the change of waveguide mode effective index with temperature and λ is the vacuum wavelength of guided light. The thermo-optic effect has been widely used in silicon guided wave components since silicon exhibits a relatively high thermo-optic coefficient ($1.86 \times 10^{-4}\,\mathrm{K}^{-1}$) at 1.55 μm wavelength.

In practical applications, the thermo-optic effect is utilized by heating the waveguide through a metal heater mostly placed on top of the waveguide or at the side of the waveguide as shown in Figure 13.2(d). The metal heaters are usually composed of materials such as Al, NiCr, etc. Position optimization of heater is the main task in thermo-optic effect-based switches, since the placement of the heater can greatly affect the optical as well as the thermal performance of the device. The thermo-optic switches exhibit switching time of the order of few microseconds, which is adequate for some switching applications. However, it is unsuitable for modern telecommunication applications. The thermo-optic switches enable the dense integration of the switches due to their low insertion losses. The main drawbacks of the thermo-optic effect are high driving power and high power dissipation.

13.4.4 All-optical Effect

All-optical effect implies the control of one light beam by another and can be used for realizing switching devices. In all-optical effect-based switches, one beam causes the change in effective index while another one is used as a probe. For all-optical switching, the interaction between the photons is realized efficiently using the nonlinear optical media, hence to a great extent, the performance of the switching device is determined by the nonlinear optical response of the optical medium. A material can exhibit various orders of non-linearity. It is the third-order optical nonlinearity that is exploited for the all-optical switching in photonic devices. Therefore, for the realization of high-performing all-optical switches, it is essential that the optical material should possess high third-order susceptibility and ultra-fast response time. However, traditional semiconductor materials have small third non-linear susceptibilities. The enhancement of the non-linearity can be achieved using the micro-structures, including micro-cavities and the slow light effect.

Another issue limiting the performance of the integrated all-optic photonic switching is the methods of material excitation. The vertical triggered method, in which the laser pulses are applied from vertically above the waveguide, is not practically applicable in the integrated photonic devices. Another method involves the on-chip excitation, where the switching is triggered by the controlled light passing through the waveguides. This method is suitable for the integration purpose, but it suffers from the faint intensity of the propagating light and requires strong third-order nonlinear susceptibility. Considering these, it is relatively difficult to enable the all-optical switching with high response time, low power consumption, and device structure suitable for the high-density integration [9]. In silicon photonics, different nonlinear effects such as the two-photon absorption (TPA), Kerr effect, free carrier absorption, and thermo-optic effect have been demonstrated to enable the all-optical switching with diverse properties. In order to further improve the performance of the all-optical switching devices, other novel designs and materials with high refractive index responses are used in addition to silicon. Some of the widely used nonlinear photonic structures are photonic crystals and nanocavities [10, 11], nonlinear dielectric microring resonators [12], and nonlinear plasmonic nanostructures [13]. A silicon microring resonator has been used to demonstrate the all-optical switching via thermal refractive index modulation with a switching time of the order of microseconds [14]. The high switching speed of 500 ps was experimentally demonstrated using undoped crystalline silicon microrings [15]. Utilizing faster electron-hole recombination and a shorter carrier lifetime provided by grain boundaries, further reduction in switching time to 135 ps in the polycrystalline silicon microring was demonstrated with a high extinction ratio of 10 dB [16].

13.5 Major Configurations

13.5.1 Directional Coupler

A directional coupler (DC) consists of two waveguides placed in closed proximity (see Figure 13.3). The light launched from one of the waveguides is coupled into the other waveguide when the separation between the two waveguides is small enough to enable evanescent interaction. In the case of two identical waveguides, all the optical power launched into the input waveguide is transferred to the other waveguide after traveling a distance called the coupling length. The coupled power returns to the original waveguide after traveling a distance equal to twice the coupling length. In this manner, the light can be transferred between the waveguides in a periodic manner. If length of the waveguides is fixed equal to one coupling length, the light launched from the input waveguide will exit from the other waveguide representing the cross state of the switch while it will exit from the same waveguide representing the bar state of the switch if the length of the waveguide is equal to twice the coupling length. The bar and cross states are represented by the schematic drawing in Figure 13.3(b) and (c). By inducing the refractive index change (and hence the optical phase) in one of the waveguides, light can be made to exit either from the same waveguide or the other. The operation principle of the directional coupler can be understood using the concept of normal modes. The normal modes are the two lowest-order modes of a single system consisting of two coupled waveguides. As these two modes propagate through the waveguides, the interference between them results in the power transfer from one waveguide to the other. When the optical power is launched into the input waveguide, the two normal modes, symmetric and antisymmetric, are excited in the two-waveguide system.

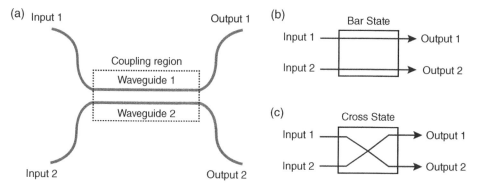

Figure 13.3 (a) Silicon photonic switch configuration based on the directional coupler, (b) Bar state, and (c) Cross state.

After traveling a length equal to the one coupling length, the phase of the two modes differ by π, and the modes interfere destructively in the same waveguide while interfering constructively in the other waveguide, giving the cross state of the switch. The power for the two output ports of the coupler is represented by the following equations:

$$P_{1-out} = 1 - F\sin^2(\gamma z) \tag{13.6}$$

$$P_{2-out} = F\sin^2(\gamma z) \tag{13.7}$$

$$L_c = \frac{\pi}{2\gamma} \tag{13.8}$$

$$\gamma^2 = \kappa^2 + \Delta^2 \tag{13.9}$$

$$F = \frac{\kappa^2}{\gamma^2} = \frac{1}{1 + \left(\dfrac{\Delta}{\kappa}\right)^2} \tag{13.10}$$

where κ is the coupling coefficient, Δ is the phase mismatch between the two supermodes, and F denotes the coupling efficiency. The power transfers maximally from one waveguide to the other at each L_c distance. Therefore, L_c is called cross-coupling length or beat length. The 3-dB coupling length L_{3-dB} of the directional coupler, at which the transfer of the power is 50%, can be estimated as:

$$L_{3dB} = \frac{L_c}{2}. \tag{13.11}$$

13.5.2 Microring Resonator

Microring resonators (MRRs) are good candidates for photonic switching applications owing to their compact footprint, strong resonance field enhancement, and narrowband wavelength selectivity. MRR-based switches have a compact active length and relatively low tuning power consumption [17]. Due to their resonant behavior, MRRs can be utilized to tune the resonance wavelength by a few nm without the need for a long active waveguide to achieve a π phase shift. A MRR typically consists of a microring formed by looping around a single silicon waveguide into a ring structure and two bus waveguides placed in close proximity to the microring. When there are two bus waveguides placed closely to MRR, the resulting configuration is called add-drop configuration and this is shown in Figure 13.4. The light that enters from the input port can leave from either drop port or through port, cf. Figure 13.4(b) and (c), depending upon whether the resonance condition is satisfied or not.

When the resonance condition given by ($2\pi R n_{eff} = m\lambda$, $m = 1, 2, 3...$) is satisfied, the light of a particular wavelength is coupled to the microring and will exit from the drop port. Otherwise, the light will exit from the through port. The transmission at the through ($T_{through}$) port and drop (T_{drop}) port can be calculated using the following expressions [17]:

$$T_{through} = \frac{r_2^2 a^2 - 2r_1 r_2 a\cos\phi + r_1^2}{1 - 2r_1 r_2 a\cos\phi + (r_1 r_2 a)^2}, \tag{13.12}$$

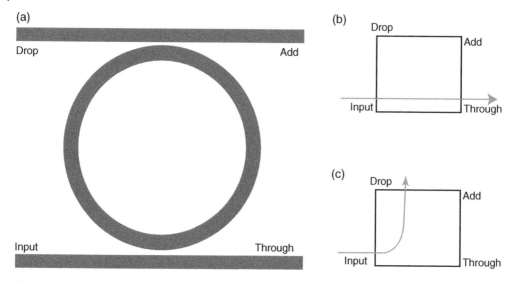

Figure 13.4 (a) Add-drop configuration of a microring resonator that can be used for switching purpose. Representation of light leaving at (b) Through port and (c) Drop port.

$$T_{drop} = \frac{\left(1 - r_1^2\right)\left(1 - r_2^2\right)a}{1 - 2r_1 r_2 a \cos\phi + \left(r_1 r_2 a\right)^2},$$ (13.13)

where r_1 and r_2 are the self coupling coefficients and $a = \exp(-\alpha 2\pi R)$ is the round trip loss, with α and R representing the absorption coefficient and radius of microring, respectively. The free spectral range (FSR) measures the separation between the two resonance wavelengths. The cavity resonances exhibit a finite linewidth called the 3-dB bandwidth or the full-width-half-maximum (FWHM). A measure of sharpness of resonances relative to their spacing (FSR) is given by the parameter finesse ($\mathcal{F}$):

$$\mathcal{F} = \frac{FSR}{FWHM}.$$ (13.14)

Another parameter of interest is the quality factor (Q), which is the measure of sharpness of resonances relative to their central frequency.

$$Q = \frac{\lambda_{resonance}}{FWHM},$$ (13.15)

By using electro-optic, thermo-optic, or hybridizing microring with other materials, one can achieve the switching functionalities in MRR. A large-scale switch fabric can also be created by cascading multiple MRR [18].

13.5.3 Mach–Zehnder Interferometer

The Mach–Zehnder Interferometer (MZI) is an interferometric device consisting of two 3-dB couplers/splitters, and a phase tuning section is often placed in one of the arms as shown in Figure 13.5. The input 3-dB coupler splits the light into two waveguide arms while the output 3-dB coupler acts as a combiner. The tuning section of the MZI modifies the phase of propagating light in such a way that it gives either constructive or destructive interference with light in the coupler arms, and the output signal exits from either the bar port or the cross port based on phase-matching condition,

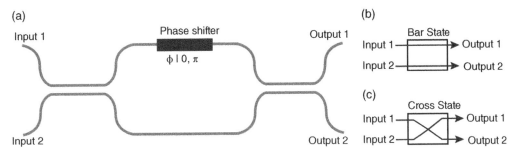

Figure 13.5 (a) Silicon photonic switch configuration based on Mach–Zehnder interferometer, (b) Bar state, and (c) Cross state.

cf. Figure 13.5(b) and (c). The output of the MZI depends on the critical 3-dB ratio of the couplers and losses inside the two arms of the MZI. The phase change introduced by one of the arms with tuning section of length (L) can be written as $\Delta\phi = \Delta\beta L$, where $\Delta\beta$ is the change in the propagation constant of light due to the tuning section. The ratio of the output power (P_{out}) to input power (P_{in}) for the bar and cross states are given by the equations

$$\frac{P_{2-out}}{P_{in}} = cos^2\left(\Delta\beta\frac{L}{2}\right) \quad (cross) \tag{13.16}$$

$$\frac{P_{1-out}}{P_{in}} = sin^2\left(\Delta\beta\frac{L}{2}\right) \quad (bar). \tag{13.17}$$

The phase shifter can be constructed in silicon waveguides by utilizing the thermo-optic effect by using a heater or exploiting the carrier effects using the PIN structure. By inducing the phase difference between the two arms as 0 or π, the output can be switched from the bar state to the cross state and vice versa.

13.5.4 Micro-Electro-Mechanical System

Micro-electro-mechanical system (MEMS)-based switches utilize the mechanical tuning method and have a stronger effect than the conventional material properties effects. They mechanically move or deform the waveguides to switch the optical signal between different output ports. MEMS-based switches have several advantages over the switches based on the conventional methods including low power consumption, high scalability, and the fabrication process being compatible with semiconductors [19]. The switching time of MEMS-based switches ranges from sub-microseconds to microseconds [19]. The four principles used in the MEMS-based switches for the manipulation of light are shown in Figure 13.6 [20].

In the first principle, the coupling between the stationary waveguides is changed using the MEMS-actuated mirrors in the gap between the waveguides as depicted in Figure 13.6(a). In the second method, as shown in Figure 13.6(b), the direction of the waveguide itself is changed to channel the light from one waveguide to another. The third method, depicted in Figure 13.6(c), is to change the propagation of the light with the help of an external element by interacting light with it. In the fourth method, the refractive index of the waveguide itself is changed by inducing the longitudinal strain, as shown in Figure 13.6(d). Using various photonic structures, MEMS-based switches have been realized in the past. In coupled waveguide configuration, the extinction ratio of 17 dB has been experimentally demonstrated for 1550 nm wavelength [21]. In [22], vertically coupled rib waveguides are used to construct a unit cell to reduce the device footprint and insertion loss. Using

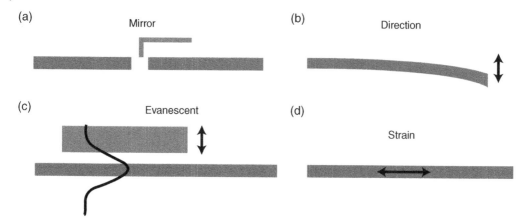

Figure 13.6 MEMS elements used to realize reconfigurable components. The reconfigurability can be achieved using principles such as (a) MEMS-actuated mirror, (b) changing the direction of the waveguide, (c) evanescent coupling of light to an external element, and (d) inducing strain in the waveguide. Modified from [20].

the multiple unit cells, a 64×64 on-chip switch with a high extinction ratio of 60 dB and fast switching time of about 0.91 μs with a low insertion loss of about 3.7 dB was demonstrated [22].

13.6 Hybrid Silicon Photonic Switches

13.6.1 III-V Materials

The III-V materials offer the opportunity to realize compact, low-cost, energy-efficient switch fabrics using a hybrid III-V/Si platform. The IIIV/Si material system leverages the low-loss guidance of silicon photonics while keeping the low power consumption and small footprint. The hybrid integration of III-V and silicon has been possible due to the advancement of wafer bonding techniques [23]. The bonding of III-V materials on silicon wafers allows the fabrication with a high-throughput process. This hybrid platform effectively combines the strengths of both platforms. With the help of III-V material, the electro-optic effect can be enhanced while keeping most of the light guided in the low-loss silicon waveguides. In addition, the hybrid III-V/Si photonic components can be fabricated using a CMOS-compatible process and this scales easily in terms of both wafer size and integration. By using a hybrid III-V/Si platform, the enhancement of the electro-optic effect has been demonstrated for phase manipulation [24]. Hybrid semiconductor-optical-amplifier (SOA) element was demonstrated by wafer bonding of III-V on a silicon waveguide where optical mode is mostly confined in the silicon waveguide and evanescently interacts with the III-V layer [25]. Such a configuration restricts the modal gain since only the evanescent tail of the propagating mode is influenced by the III-V layer. An MZI-based switch was demonstrated on the hybrid silicon platform with a power consumption of about 1 mW [24]. However, this switch requires optical amplifiers for the large-scale switch fabrics to compensate for the optical loss of each element.

13.6.2 2D Materials

An efficient switching device would require a simple heating mechanism providing efficient heating and be scalable at a large scale for the switching application. Recently 2D materials have attracted great attention in this regard for photonic applications due to their excellent portfolio of material properties. 2D materials have novel electronic and optical properties distinctively different from the bulk parental materials. Various emerging 2D materials include molybdenum disulfide (MoS_2), tungsten diselenide (WSe_2), hexagonal boron nitride (hBN), and graphene [26]. Among these, graphene has attracted great attention for integrated photonic devices and complex nanostructures due to its high scalability [26–29]. The surface of the 2D materials is naturally passivated and does not show the lattice mismatch since it has no dangling bonds. Therefore, it is easy to integrate or place 2D materials on silicon waveguides. The room temperature values of the thermal conductivity of graphene can go up to 5.30×10^3 W/mK. The ultra-high thermal conductivity and low loss of the graphene is beneficial for the thermal management of the thermo-optic devices [30]. Graphene as a heat conductor has been used to tune one of the arms of the MZI thermally. The graphene acts as a heat conductor and transports heat from the metal heater to the MZI arm. An spectral shift of 7 nm was shown for an applied heating power of 110 mW [27]. A microring resonator-based device by coating the microring with the graphene has been demonstrated to tune the resonance wavelength. By electrically heating the graphene, the thermal energy generated can shift the resonance wavelength by 2.9 nm with an electrical power consumption of 28 mW. It also exhibited a modulation depth of 7 dB with switching time of 750 ns [28]. Using a graphene heater placed at a small (240 nm) distance from the silicon microring, without any light absorption, demonstrated a tuning power of 22 mW per spectral range with a response time of 3 μs [29]. By optimizing the coverage length of graphene on a microring resonator, an on-off microring switch exhibiting an extinction ratio of 12.8 dB with a voltage of 8.8 V has been demonstrated [31].

13.6.3 Phase Change Materials

Photonic switches based on conventional methods have limitations in terms of speed, optical loss, and footprint. The electro-optic effect-based switches can enable high-speed but suffer from high power consumption and high insertion losses. The thermo-optic effect-based switches have a slow response time of the microsecond order, making them unsuitable for modern telecommunication and data center-related applications. In addition, these conventional effects need a continuous static power bias to sustain the device state, adding heavily to overall power consumption of the device.

To surmount the above-mentioned limitations, high-performance, reconfigurable switches having low static power consumption, high optical contrast, and ultrafast response time are required. Integration of phase change materials (PCMs) represents a promising way for realizing nonvolatile photonic switching davices on a SOI platform. Apart from the nonvolatility, there are many advantages of using PCM material for photonic switching applications [32]. First, PCMs exhibit a high refractive index contrast when switched from one physical state to another. Second, the transition between the phases proceeds with a fast switching time ranging from sub-nanosecond to a few picoseconds. Third, switching between the physical states can be achieved by thermal, electrical, or optical means. Fourth, PCM can be easily deposited, with the silicon material using the CMOS-compatible fabrication process, scaled down to the nanoscale, making it possible to realize miniaturized photonic switching devices. Fifth, a high cyclability of the devices up to 10^{10} has been demonstrated in the switching devices [33]. Therefore, PCMs are an ideal candidate for realizing reconfigurable hybrid PCM-silicon photonic switching devices. Until now, extensive efforts have been made to realize non-volatile, ultra-compact, CMOS-compatible photonic switching devices by employing $Ge_2Sb_2Te_5$ or GST. The GST has given the most promising results and is extensively used in various photonic devices. In most devices, hybrid GST-Si waveguides are formed by placing GST on top of the silicon waveguide, which manipulates the absorption and phase of the propagating light through evanescent interaction. The optical power is absorbed in the hybrid waveguide region and depending on the structural phase of the GST layer, the output differs. In the amorphous phase with low optical loss, the light passes through waveguides without much appreciable loss. This gives the high or the ON state of the switch. On the other hand, the crystalline phase with high optical loss absorbs most of the light and produces a low or OFF state of the switch. The switch state can be changed by switching the phase of the GST layer from amorphous to crystalline and vice versa. The crystalline state is obtained by raising the GST temperature above the crystallization temperature ($\sim$ 150 °C) but keeping the maximum temperature below the amorphization temperature ($\sim$ 600 °C) [34]. For amorphization, the GST temperature should be raised above the melting temperature and then rapidly quenched down to form amorphous phase.

Optical and electrical heating mechanisms have been used to induce the phase change behavior in GST-based photonic switches. Free space laser heating was demonstrated in a hybrid GST-Si waveguide device [35]. Improvement in the insertion losses and switching time was achieved using a silicon multimode interference waveguide with optical light focused on a circular GST cell placed on top of the silicon waveguide [36]. This heating approach was also used in the silicon microring ring resonator integrated with a GST thickness of 20 nm and 12 dB extinction ratio was obtained [37]. However, this approach is not viable for realizing practical integrated photonic swiching devices. Many researchers have used the on-chip optical pumping technique to heat the GST-Si waveguides e.g. [38, 39]. In this case, a pump is coupled to the waveguide (by an end-fire or grating coupler scheme) and is made to pass through the hybrid waveguide region. In [39], a switching contrast of 12% with energy consumption of 9.5 nJ and response time of 3.8 μs was achieved using a 4 μm-long hybrid GST cell placed on top of a rib waveguide. This on-chip pumping approach is energy-inefficient since only the evanescent wave of the propagating mode interacts with the GST layer placed on top of the waveguide.

Another approach [40–42], in which GST can be placed in the trench created by partially etching the silicon waveguide, is a better alternate and enhances the light-matter interaction, enabling high optical readout contrast with a compact active volume. In a 1 × 1 switch operating at 1.55 μm, high extinction ratio of 43 dB with a moderately low insertion loss of 2.7 dB was achieved [40]. This switch was able to maintain an extinction ratio above 30 dB for a wavelength span of 1500–1600 nm. This high extinction ratio was obtained with a very compact active volume of $400 \times 180 \times 450$ nm^3 (length × height × width). The phase change analysis of the embedded GST confirmed that during the phase change process, the phase change occured in most of the GST regions. The switch states can be altered with 1.6 mW and 7.2 mW of power for the process of crystallization (ON to OFF) and amorphization (OFF to ON), respectively [40].

Electrically induced phase change in the hybrid GST-Si waveguide has been demonstrated using Joule heating. Low-loss indium-tin-oxide (ITO) material as an electrode allows for electrical heating with negligible optical losses and provided localized heating of the active region. In electrical heating, a single electrical pulse is able to induce crystallization and amorphization. The heating of GST placed on top of a silicon waveguide was demonstrated [43] with an extinction ratio of 1.2 dB. The low extinction ratio is due to the weak interaction with the optical mode and small active volume.

Taking the embedded approach 1 × 1 waveguide switch and 1 × 2 directional coupler switch for operation at 2.1 μm were also investigated. The GST phase was transformed by applying electrical pulses through ITO electrodes, which alters the device state [41]. Both amorphization and crystallization processes were studied to confirm whether the phase transition occurs in the complete GST region. In a 1 × 1 waveguide switch, an extinction ratio of 33.79 dB was obtained with 0.52 dB insertion loss for an optimized GST length of only 0.92 μm. The crystallization was induced by a 5 V pulse which corresponds to the energy consumption of 0.9 nJ, while amorphization was achieved with a 7.5 V pulse which increases the GST temperature above the melting point, and energy consumption of 62.21 nJ.

In a 1×2 directional coupler switch, reversible switching was achieved with an extinction ratio of 10.33 dB and 5.23 dB in the cross and bar state, respectively. The optimized coupler has an active length of only 52 μm with a coupling gap of 100 nm. The crystallization and amorphization of GST was achieved with 6 V and 7.5 V electrical pulse, respectively. The phase transition of the GST was achieved with an energy of 44.67 nJ per cycle.

Design and analysis of a 1×2 tunable switch based on a hybrid GST-silicon microring resonator as shown in Figure 13.7 [42] was carried out. The overview of the devices is shown in Figure 13.7(a). The hybrid part of the microring consists of 20 nm thick GST layer placed on top of the partially etched silicon waveguide. The ITO layers of 50 nm thickness are placed on top of the GST and the side of the hybrid waveguide for electrically induced Joule heating as shown in Figure 13.7(b). The output of the microring is switched between through and drop port by electrically inducing the phase change in the embedded GST layer. Through port transmission spectrum of the device, covering one of the resonance wavelengths, is shown in Figure 13.7(c). The switch exhibited high extinction ratio of 18.75 dB at the through port as shown in Figure 13.7(c), while for the drop port (not shown here), the extinction ratio was 25.57 dB. The switch states were interchangeable by applying 5 V and 8 V pulses, and energy consumption for one switching cycle of the switch was 108.11 mW. High value of amorphous GST thermo-optic coefficient (1.1×10^{-3} K^{-1}) was exploited for tuning of the resonance wavelength of microring switch. Tuning curve of resonance wavelength shift as a function of applied power is shown in Figure 13.7(d). The switch exhibits a wavelength tuning efficiency of 1.16 nm/mW. For a GST length of 7 μm, we obtained a maximum wavelength tuning range of 4.63 nm [42].

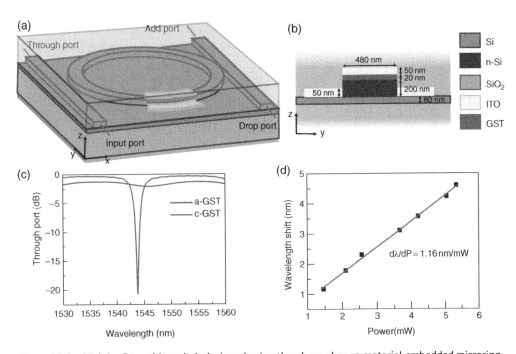

Figure 13.7 (a) A 1×2 tunable switch designed using the phase change material embedded microring resonator. (b) Side-view cross-section of the microring showing the hybrid waveguide region. (c) Transmission spectrum at through port for the amorphous and crystalline phase of GST. (d) Tuning curve of switch in the amorphous phase of GST. Reprinted with permission from [42] © The Optical Society.

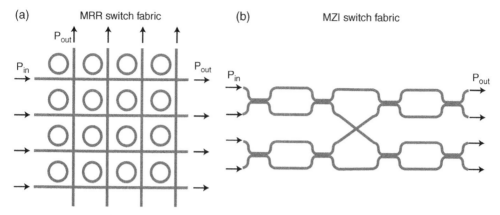

Figure 13.8 Representation of (a) MRR- and (b) MZI-based switch fabrics.

13.7 Switch Fabrics Using MRR and MZI

Several large-scale silicon switch fabrics based on the switching building blocks have been demonstrated. The MZI and MRR are more suitable for large-scale switch fabrics as they are highly scalable. A representative MRR-based switch fabric is shown in Figure 13.8(a). Various types of MRR based switch fabrics with different port counts have been reported. These switch fabrics include a 5×5 electro-optic switch fabric [44], a modular 8×8 switch fabric [45], and a 48×8 thermo-optic switch fabric [46]. The MZI-based switch fabrics are represented in Figure 13.8(b). Large-scale switch matrices based on the MZI elements have been demonstrated in recent years. Examples include 16×16 electro-optic MZI switch fabric [47], 32×32 electro-optic switch fabric [48], 64×64 thermo-optic switch fabric [49], and 32×32 thermo-optic switch fabric [50].

13.8 Summary

The ever-increasing demand of interconnects in data centers and telecommunication networks motivates the deployment of high-performance switching technologies. Photonic switching has received great attention to potentially address the challenges regarding bandwidth, cost, and power consumption in data centers. In this chapter, photonics-based switching relying on various physical effects such as electro-optic, thermo-optic, free carrier injection/extraction, and all-optical effects has been presented. Major configurations of the switching devices viz. MRR, DC, MZI, and MEMS technologies are covered. The need for hybrid silicon photonics with materials such as III-V/Si, 2D materials, and phase change materials has been presented. Furthermore, the emerging nonvolatile switches based on the phase change material are covered and discussed in detail.

Bibliography

1 G.I. Papadimitriou, C. Papazoglou, and A.S. Pomportsis. Optical switching: switch fabrics, techniques, and architectures. *Journal of Lightwave Technology*, 21(2):384–405, 2003.

2 R. Soref and J. Larenzo. All-silicon active and passive guided-wave components for $\lambda = 1.3$ and 1.6 μm. *IEEE Journal of Quantum Electronics*, 22(6):873–879, 1986.

3 B. Jalali and S. Fathpour. Silicon photonics. *Journal of Lightwave Technology*, 24(12): 4600–4615, 2006.

4 D. Inniss and R. Rubenstein. *Silicon Photonics: Fueling the Next Information Revolution*. Morgan Kaufmann, Cambridge, MA; Singapore, 2017.

5 R. Kumar, N. Ali, and S. Singh. High performance and cmos compatible photonic switches based on phase change materials. In *Optoelectronic Devices and Integration VIII*, volume 11184, page 111840C. International Society for Optics and Photonics, 2019.

6 B. Chmielak, M. Waldow, C. Matheisen, C. Ripperda, J. Bolten, T. Wahlbrink, M. Nagel, F. Merget, and H. Kurz. Pockels effect based fully integrated, strained silicon electro-optic modulator. *Optics Express*, 19(18):17212–17219, 2011.

7 R. Soref and B. Bennett. Electrooptical effects in silicon. *IEEE Journal of Quantum Electronics*, 23(1):123–129, 1987.

8 X. Xiao, H. Xu, X. Li, Z. Li, T. Chu, Y. Yu, and J. Yu. High-speed, low-loss silicon mach–zehnder modulators with doping optimization. *Optics Express*, 21(4): 4116–4125, 2013.

9 Z. Chai, X. Hu, F. Wang, X. Niu, J. Xie, and Q. Gong. Ultrafast all-optical switching. *Advanced Optical Materials*, 5(7):1600665, 2017.

10 S. Fan. Sharp asymmetric line shapes in side-coupled waveguide-cavity systems. *Applied Physics Letters*, 80(6):908–910, 2002.

11 M.F. Yanik, S. Fan, and M. Soljačić. High-contrast all-optical bistable switching in photonic crystal microcavities. *Applied Physics Letters*, 83(14):2739–2741, 2003.

12 O. Wada. Femtosecond all-optical devices for ultrafast communication and signal processing. *New Journal of Physics*, 6(1):183, 2004.

13 A. Pasquazi, S. Stivala, G. Assanto, V. Amendola, M. Meneghetti, M. Cucini, and D. Comoretto. In situ tuning of a photonic band gap with laser pulses. *Applied Physics Letters*, 93(9):091111, 2008.

14 V.R. Almeida and M. Lipson. Optical bistability on a silicon chip. *Optics Letters*, 29(20):2387–2389, 2004.

15 V.R. Almeida, C.A. Barrios, R.R. Panepucci, and M. Lipson. All-optical control of light on a silicon chip. *Nature*, 431(7012):1081–1084, 2004.

16 K. Preston, P. Dong, B. Schmidt, and M. Lipson. High-speed all-optical modulation using polycrystalline silicon microring resonators. *Applied Physics Letters*, 92(15):151104, 2008.

17 W. Bogaerts, P. De Heyn, T. Van Vaerenbergh, K. De Vos, S.K. Selvaraja, T. Claes, P. Dumon, P. Bienstman, D. Van Thourhout, and R. Baets. Silicon microring resonators. *Laser & Photonics Reviews*, 6(1):47–73, 2012.

18 H. Jia, Y. Zhao, L. Zhang, Q. Chen, J. Ding, X. Fu, and L. Yang. Five-port optical router based on silicon microring optical switches for photonic networks-on-chip. *IEEE Photonics Technology Letters*, 28(9):947–950, 2016.

19 X. Tu, C. Song, T. Huang, Z. Chen, and H. Fu. State of the art and perspectives on silicon photonic switches. *Micromachines*, 10(1):51, 2019.

20 F. Chollet. Devices based on co-integrated mems actuators and optical waveguide: A review. *Micromachines*, 7(2):18, 2016.

21 Y. Akihama, Y. Kanamori, and K. Hane. Ultra-small silicon waveguide coupler switch using gap-variable mechanism. *Optics Express*, 19(24):23658–23663, 2011.

22 T.J. Seok, N. Quack, S. Han, R.S. Muller, and M.C. Wu. Large-scale broadband digital silicon photonic switches with vertical adiabatic couplers. *Optica*, 3(1):64–70, 2016.

23 M.J.R. Heck, J.F. Bauters, M.L. Davenport, J.K. Doylend, S. Jain, G. Kurczveil, S. Srinivasan, Y. Tang, and J.E. Bowers. Hybrid silicon photonic integrated circuit technology. *IEEE Journal of Selected Topics in Quantum Electronics*, 19(4):6100117, 2012.

24 L. Chen, E. Hall, L. Theogarajan, and J. Bowers. Photonic switching for data center applications. *IEEE Photonics Journal*, 3(5):834–844, 2011.

25 H. Park, A.W. Fang, O. Cohen, R. Jones, M.J. Paniccia, and J.E. Bowers. A hybrid algainas–silicon evanescent amplifier. *IEEE Photonics Technology Letters*, 19(4):230–232, 2007.

26 F. Xia, H. Wang, D. Xiao, M. Dubey, and A. Ramasubramaniam. Two-dimensional material nanophotonics. *Nature Photonics*, 8(12):899–907, 2014.

27 L. Yu, D. Dai, and S. He. Graphene-based transparent flexible heat conductor for thermally tuning nanophotonic integrated devices. *Applied Physics Letters*, 105(25):251104, 2014.

28 S. Gan, C. Cheng, Y. Zhan, B. Huang, X. Gan, S. Li, S. Lin, X. Li, J. Zhao, H. Chen, et al. A highly efficient thermo-optic microring modulator assisted by graphene. *Nanoscale*, 7(47):20249–20255, 2015.

29 D. Schall, M. Mohsin, A.A. Sagade, M. Otto, B. Chmielak, S. Suckow, A.L. Giesecke, D. Neumaier, and H. Kurz. Infrared transparent graphene heater for silicon photonic integrated circuits. *Optics Express*, 24(8):7871–7878, 2016.

30 A.A. Balandin, S. Ghosh, W. Bao, I. Calizo, D. Teweldebrhan, F. Miao, and C.N. Lau. Superior thermal conductivity of single-layer graphene. *Nano Letters*, 8(3):902–907, 2008.

31 Y. Ding, X. Zhu, S. Xiao, H. Hu, L.H. Frandsen, N.A. Mortensen, and K. Yvind. Effective electro-optical modulation with high extinction ratio by a graphene–silicon microring resonator. *Nano Letters*, 15(7):4393–4400, 2015.

32 M. Wuttig, H. Bhaskaran, and T. Taubner. Phase-change materials for non-volatile photonic applications. *Nature Photonics*, 11(8):465–476, 2017.

33 S.B. Kim, G.W. Burr, W. Kim, and S.-W. Nam. Phase-change memory cycling endurance. *MRS Bulletin*, 44(9):710–714, 2019.

34 P. Guo, A.M. Sarangan, and I. Agha. A review of germanium-antimony-telluride phase change materials for non-volatile memories and optical modulators. *Applied Sciences*, 9(3):530, 2019.

35 Y. Ikuma, Y. Shoji, M. Kuwahara, X. Wang, K. Kintaka, H. Kawashima, D. Tanaka, and H. Tsuda. Small-sized optical gate switch using $Ge_2Sb_2Te_5$ phase-change material integrated with silicon waveguide. *Electronics Letters*, 46(5):368–369, 2010.

36 D. Tanaka, Y. Shoji, M. Kuwahara, X. Wang, K. Kintaka, H. Kawashima, T. Toyosaki, Y. Ikuma, and H. Tsuda. Ultra-small, self-holding, optical gate switch using $Ge_2Sb_2Te_5$ with a multi-mode si waveguide. *Optics Express*, 20(9):10283–10294, 2012.

37 M. Rudé, J. Pello, R.E. Simpson, J. Osmond, G. Roelkens, J.J.G.M. van der Tol, and V. Pruneri. Optical switching at $1.55\,\mu$ m in silicon racetrack resonators using phase change materials. *Applied Physics Letters*, 103(14):141119, 2013.

38 H. Zhang, L. Zhou, J. Xu, L. Lu, J. Chen, and B.M.A. Rahman. All-optical non-volatile tuning of an amzi-coupled ring resonator with gst phase-change material. *Optics Letters*, 43(22):5539–5542, 2018.

39 X. Li, N. Youngblood, Z. Cheng, S.G.-C. Carrillo, E. Gemo, W.H.P. Pernice, C.D. Wright, and H. Bhaskaran. Experimental investigation of silicon and silicon nitride platforms for phase-change photonic in-memory computing. *Optica*, 7(3):218–225, 2020.

40 N. Ali and R. Kumar. Design of a novel nanoscale high-performance phase-change silicon photonic switch. *Photonics and Nanostructures-Fundamentals and Applications*, 32:81–85, 2018.

41 N. Ali and R. Kumar. Mid-infrared non-volatile silicon photonic switches using nanoscale $Ge_2Sb_2Te_5$ embedded in silicon-on-insulator waveguides. *Nanotechnology*, 31(11): 115207, 2020.

42 N. Ali, R.R. Panepucci, Y. Xie, D. Dai, and R. Kumar. Electrically controlled 1× 2 tunable switch using a phase change material embedded silicon microring. *Applied Optics*, 60(13):3559–3568, 2021.

43 K. Kato, M. Kuwahara, H. Kawashima, T. Tsuruoka, and H. Tsuda. Current-driven phase-change optical gate switch using indium–tin-oxide heater. *Applied Physics Express*, 10(7):072201, 2017.

44 A.W. Poon, X. Luo, F. Xu, and H. Chen. Cascaded microresonator-based matrix switch for silicon on-chip optical interconnection. *Proceedings of the IEEE*, 97(7):1216–1238, 2009.

45 D. Nikolova, D.M. Calhoun, Y. Liu, S. Rumley, A. Novack, T. Baehr-Jones, M. Hochberg, and K. Bergman. Modular architecture for fully non-blocking silicon photonic switch fabric. *Microsystems & Nanoengineering*, 3(1):1–9, 2017.

46 F. Testa, C.J. Oton, C. Kopp, J.-M. Lee, R. Ortuno, R. Enne, S. Tondini, G. Chiaretti, A. Bianchi, P. Pintus, et al. Design and implementation of an integrated reconfigurable silicon photonics switch matrix in IRIS project. *IEEE Journal of Selected Topics in Quantum Electronics*, 22(6):155–168, 2016.

47 L. Lu, S. Zhao, L. Zhou, D. Li, Z. Li, M. Wang, X. Li, and J. Chen. 16× 16 non-blocking silicon optical switch based on electro-optic mach-zehnder interferometers. *Optics Express*, 24(9):9295–9307, 2016.

48 L. Qiao, W. Tang, and T. Chu. 32× 32 silicon electro-optic switch with built-in monitors and balanced-status units. *Scientific Reports*, 7(1):1–7, 2017.

49 L. Qiao, W. Tang, and T. Chu. Ultra-large-scale silicon optical switches. In *2016 IEEE 13th International Conference on Group IV Photonics (GFP)*, pages 1–2. IEEE, 2016.

50 P. Dumais, D.J. Goodwill, D. Celo, J. Jiang, C. Zhang, F. Zhao, X. Tu, C. Zhang, S. Yan, J. He, et al. Silicon photonic switch subsystem with 900 monolithically integrated calibration photodiodes and 64-fiber package. *Journal of Lightwave Technology*, 36(2):233–238, 2017.

Part C

Application of Optical Switches in Networks

14

Switch Control: Bridging the Last Mile for Optical Data Centers

Nicola Calabretta[1] and Xuwei Xue[2]

[1] *Electro-Optical Communication, Eindhoven University of Technology, Eindhoven, Netherlands*
[2] *State Key Laboratory of Information Photonics and Optical Communications (IPOC), Beijing University of Posts and Telecommunications, Beijing, China*

14.1 Introduction

With the escalation of traffic-boosting applications, such as cloud computing, Internet of Things, and high-definition streaming, the bandwidth growth in data centers (DCs) exceeds that of wide-area telecom networks and even outpaces the bandwidth growth rate of electrical switch application-specific integrated circuits (ASICs). Due to the technical challenge involved in increasing the pin-density on the Ball Grid Array (BGA) packaging technique, current electrical switches are expected to hit the bandwidth bottleneck in two generations from now. To overcome the bandwidth bottleneck of electrical switches, switching the traffic into the optical domain has been considerably investigated as a future-proof solution supplying ultra-bandwidth. Benefiting from the optical transparency, the optical switch with high bandwidth is independent of the bit rate and data format of the traffic. Moreover, migrating the switching functionality from the electrical to the optical domain removes the power-consuming optical-electrical-optical (O-E-O) conversions and eliminates the dedicated electronics circuits for various-format modulation, hence significantly decreasing cost and processing delay.

Despite the promises held by the optical switching technique, there are still several challenges that need to be addressed to practically deploy optical switches in DC networks (DCNs). First, to fully utilize the nanoseconds-level hardware switching time, a fast control mechanism is required to configure the switch on the nanoseconds time scale to fast forward the packets. Besides this, the controlling overhead should be independent of the network scale. Second, as no effective buffer exists in the optical domain, the conflicted packets at the optical switch would be dropped and this results in high packet loss. Thus, packet contention resolution is another unsolved challenge to complete the fast switch control. Third, in optical switched network, new physical connections are created every time the switch configuration changes. This implies that the receiver has to continuously adjust the local clock to properly sample the incoming packets and recover the data. The longer this process takes, the lower the network throughput will be, particularly for the intra-data

Optical Switching: Device Technology and Applications in Networks, First Edition. Edited by Dalia Nandi, Sandip Nandi, Angsuman Sarkar, and Chandan Kumar Sarkar.
© 2022 John Wiley & Sons, Inc. Published 2022 by John Wiley & Sons, Inc.

center scenarios where many applications produce short traffic packets. To overcome the aforementioned challenges, the optics and networking communities have been working on optical switch and control systems for many years but each community just tried to solve these problems from their own perspective. The optical switch control, from a global perspective, has the potential to overcome these challenges with less resource usage.

14.2 Switch Control Classification

To date, two main types of switches classified by the switch reconfiguration time have been investigated: slow micro-electrical mechanical system (MEMS) optical switches with milliseconds switching magnitude, and fast arrayed waveguide grating routers (AWGRs) based optical switches with nanoseconds switching configuration time. Traditional electrical switch schemes like the Ethernet switch, even with limited bandwidth, also have referring significance for the design of counterpart optical switch mechanisms. Thus, determined by the switches exploited, switch control can vary three types of class from electrical to optical with switching magnitude of milliseconds to nanoseconds. The following sections elaborate on the electrical switch control, slow optical switch control, and fast optical switch control.

14.2.1 Electrical Switch Control

Ethernet switches are the most representative and widely used switches among electrical switches [1]. An Ethernet switch generally consists of a certain number of input and output ports, memory or buffer, microprocessors, and the switching hardware [2]. According to the ISO/OSI model, an Ethernet switch is a link layer device and implements the function of the message receiving and forwarding related to the MAC addresses and certain scheduling algorithms. In addition to avoiding packet collisions, the switch is also employed as a flow controller by sending queue status back to the transmitter to suspend the packet transmission.

Flow characteristics and the topology of network around the switch are two vital factors for the performance of an Ethernet switch. In a traditional Ethernet network, a switch dynamically learns the topology of the adjacency network and constructs a mapping table. The dynamic mapping table and the packets schedule are implemented by exploiting algorithms which can decrease the processing latency and increase the performance of switch. Therefore, to control the switch, the design of the switching technology should be based on the application of transmitting packets to minimize the latency of transmission and optimize system performance.

In order to ensure the stability and robustness of the real-time control system, the most important control performance indicators can be characterized by the following three parameters: the maximum allowable transmission time, the sampling rate, and the richness of control data. The maximum allowable transmission time is a key factor for the stability of the control system. If the packet transmission time is larger than the maximum allowable transmission time, the performance of the entire system will be degraded, and a massive or possibly unbounded output will be generated. The sampling rate is the frequency at which the sensing device of the control system collects data. The value of the sampling rate depends on the performance of the processing unit and the communication medium. The abundance of control data directly determines the resolution of sampling and transmission data. The richness of the control data is determined by the quantizable packet size in bits.

In a real-time network, when a node transmits control data from a transmitter to a receiver or from a control to an actuator, it will send a request packet containing the source and destination node MAC and IP addresses and the control parameter sets to the switch. The feasibility of scheduling request transmission is then calculated based on the earliest deadline priority scheduling algorithm. When the transmission request is not feasible, the switch will recommend that the transmitting node adjust the transmission rate according to the active queue control. In addition, the switch will send a packet with an acknowledgement or notify the transmitter.

The earliest deadline first (EDF) algorithm is employed as the scheduling algorithm for all incoming packages, which is a dynamic priority scheduling algorithm [3]. A switch scheduler exploiting the EDF algorithm always forwards the package whose absolute deadline is the earliest. Then package priorities are not fixed but change depending on the closeness of their deadline. The schedulable condition is checked by computing the total utilization of the requested packages. When receiving a request package, the switch calculates the total utilization of all the request packages. If the schedule is feasible, the switch acknowledges the transmitter with network schedule parameters. If the schedule is not feasible, the switch sends out a set of recommended control parameters for the request node. These control parameters are suggested based on the status of switch queue and the active queue control scheme. When the requesting network traffic is larger than the switch capacity, a request will not be feasibly scheduled. Under this situation, the switch will adopt the active queue control (AQC) to advise the transmitters of further incoming packages. Otherwise, the network payload will be overloaded and the performance of control will be degraded [3].

In general, the Ethernet switch controller implement a scheduling algorithm for the incoming network traffic by utilizing related switch variables such as queue length and throughput to allocate bandwidth for each connected node. When the network traffic load is low, the switch adopts the EDF algorithm to verify the feasibility of scheduling and transmit the incoming packets. When the network traffic load is too high, the AQC is adopted to allocate the available network bandwidth properly.

14.2.2 Slow Optical Switch Control

Bandwidth is limited by current optoelectronic optical switches being used for signal switching, which require conversion of signals from optical to electronic to process signal, and back to optical for transmitting. In order to eliminate this bottleneck, various approaches have been employed. The optical switch is a device that switches an optical signal from one optical port to another, without having to first convert the optical signal into an electrical signal. Optical switches became important because of the telecommunications industry's desire to focus on all-optical networks (AON), meaning total exclusion of electronics [4, 5].

A slow optical switch is a device that enables signals in optical fibers or integrated optical circuits to be selectively switched from one port to another within milliseconds and microseconds, such as those using moving fibers. Mechanical optical switches are typical switches that rely on the movement of optical fibers or optical components to switch the optical path, such as a mobile optical fiber type, moving the sleeve to move the lens (including mirrors, prisms, and self-focusing lens) types. The biggest advantage of this kind of optical switch is the low insertion loss and low crosstalk. Its disadvantage is slow which within milliseconds and microseconds and easy to wear, easy to vibration, impact shocks.

One approach to optical switching has been the use of micro-electrical mechanical system (MEMS) technology to fabricate tiny mirrors that perform the switching function. These tiny mirrors steer optical signals directly, without O/E conversion operation. MEMS optical switch are micrometer-scale devices that rely on mechanical moving micro-mirrors to switch the optical signal from input ports to output ports. In general, the core of the MEMS optical switch control is how to fast and stably move micro-mirrors to satisfy the switching function. Meanwhile, considering the wear of the systems, a considerable number of control mechanisms have been employed.

The structure of MEMS consists of mechanical moving parts controlled by electronics, and hence usually show oscillating transitions between on and off states. This reduces switching speed and reliability, as the switch tends to wear out faster. Application of proper control system design emerges as a solution to these problems. The control system generally has two parts, a feed-forward portion responsible for reaching proximity to the desired position and a feedback portion responsible for shaping the system dynamics for transient response conditioning [6].

Another approach to optical switching has been the use of Liquid Crystal on Silicon (LCoS) technology to design wavelength-selective switch (WSS) systems for WDM applications. The LCoS device implement the amplitude, phase, or polarization modulation of the incident light by utilizing the electrically modulated optical properties of liquid crystals (LCs). The LCoS devices provided in the commercial market are reflective and composed of pixels coated with aluminum mirrors on the silicon backplane. The applied voltage on each pixel is individually controlled by the integrated driving circuitry underneath the aluminum mirrors on the silicon backplane [7].

LCoS is a very versatile switching element, although it is capable of easily switching between multiple ports there is no requirement to lock the spectrum of the switching to any predetermined channel plan. LCoS can be used for its polarization switching capabilities in a similar fashion to LC switches but the true flexibility of LCoS is unleashed by employing the LCoS in a variable phase spatial light modulator (SLM) which can be used to create from an incoming phase front the required outgoing phase front to couple efficiently to a configured output port selected by a applying a simple mathematical relationship [8–10].

It is simplest to consider the LCOS as being divided into a switching plane and a wavelength dispersive plane. The wavelengths of the optical spectrum are dispersed along the wavelength axis so that each wavelength can be operated upon separately and directed to the desired port. As such it is straightforward to have contiguous switching elements of variable width without any in-band artifacts or glitches that could be detrimental to cascaded performance without sacrificing the spectral resolution of the optical switch. LCOS is now a mature wavelength switching technology which has proven itself in current applications that do not fully utilize the full flexibility such as 50 GHz multiport WSS [11].

14.2.3 Fast Optical Switch Control

Compare to slow optical switch, a fast optical switch is a device that enables signals in optical fibers or integrated optical circuits to be selectively switched from one circuit to another within nanoseconds and sub-nanoseconds. Although lowering the time order of magnitude, it brings much challenges such as fast reconfiguration, precise time synchronization, fast clock data recovery, and scalability. Mach–Zehnder interferometers based optical switch, semiconductor optical amplifier based optical switch and AWGRs with tunable lasers are three typical fast optical switches.

The Mach–Zehnder interferometer (MZI) based switch consists of two 3 dB couplers, connected by two interferometer arms and shown in Figure 14.1. By controlling the effective refractive index of one of the arms, the phase difference at the beginning of the coupler can be changed, such that the light switches from one output port to the other. This switch has the advantage that the phase shifting part and the mode coupling part are separated, such that both can be optimized separately. Photonic switch networks constructed with MZI as switching elements (SE) can achieve nanosecond switching speed and has been demonstrated to scale up to 32-by-32 fabric port count [12]. Further scalability of the silicon integrated switch fabrics is limited by insertion loss and switching crosstalk [13].

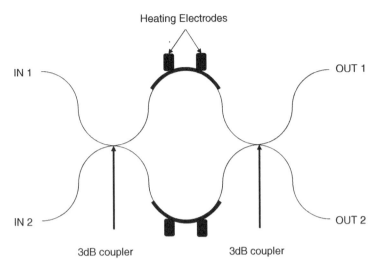

Figure 14.1 The structure of a MZI.

SOA is an active optoelectronic device with optical gain similar in structure to the semiconductor laser. The SOA is of small size and electrically pumped. It can be potentially less expensive than the EDFA and can be integrated with semiconductor lasers, modulators, etc. With the increase of the intensity of the current injected into the device, when the number of particle inversion in the SOA reaches a certain degree, the optical gain begins to appear in the SOA, and the injection current corresponding to the transparent medium is the threshold current of the SOA. Beyond this threshold, SOA emerges the light amplification capabilities. When the SOA is turned on, the input optical signal passes through the device and the optical signal power is amplified. After the SOA is closed, the input optical signal is absorbed by the SOA. According to the wavelength and input port of the input optical signal, the switching state of the corresponding SOA can be adjusted and different optical signal paths can be selected to achieve wavelength switching. The combination of amplification in the on-state and absorption in the off state makes this device capable of achieving very high extinction ratios. High-radix switches can be fabricated by integrating SOAs with passive components [14].

Flow control and software-defined networking (SDN) based control are typical switch control schemes among SOA-based optical switches. Flow control technique is employed to resolve the packet contentions when multiple optical data packets have the same destination. Once contentions occur, the data packets with higher priority will be forwarded to the destination Top of Racks (ToRs) while the conflicted packets with lower priority will be forwarded to the ToRs with no destination request. This kind of packet forwarding mechanism guarantees that the receivers at each ToR receive a continuous traffic flow at every time slot. SDN-enabled control and orchestration plane with extended OpenFlow (OF) protocol has been developed and implemented for the prototyped flow-controlled and clock-distributed optical data center network. With the abstracted information and translation offered by the OF-agent, the SDN controller can flexibly slice the optical networks by updating the look-up table and monitoring the stored statistics of the data plane. Such SDN-enabled control interface is the key optimizer to implement the SDN control scheme for a programmable optical data center network.

AWGR, which is a passive and lossless optical interconnect element, basically, as shown in Figure 14.2, is a fully-connected structure [15]. AWGR-based switching fabric is power-efficient as

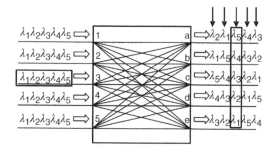

Figure 14.2 Fully-connected topology of N-by-N AWGR. Source: Modified from [15].

the signal is only switched to the desired output via the appropriate wavelength, instead of employing the broadcast-and-select mechanism that accompanies excessive power splitting losses. The cyclic wavelength routing characteristic of the AWGR allows different inputs to reach the same output simultaneously by using different wavelengths. Non-blocking switching from any input port to any output port can easily be achieved by simply tuning the wavelength at each input. The power consumed in tunable wavelength converters (TWC) [16], the loopback shared buffer, and the control plane logic scale linearly with the number of ports, unlike other switches. In contrast to SOA, AWGR is a passive device depending on the control of input and output node. Predefined control and label control are employed to implement AWGR-based optical switch control.

14.3 Challenges for Switch Fabric Control

The optics and networking communities have been extensively investigating optical switching control techniques, especial for fast (nanoseconds) switching control, for many years. Nevertheless, each community address the challenges and problems from their own perspective. The optics community focuses on developing technologies for individual devices and components [17–19] that achieve nanoseconds-level optical switching while devoting little attention to solving the switch fabric control challenges, e.g., scalable control plane, precise time synchronization, fast burst clock data recovery, lack of optical buffer, and reliability.

14.3.1 Scalable Control Plane

The control function is implemented by the switching control mechanism. In a simple end-to-end network topology, control plane executing control function can easily implement dual communication with low delay and high bandwidth. However, when the scale of network is expanded exponentially, which means the number of nodes in the network is increased to tens of thousands of nodes, whether the switching controller achieves the previous effect is a huge challenge. The increasing number of controlled nodes increases the burden of the switch, and brings about power consumption increase and blocking problems. One feasible solution to reduce the burden is to cascade or expand the use scale of the switch on the basis of the original structure. Whether original switch control is applicable to the new condition or meets the same QoS requirement are factors to consider in scalable control plane.

On the other hand, the increased nodes require the increasing radix of switch. Under the current existing technology, the radix of SOA of 32-by-32 has been achieved, which can support up to 10,000 nodes, but the increase of the insertion loss, which can be attributed to the increase of radix, has affected the design scheme of the switch. The radix of the AWGR is an important consideration in determining the scalability of the interconnect networks. An AWGR-based switch has the potential ability to implement a high radix for the switch, up to 128 [18]. While high port count silicon photonic AWGRs with 512 ports have been experimentally demonstrated, in-band crosstalk and the number of wavelengths become challenging. Noting that 32-port AWGRs are commercially available and that 64-port AWGRs were demonstrated with < −40 dB crosstalk and ~6 dB insertion loss, the use of AWGRs with a port count value up to 64 is a viable solution both for passive AWGR interconnection and for active AWGR switches. Therefore, the applicability of the simple model is the first step towards implementation and the feasibility of the large-scale network should be considered in the design of the controller.

14.3.2 Precise Time Synchronization

In pace with the scalability improvement of the switch structure, the demand for precision in time synchronization necessitates a key factor in the performance of switching [20]. Above all, the significance of precise time synchronization is proposed by the utilization rate of the bandwidth. On the basis of switches, to guarantee the validity of signals, the switch window will remain open before the transmission is completed under control. However, the throughput of the switches is ascertained with the guard bandwidth, in which the time synchronization can connect both switch and the endpoints at an appropriate granularity to increase the bandwidth utilization. Ideally, the permission of the lowest interval can reach the nanosecond in recent researches. Not only this, but the precise time-slot alignment and the ultrafast control of the pass window based on time synchronization are of paramount importance in the time division multiplexing (TDM) and wavelength division multiplexing (WDM) system etc.

Corresponding to the different scenarios, there are several schemes of precise time synchronization nevertheless contributing various accuracy and precision to the switching process. Moreover, the precision of time synchronization protocol varies from millisecond to sub-nanosecond in experiments. In a variety of such time synchronization protocols, network time protocol (NTP), the Global Positioning System (GPS), BeiDou Navigation Satellite System (BDS), IEEE 1588, precision time protocol (PTP, IEEE 1588V2), and Datacenter Time Protocol (DTP) are indicated in the order of the average precision elevating in systems. The introduction of techniques mentioned is shown in Table 14.1 with scalability and synchronization precision.

In the paper by Patel et al. [21], the stability of the technology-based NTP depends on stable data center network (DCN). Imagine with such a scene, the effect of network fluctuations will contribute the disconnection of the client to the clock source server, which results in the loss of capturing reference time. This scheme can merely reach the accuracy of synchronization at the 50 ms level.

Table 14.1 The comparison of different precise time synchronization technologies.

Technique	Synchronized Precision	Scalability	Summary
NTP	Milliseconds	Sufficient	Does not reach the order of fast switch.
GPS/ BDS	Tens of nanoseconds	Insufficient	In the ultrafast switching of DCN, the signal intensity is unable to support the nanosecond switch.
PTP	Sub-microseconds	Insufficient	The sophisticated devices and the high cost are confined.
DTP	Nanoseconds	Insufficient	The difficulty in cascading limits its application in scalable networks.

Moreover, according to the main research in industry, the global navigation satellite system (GNSS) is a feasible scheme to acquire the reference time and clock frequency. Nevertheless, extra receivers of GNSS and cables deploying have increased the cost in a considerable scale. Furthermore, the sub-microseconds accuracy of PTP implemented with dedicated devices of PTP and specialized cables, dramatically taking an additional cost in the system [22, 23]. The researchers of Cornell University [24] proposed the time synchronization mechanism on the basis of the physical layer, Datacenter Time Protocol (DTP). However, to synchronize the time and frequency, the master and slave will deploy dedicated clock channel and devices on the physical layer, while the complexity of link expanded with the application in datacenters.

With the problems in time synchronization, the researchers also proposed many other protocols to comprise, for example, the enhanced Just-in-time protocol [25], which reserves the resources to compensate the optical burst switching (OBS). Despite the challenges for the precise time synchronization, there may be a potential method to implement the accuracy, stability, and scalability simultaneously.

14.3.3 Fast Burst Clock Data Recovery

Before the introduction of the demands based on clock and data recovery (CDR), we will first explain the concept of CDR. While the optical packets are after transmission, the clock of the same frequency is required to recover the data carried on the optical signals in the receiver. However, the simplified method is the distribution of the frequency of the transmitter with an extra cost on the specialized cable. To this end, the economic feasibility of optical transmission is biased toward recovering the clock frequency from the injecting optical signals.

In the switching scenario, according to the research of Clark et al. [26], 97.8% of packets are small packets sized 576 bytes or less. However, compared to the electrical switches, optical switching requires the reconfiguration of the link every time in the instant of the dynamical state of the laser for transmission. The momentary process introduced undesirable burst-CDR locking time for off-the-shelf nanosecond-optical switches that can make small packets transmit practically. Moreover, the minimum guard-band for the packets should meet the sum of optical switching time and CDR locked time, which confines the throughput and the bandwidth utilization of the switches. Additionally, the burst signals meet the recovery time sum-up increasing in repetitive operations, which also reduces the utilization rate for the inducement of long guard-band, shown in Figure 14.3.

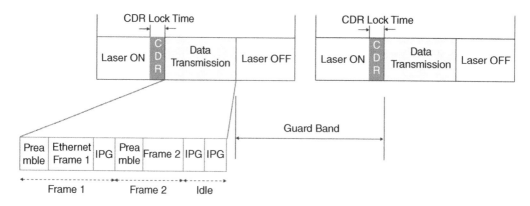

Figure 14.3 The architecture of clock and data recovery in optical packets.

The preamble block consists of some symbols that claim the subsequent block is the valid information. Additionally, the IPG block is the acronym for Inter-Packet Gap, forbidding jitter and other influences in the packets. Moreover, the fundamental reason for the burst characteristic of optical signals is the lock loss of the phase locked loop (PLL) while there is no signal in the optical channel with a long gap. However, the procedure of CDR will be required, ensuring the optical signal injection. The process of CDR is divided in three parts: frequency recovery, phase recovery, and data recovery. The schemes for a burst-mode receiver can basically be categorized into three main architectures: PLL-based burst-mode CDR (PLL-based BMCDR), Oversampling-based burst-mode CDR (Oversampling-based BMCDR), and Gated-Voltage-Oscillator-based burst-mode CDR (GVCO-based BMCDR). In the experiment of Verbeke et al. [27], a 25 Gb/s all-digital CDR circuit for burst mode achieved 37.5 ns recovery without the start-of-burst signal. Furthermore, in recent research of Zheng et al. [28], a DSP-assisted 25 Gb/s burst-mode receiver for 50G-PON upstream transmission has been validated experimentally with approximately 200 ns recovery time.

However, with the fast CDR demands of both point-to-point and multipoint-to-point (MP2P) systems, the process of CDR should be implemented in nanoseconds or sub-nanoseconds to match the development of transmission rates, which brings challenges for system control. Not focusing on the circuit or techniques to reduce the burst-CDR locking time, in the previously mentioned experiment of [26], a feasible method proposed from the control plane of the system to relieve burst-CDR locking time has been demonstrated. It exploits the distribution of reference clock in the system and the phase pre-shifting based on the last value simultaneously to compensate the fluctuation of phase in transmission to achieve the lower CDR-locking time.

14.3.4 Lack of Optical Buffer

The lack of optical buffer is one of the main architectural differences between electrical switches and optical switches, where the electrical switches employ random access memories (RAM) to buffer data packets that lost contention. Because no effective RAM exists in the optical domain, conflicted packets at the optical switch would be dropped and this results in high packet loss. Despite several approaches having been proposed to overcome this issue, based either on optical fiber delay lines (FDLs), wavelength conversion, or deflection routing, none of them is practical for large-scale data center networks, due to the fixed buffering time (FDLs), extra hardware deployment (wavelength conversion), and management complexity (deflection routing).

Although there is a lack of optical buffer to overcome the bottleneck, it is the key part of optical packet switching, optical routing, and optical computing. Optical packet switching provides an almost arbitrary fine granularity but faces significant challenges in the processing and buffering of bits at high speeds. The simplest solution to overcome the contention problem is to buffer contending packets, thus exploiting the time domain. This technique is widely used in traditional electronic packet switches, where packets are stored in the switch RAM until the switch is ready to forward them. Electronic RAM is cheap and fast. On the contrary, photons are bosons, and they are theoretically impossible to stop without converting them into other forms of energy, so optical RAM does not exist. The only way out is to delay the light signal for a period of time so that it can be processed at high speeds. FDLs are the only way to "buffer" a packet in the optical domain. Contending packets are sent to travel over an additional fiber length and are thus delayed for a specific amount of time [29].

To highlight the scaling problem in large-scale OPSs, consider an OPS or router with 1000 incoming and outgoing channels, each at a data rate of 40 Gb/s. In electronic routers, the buffering capacity per port is usually equal to around 250 ms of delay per port. At the 40 Gb/s data rate, this corresponds to a buffer capacity of around 10 Gb. If single-wavelength FDLs were utilized in place of electronics for this buffering, then the total length of fiber needed for buffering all ports in the router would be approximately 40 Gm, or about 150 times the distance from the earth to the moon. Even if we set aside issues of signal distortion by dispersion and the very significant problem of power consumption by the necessary inline amplifiers, these lengths of fiber are quite unrealistic just on the basis of the physical space required to house the delay lines.

Several design ideas for optical buffer are presented here. The first is to increase the transmission distance, which is similar to the optical delay line, the second is to reduce the group speed, and the last is to design the optical buffer according to the specific application scenarios. By the way, fiber-optic loop and slow light combination is the future of buffer [29, 30].

14.3.5   Reliability

Reliability is a standard concept for evaluating the effectiveness of a network system. Unlike the reliability of optical switching technology, the QoS after long-time link construction has mainly been investigated. A well-designed switching control scheme can control the optical switch continuously for a long time with low latency and low packet loss. In order to verify that the reliability requirements are met, it is necessary to carry out long-term chain building tests in the experimental scenarios when designing the switching control mechanism.

14.4   Switch Fabric Control: State of the Art

14.4.1   Predefined Control

Predefined control is a control scheme in which data is deployed on the switch path before being transmitted to the switch input node. Fast optical switches based on passive optical device AWGR are widely used due to the interference of insertion loss and crosstalk. An optical switch based on AWGR can realize the design of a high-radix situation, which can reach 128×128. The switch principle of a passive optical device is fixed by the device, not directly to reconstruct or control switches switching path, so need incoming ports to see predefined data in advance, after the predefined optical data can be carried out in accordance with the scheduled way switch, and finally complete the end-to-end communication. Precise time synchronization is the key to implement predefined control. It requires the synchronization of all sub-clocks, also called the slave. The operation is effective only when all parts of the clock are synchronized in the same time domain. A White Rabbit (WR) switch is an Ethernet switch based on high-precision clock distribution with performance much better than that of the PTP 1588 synchronization protocol. Generally, predefined control technology is a new control scheme proposed by the device itself, which needs continuous development.

14.4.2   SDN Control

To fulfill the promises of facilitating virtualization and enhancing the network performance by providing simplicity, programmability, and flexibility, SDN is penetrating the optical DCNs. OpenFlow (OF), the core technology of SDN, achieves centralized control of large-scale network traffic by separating the control plane of network devices from the data plane and makes network management more convenient. After SDN deployment, the connectivity of routers of each node in the network is completely controlled by the upper layer automatically, without manual adjustment, and only corresponding network rules need to be defined in advance. In addition, users are allowed to modify the built-in protocols of the device according to their needs to achieve better data switch performance.

To elaborate the control scheme, as is shown in Figure 14.4, the OpenDaylight (ODL) and OpenStack platforms are deployed as the base SDN controller and orchestrator, connecting the data plane by means of integrated OF Agents implementing an extended OF protocol. The OF Agents deployed on top of FPGA-based ToRs and switch controllers are mediation entities between the control plane and data plane. Cooperating with OF protocol, these Agents enable the communications between the southbound interface (SBI) of ODL controller and the Peripheral Component Interconnect Express (PCIE) interfaces of data plane.

Apart from the forwarding control of optical packets, the FPGA-based switch controllers report the counts of NACK signals indicating packet retransmissions to the SDN controller which, in turn, forwards them to the orchestrator for the full automation of network slicing (NS) deployment, featuring dynamic flow priority assignment and automatic load balancing to decrease the transmission latency and packet loss. To fulfil the promises of facilitating virtualization and enhance the network performance by providing simplicity, programmability, and flexibility, SDN is penetrating optical DCNs. In light of this, an SDN-enabled control and orchestration plane with extended OF protocol has been developed and implemented for the prototyped flow-controlled and clock-distributed optical DCN. With the abstracted information and translation offered by the OF-agent, the SDN controller can flexibly slice the optical networks by updating the look-up table and monitoring the stored statistics of the data plane. Such SDN-enabled control interface is

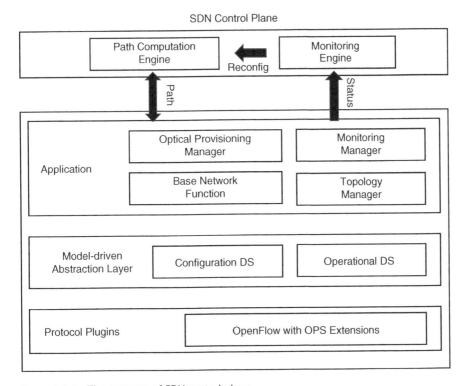

Figure 14.4 The structure of SDN control plane.

the key enabler to realize the SDN control framework for a programmable optical data center network. By exploiting the nanoseconds forwarding of hardware control and the decoupled SDN control, assessments demonstrate the QoS guaranteed operation for the applications running in the optical switch-based sliced networks [31, 32].

14.4.3 Label Control

Label control with a packet retransmission mechanism provides a promising solution to address the optical packet contention by pushing the buffer to the edge nodes in the electrical domain. In the fast label control system, each of the bidirectional label channels is a continuous link not only used to send the label requests from ToRs to switch controller and the flow control signals from the switch controller to the ToRs, but also used to distribute the clock from the switch controller to all the connected ToRs. In this way, all the transceivers of the ToRs are clocked with the same frequency and thereby the receivers eliminate the time-consuming clock frequency recovery step. Moreover, to guarantee that the CDR circuits are active, the switch controller, which has the full vision of the traffic from the ToRs, exploits the multicast capability of the optical switch to forward packets to the non-destined ToRs to fill the empty slots. It shows that this clock distribution technique achieves a constant 3.1 ns data recovery time regardless of the IPG length and with no deployment of the high-cost burst-mode CDR receivers [33].

The bidirectional label channels carrying the label signals and flow control signals are exploited in the nanoseconds optical switch and control system to implement the fast switch control, time allocation, and flow control as well as the clock distribution. To further improve the utilization of the label channels and to provision differentiated quality of service (QoS), the label channels can be utilized to deliver more useful information apart from the destination request and priority. For instance, one lower speed (that means lower cost) label channel can be used to carry several (current and following) destination requests for data packets from multiple data channels in an asynchronous mode. This could boost the label processing capability at the switch controller since the controller knows more destination requests. This could also improve the transmission throughput since the coming packets can be transmitted immediately, and do not need to follow the synchronous time slot. Moreover, heterogeneous requests can be delivered by the label signals to the switch controller to enable the dynamic QoS provisioning for data center applications with different requirements [34].

To elaborate the control scheme, as is shown in Figure 14.5, the LIONS [35, 36] consists of an AWGR, tunable wavelength converters (TWCs), an electrical control plane (CP), electrical loopback buffers, label extractors (LEs), and fiber delay lines. Between each terminal node and the switch, there is an optical channel adapter that serves as the media interface [37]. In general, the AWGR-based switching fabric can easily realize the output queue, provided that a 1:N optical DEMUX with N receivers is available at each AWGR output. However, N receivers at each output may not be practical or scalable since this requires a total of N^2 receivers for the whole system.

When the optical labels arrive at the label extractor (LE), the optical labels are separated from the optical payloads by it and then transmitted to the control plane. The control plane is an electrical processing plane. After the optical to electrical (O-E) conversion, the converted labels from the optical label packet first enter the label processor. The label processor has the preamble detector that implements the pre-process operation of the inputs. Once the detector detects a valid input, the label processor will start to record the label contents which contains destination address and packet length following the preamble. The label processor then maps the destination address to the desired output port, and sends a request to the proper arbiter for contention resolution. After the

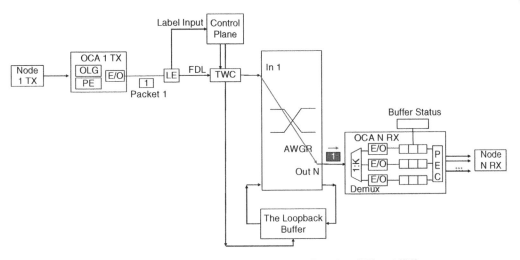

Figure 14.5 The system diagram of the optical switch. Source: Based on [35] and [36].

arbiter processing, the control plane generates control signals for TWCs, setting their outputs to the proper wavelengths. For the winning inputs, the control plane assigns wavelengths that enable them to send packets to the desired AWGR output, while for inputs that do not get grants, the control plane assigns wavelengths that force them to send packets to the AWGR output connecting with the loopback shared buffer. While the control plane receives the label and makes the decision, the corresponding packet payload travels through the fixed length fiber delay line (FDL) to compensate for the control plane latency. The packets arrive at the inputs of TWCs after the TWCs set their output to the proper wavelengths. The control plane latency is measured as the time between the first bit of the optical label arrives at the O-E converter and the TWC finishes the output wavelength tuning. Since the arbiter can distribute the TWC control signals in the same cycle for the requests that arrive on the same arbitration cycle, the control latency is identical for any packet arriving at any input [38, 39].

As shown in Figure 14.6, the SOA-based FOS prototype includes an FPGA-based switch controller and a switch fabric [40]. The switch fabric is a wavelength and space switch based on strictly non-blocking architecture. The switch fabric consists of N identical modules and each of them handles the packets from the corresponding TORs. The label channels carry the packet destination and are processed by the switch control of each module. Meanwhile, the optical packets are fed into the SOA-based 1×N switch. The switch controllers retrieve the label bits and check possible packets contentions, thereby configuring the 1×N switch to forward the optical packets. Moreover, the switch controllers also generate the ACK/NACK flow control messages, which are sent back to the ToRs to release or to ask for retransmission of the packets in case of contention, respectively.

14.4.4 AI Control

To date, AI control, employed in flow control, topology reconfiguration, and control prediction, is a new-type control scheme for high-layer network structure. The high-layer network control is primarily for virtualization and cloud service. An intelligent engine that is easy to configure and use with high efficiency, combined with diverse domain knowledge and models, is needed for modern data centers to quickly learn and extract targeted valuable information and strategies from the massive amounts of data generated by various applications. The intelligent engine enables the

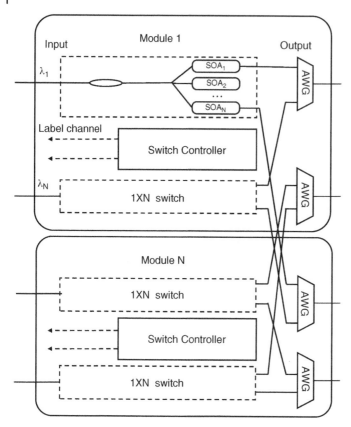

Figure 14.6 SOA-based FOS prototype. Source: Modified from [40].

data centers to provide rich platform services and application programming interfaces (APIs) with pre-integrated machine learning, graphics engine, and search capabilities, as well as artificial intelligence (AI) services and APIs in common fields such as visual, voice, and language processing. These intelligent platforms and general AI/machine learning/deep learning services could work closely with the heterogeneous computing hardware such as graphics processing units (GPUs) and field-programmable gate arrays (FPGAs) to implement the in-depth optimization of application performance.

Bibliography

1 K.C. Lee and S. Lee. Performance evaluation of switched Ethernet for networked control systems. In *IECON 02 – 28th Annual Conference of the Industrial Electronics Society*, 2002.

2 C.S. Hsu and K. Kase. *Industrial ethernet switch in factories and control applications.* 39:51–56, 2016.

3 F.L. Lian, Y.C. Tu, and C.W. Li, editors. Ethernet switch controller design for real-time control applications. In *IEEE International Conference on Control Applications*, 2005.

4 L. Xu, H.G. Perros, G. Rouskas. Techniques for optical packet switching and optical burst switching. *IEEE Communications Magazine*, 39(1):136–142, 2001.

5 G.I. Papadimitriou. Optical switching: switch fabrics, techniques, and architectures. *Journal of Lightwave Technology*, 21(2):384–405, 2003.

6 B. Borovic, C. Hong, A.Q. Liu, L. Xie, and F.L. Lewis, editors. Control of a MEMS optical switch. In *IEEE Conference on Decision & Control*, 2004.

7 J.M. Harris, R. Lindquist, J.K. Rhee, and J.E. Webb. Liquid-crystal based optical switching. In Tarek S. El-Bawab, editor, *Optical Switching*, Springer US, 141–167, 2006.

8 Y. Sakurai, M. Kawasugi, Y. Hotta, S. Khan, and N. Uehara. LCOS-based 4×4 wavelength cross-connect switch for flexible channel management in ROADMs. In *Optical Fiber Communication Conference/National Fiber Optic Engineers Conference*, 2011.

9 M. Iwama, et al. LCOS-based flexible grid 1×40 wavelength selective switch using planar lightwave circuit as spot size converter. In *Optical Fiber Communications Conference & Exhibition IEEE*, 2015.

10 J. Lee, Y. Chung, and C.G. Oh. ASIC design of color control driver for LCOS (liquid crystal on silicon) micro display. *IEEE Transactions on Consumer Electronics*, 47(3):278–282.

11 W. Mi, L. Zong, M. Lei, M. Andres, Y. Ye, H. Zhao, et al. LCoS SLM study and its application in wavelength selective switch. *Photonics*, 4(2):22, 2017.

12 A. Kumar, A. Ashish, and A. Kumar. Optical fiber communication system performance using MZI switching. *International Journal of Soft Computing & Engineering*, 2(3):98–107, 2012.

13 F. Shokraneh and M.S. Nezami, and O. Liboiron-Ladouceur. Theoretical and experimental analysis of a 4×4 reconfigurable MZI-based linear optical processor. *Journal of Lightwave Technology*, 38(6):1258–1267, 2020.

14 N. Boudriga and M. Sliti. All optical switching control. In *16th International Conference on Transparent Optical Networks*, 2014.

15 M. Imran, P. Landais, M. Collier, and K. Katrinis, editors. Performance analysis of optical burst switching with fast optical switches for data center networks. *2015 17th International Conference on Transparent Optical Networks (ICTON)*, 2015.

16 R. Proietti, Z. Cao, C.J. Nitta, and Y.A. Li. A scalable, low-latency, high-throughput, optical interconnect architecture based on arrayed waveguide grating routers. *Journal of Lightwave Technology*, 33(4):911–920, 2015.

17 S. Cheung, T. Su, K. Okamoto, and S. Yoo. Ultra-compact silicon photonic 512×512 25 GHz arrayed waveguide grating router. *IEEE Journal of Selected Topics in Quantum Electronics*, 20(4): 310–316, 2013.

18 A. Wonfor, H. Wang, R. Penty, and I. White. Large port count high-speed optical switch fabric for use within datacenters. *Journal of Optical Communications Networking*, 3(8):32–39, 2011.

19 R. Stabile, A. Albores-Mejia, and K. Williams. Monolithic active-passive 16×16 optoelectronic switches. *Optics Letters*, 37(22):4666–4668, 2012.

20 H. Ballani, P. Costa, I. Haller, K. Jozwik, K. Shi, B. Thomsen, et al., editors. Bridging the last mile for optical switching in data centers. In *2018 Optical Fiber Communications Conference and Exposition (OFC)*, 11–15 March 2018.

21 Y.S. Patel, A. Page, M. Nagdev, A. Choubey, R. Misra, and S.K. Das. On demand clock synchronization for live VM migration in distributed cloud data centers. *Journal of Parallel and Distributed Computing*, 138:15–31, 2020.

22 D.A. Popescu, A.W. Moore, editors. PTPmesh: data center network latency measurements using PTP. In *2017 IEEE 25th International Symposium on Modeling, Analysis, and Simulation of Computer and Telecommunication Systems (MASCOTS)*, 20–22 September, 2017.

23 B. Guo, Y. Shang, Y. Zhang, W. Li, S. Yin, Y. Zhang, et al. Timeslot switching-based optical bypass in data center for intrarack elephant flow with an ultrafast DPDK-enabled timeslot allocator. *Journal of Lightwave Technology*, 37(10):2253–2260, 2019.

24 V. Shrivastav, K.S. Lee, H. Wang, and H. Weatherspoon. Globally synchronized time via datacenter networks. *IEEE/ACM Transactions on Networking*, 27(4):1401–1416, 2019.

25 J.J.P.C. Rodrigues and M.M. Freire, editors. Performance assessment of enhanced just-in-time protocol in OBS networks taking into account control packet processing and optical switch configuration times. *22nd International Conference on Advanced Information Networking and Applications - Workshops (AINA Workshops – 2008)*, 25–28 March 2008.

26 K. Clark, H. Ballani, P. Bayvel, D. Cletheroe, T. Gerard, I. Haller, K. Jozwik, K. Shi, B. Thomsen, P. Watts, H. Williams, G. Zervas, P. Costa, and Z. Liu. Sub-nanosecond clock and data recovery in an optically-switched data centre network. In *2018 European Conference on Optical Communication (ECOC)*, 2018.

27 M. Verbeke, P. Rombouts, H. Ramon, J. Verbist, J. Bauwelinck, X. Yin, et al. A 25 Gb/s All-digital clock and data recovery circuit for burst-mode applications in PONs. *Journal of Lightwave Technology*, 36(8):1503–1509, 2018.

28 H. Zheng, A. Shen, N. Cheng, N. Chand, F. Effenberger, X. Liu, editors. High-performance 50G-PON burst-mode upstream transmission at 25 Gb/s with DSP-assisted fast burst synchronization and recovery. In *2019 Asia Communications and Photonics Conference (ACP)*, 2–5 November, 2019.

29 R.S. Tucker, Ku Pei-Cheng, and C.J. Chang-Hasnain. Slow-light optical buffers: capabilities and fundamental limitations. *Journal of Lightwave Technology*, 23(12), 2005.

30 H. Liu, L. Yuan, P. Han, J. Huang, and D. Kong. Multicast contention resolution based on time-frequency joint scheduling in elastic optical switching networks. *Optics Communications*, 383: 441–445, 2017.

31 I.H. White, K.A. Williams, R.V. Penty, et al. Control architecture for high capacity multistage photonic switch circuits. *Journal of Optical Networking*, 6(2):180–188, 2007.

32 X. Xue, K. Prifti, F. Wang, F. Yan, N. Calabretta, editors. SDN-enabled reconfigurable optical data center networks based on nanoseconds WDM photonics integrated switches. In *2019 21st International Conference on Transparent Optical Networks (ICTON)*, 2019.

33 F.L. Yan, G. Guelbenzu, N. Calabretta, editors. A novel scalable and low latency hybrid data center network architecture based on flow controlled fast optical switches. In *Optical Fiber Communication Conference*, 2018.

34 M. Leenheer, C. Develder, J. Vermeir, J. Buysse, P. Demeester, editors. Performance analysis of a hybrid optical switch. In *International Conference on Optical Network Design & Modeling*, 2008.

35 X. Ye, Y. Yin, S. Yoo, P.V. Mejia, V. Akella, editors. DOS – A scalable optical switch for datacenters. In *ACM/IEEE Symposium on Architectures for Networking & Communications Systems*, 2010.

36 Y. Yin, R. Proietti, X. Ye, C.J. Nitta, V. Akella, and S. Yoo. LIONS: An AWGR-based low-latency optical switch for high-performance computing and data centers. *IEEE Journal of Selected Topics in Quantum Electronics*, 19(2):3600409, 2013.

37 X. Xue, K. Prifti, B. Pan, F. Yan, and N. Calabretta, editors. Fast dynamic control of optical data center networks based on nanoseconds WDM photonics integrated switches. In *2019 24th OptoElectronics and Communications Conference (OECC) and 2019 International Conference on Photonics in Switching and Computing (PSC)*, 2019.

38 R. Proietti, Y. Yin, R. Yu, C.J. Nitta, V. Akella, C. Mineo, S.J.B. Yoo. Scalable optical interconnect architecture using AWGR-based TONAK LION switch with limited number of wavelengths. *Journal of Lightwave Technology*, 31(24):4087–4097, 2013.

39 R. Proietti, Z. Cao, C.J. Nitta, Y. Li, and S.J.B. Yoo. A scalable, low-latency, high-throughput, optical interconnect architecture based on arrayed waveguide grating routers. *Journal of Lightwave Technology*, 33(4):911–920, 2015.

40 F. Yan, X. Xue, B. Pan, X. Guo, and N. Calabretta, editors. FOScube: a scalable data center network architecture based on multiple parallel networks and fast optical switches. In *2018 European Conference on Optical Communication (ECOC)*, 2018.

15

Reliability in Optical Networks

Antony Gratus Varuvel and Rajendra Prasath

Indian Institute of Information Technology (IIIT) Sri City, Chittoor, Andhra Pradesh, India

15.1 Introduction

Reliability, Availability, Maintainability, and System Safety (RAMS) are the driving factors in modern scenarios for sustenance and market share, wherein an equal amount of technological expertise is assumed to be available within competing concerns/firms. Notwithstanding this, RAMS in military, nuclear, medical, and aerospace sectors are always treated as equally important to functionality [1]. A design primarily relies on the technical expertise of the people involved. The inherent reliability and safety of the system are both largely determined during early design phase of a capital project. Though the RAMS requirements are not stated explicitly by the customer/end user, in most cases, the expectation is to incorporate built-in RAMS in the products by design. Involvement of RAMS at the early stage of the design would support the organization in achieving strategic goals, ensuring core competency, maintaining excellence in service and improving cost-effectiveness [2]. On the other hand, not having the RAMS concept in place in the early stages of design could lead to overloading in terms of the following factors: a) rejection by customer; b) requiring upgrades to meet RAMS demand; c) failing to comply with certification norms; d) loss of market share and competitiveness; e) higher modification cost; f) loss of trust and loyalty among customers; g) time overrun; h) cost overrun; i) inability to attain break-even point; and j) increased warranty and support cost.

Hence, it is very much essential to cater for RAMS, as part of design, in the early stages of design phase through development to production, and even during phase-out of the products. Extensive usage and involvement of RAMS as a part of the design process would have a positive effect, which would be much more than one can imagine and anticipate. With this perspective, details of RAMS in the area of optical system and networks are elaborated in the following sections. An extensive outline of the usage of optical networks for very high speed transfer of data is provided in [3]. This chapter aims to incorporate RAMS capabilities within the system, so as to ensure that the optical network/system is highly dependable.

15.2 RAMS in Optical Networks

Need for speed in the domain of data transfer with the highest possible accuracy and without any data dropout has necessitated deep research and development in the digital communication arena. While the improvement of existing physical layers with more secure and deterministic data is undertaken (e.g., Ethernet media), the requirement for transfer of extremely high density data mandated a new physical communication medium, which can transfer data at the speed of electromagnetic waves. With these demanding requirements, the exploration of the optical medium was undertaken and is being standardized. Many technological advancements have been achieved by researchers in this domain. Irrespective of the fact that the speed and bandwidth of optical communication system is higher when compared to other demonstrated data transfer protocols and media such as High Speed Deterministic Time-Sensitive Ethernet, the reliability of optical networks ensures compliance with user/usage requirements. Unlike other physical media, a single strand of optical fiber can cater for many numbers of connected terminals/users. There are many technologies in the optical domain, from the perspective of physical components and communication protocols. This chapter is dedicated to analysing all such technologies of optical networks through the RAMS lens. More attention is paid to Reliability, which is the critical determinant for other disciplines. This means that realisation of a product with higher achievable reliability would bear significant advantages in other Availability, Maintainability and Safety aspects. A trade-off study between energy efficiency and reliability for optical networks has been carried out in [4]. Detailed emphasis on RAMS attributes are given in this chapter.

15.3 Objectives

As the lack of adherence to, and ignorance of, the tasks which are necessarily required to be carried out during the different stages of a product life cycle with RAMS capabilities/features will result in a reduction in the usefulness of the products, it is essential to identify clearly the scope of any RAMS project, in line with the other functional requirements. This chapter focuses on the following objectives:

a) to list basic RAMS tasks applicable for optical networks
b) to guide designers in realizing products with RAMS concepts, by means of design feedback from RAMS perspective
c) to identify applicable RAMS analyses during various phases/stages of product life cycle of optical networks and components, and
d) to comply with RAMS requirements.

15.4 Life Cycle of a Product/Project

Most commercially available products are designed for a short life span, whereas in the case of optical networks and components, the entire life cycle of the product is treated with the utmost importance owing to its critical functionality and the impacts resulting from downtime. Phasing out of products is also a prime concern, for products whose constituents are hazardous and/or not environmental friendly in nature. RAMS aspects are applicable and to be considered in every phase of the product or project life cycle.

15.5   Preamble to RAMS

Building a product with the RAMS concept is found to be a challenge in comparison to building a product with functionality and features. Hence, this chapter does not address functionality features and their incorporation as part of the product life cycle. Instead, detailed study and analysis of the RAMS concepts during the life cycle of the product is given importance and supplemented. Treatment of RAMS problems at a later stage of a project calls for exorbitant resources. The curve shown in Figure 15.2 depicts the cost associated with any minor change in RAMS program at a later stage of a project. Typical cost to be incurred to achieve high reliability is discussed in [5]. A flow-chart showing the interrelations among the RAMS and the realization process is given in Figure 15.1.

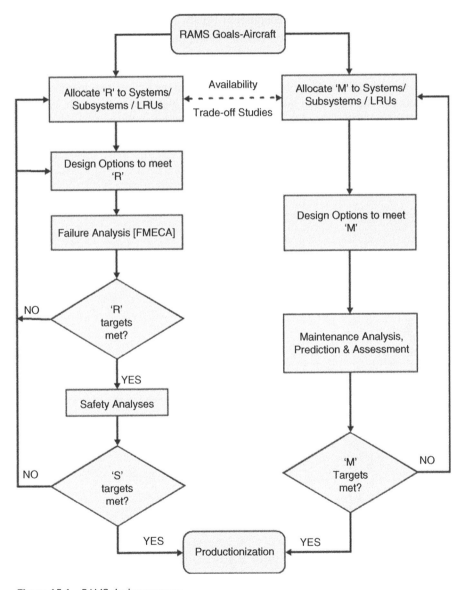

Figure 15.1   RAMS design process.

Where the objective is to improve the cost-effectiveness of any program which is constrained by cost and timeline, the adaptation of the RAMS concept will find the way out. The positive advantages of carrying out RAMS activities in parallel with design activities are: a) uncovering design deficiencies; b) improving cost-effectiveness; c) increasing system availability; d) achieving optimal design; e) reducing system down time; f) meeting or exceeding user requirements; and g) complying with certification norms. The subsections of this section deal with the attributes of RAMS and their importance.

15.5.1 Reliability

By definition, reliability is the probability that an item will perform its intended function, under the stated environmental conditions, for a specified period of time. One has to consider all the variables in the above definition, such as environmental conditions (usually, a combination of variables) and duration. Achieving the reliability goals requires strategic vision, proper planning, sufficient organizational resource allocation and the integration of reliability practices into development projects.

Figure 15.2 shows the relationship between Reliability and the associated Cost at various phases of a product timeline, extending from T_0 to product maturity. From the figure, it is clearly evident that the difference between the additional budget is exorbitantly higher in comparison with the smaller increase in reliability, as the cost of improvement in terms of reliability increases exponentially over time. Also, it could be noted that the reliability could not be improved to 1, which is

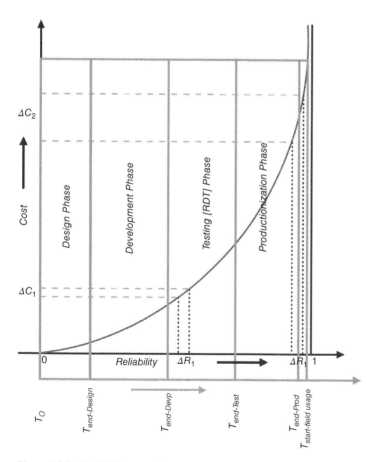

Figure 15.2 Reliability vs cost.

idealistic in nature. This is pictorially represented in Figure 15.2. The total cost of improvement in reliability is higher and higher as the present system reliability increases from 0. From the curve it is clearly evident that the incremental change in reliability will be met with an additional budget of C_1 initially and C_2 at a later stage in the program. Hence the reliability improvement could be initiated, studied and analyzed only by considering the following critical factors: a) present system reliability; b) economic zone of operation; c) minimum essential reliability required; d) maximum desirable reliability feasible; and e) optimum achievable reliability.

15.5.2 Availability

Availability is often signifies the state of the system "On Demand". In a simpler form, availability will always be higher if the reliability of the component is higher. Availability is directly proportional and linearly related to reliability if the failure distribution is exponential. But reliability is not the only factor which will ensure the availability of the system. Maintainability is the other important factor which is to be considered and accounted for when predicting/estimating/assessing system availability. Availability could be pulled downwards to "0" if maintainability is not ensured, resulting in very high Mean Down Time (MDT). Availability is always greater than or equal to reliability. For items which show more failures, availability can be improved by reducing the repair/replacement time.

15.5.3 Maintainability

The third parameter of RAMS is Maintainability. Maintainability and Reliability are the basic design drivers in addition to the functionality. All the three parameters (Functionality, Reliability & Maintainability) are considered to be in parallel. Ignoring Maintenance factors in the initial phases of design will result in time and cost overrun at a later stage, which would not be acceptable to customers beyond a certain threshold. As a rule of thumb, a lesser MTTR could be allocated to those components whose failure rates are very much higher and vice versa, implying that components with a higher failure rate (lesser MTBF, in general) are most likely to fail frequently statistically, and hence those components are to be repaired/replaced within no time/shorter time to maintain availability of the system. For an optical system, maintenance would pose a major challenge and hence enough design features are to be builtin to identify, locate, confine and report the fault, preferably well in advance of the onset of failure.

15.5.4 System Safety

From the perspective of reliability analyses, proper functioning of any system should be ensured. However, improper/inadvertent functioning of the components should also be eliminated/mitigated to ensure the effectiveness of the system. Safety analysis/assessment plays a vital role in this context. To ensure safe operation, all the unsafe conditions are to be mitigated by implementing various safety features such as serial or parallel redundancy of components, fault tolerance techniques, safety interlocks etc. As a result, reliability will be higher if the safety is higher, whereas the reverse case is not true. In any safety-critical systems/applications, where the risk and causalities involved are higher, it is often the case that a risk-based decision would be considered for the implementation of the proposed design.

Risk treatment is an important aspect as regards the acceptability of the system design, which is safety-critical in nature. Detailed exploration in terms of severity (consequences) and estimation/prediction of likelihood of occurrence is necessary to quantify the risk. With the increasing trend towards use of optical networks, even for safety-critical applications, aspects related to risk should not be ignored.

15.6   Significance of Reliability in Optical Interconnect Systems

Faults and failures downgrade reliability and hence the dependability measure of optical switches and networks [6]. There are many physical components utilized in an optical network, depending on the technology adopted. Irrespective of the technology adopted, the basic circuitry at the source and destination of an optical network is assumed to be same for all the types of optical network.

Physical or optical damage to optical devices and components is very important. Due to the very nature of the sensitivity of light to the defects, repair and/or replacement of devices is a costly and time-consuming depending on the location of the fault. Any mechanical break in the circuitry is generally easily detected using supervisory techniques (automated) and repairs can be initiated quickly, if the fault is localized. When extensive damage to a link has occurred in long lengths of optical fibre, this will be unfit for carrying traffic and the repairs could be almost impossible. Under these conditions, replacement is often undertaken. This would result in major downtime of the system, and severe economic impact, based on the application.

Propagation delay, power dissipation, and crosstalk noise are affected due to less reliable interconnects as discussed in [7]. The reliability of optical fiber and waveguide sensors becomes increasingly important as they are more frequently used in applications where a failure of the (often inaccessible) sensor might have greater consequences on cost and/or safety. The reliability of the system is directly related to the operational effectiveness of the system. System/cost-effectiveness is a measure of the ability of an item to meet service requirements of defined quantitative characteristics.

Optical networks are usually chosen for applications where high bandwidth and fast communication of data are required. The fields adopting to optical switches and system will usually demand high reliability. The initial cost of optical switches and system will be comparably higher than the conventional method of communication and switching. Due to relatively more costly investment, it is expected that the operational effectiveness and failures will be greatly enhanced compared to conventional switches and transmission media. As regards the bandwidth and users/systems connected, the optical network and switches are expected to be more reliable.

In summary, reliability is highly significant in optical networks and switches due to the following aspects:

- An optical network carries a voluminous amount of data, so the loss of a single communication medium would result in total loss of data.
- Number of users/systems connected is much higher.
- Interconnection of optical system is highly complex and needs to be highly accurate. Inaccuracies affect the throughput.
- A minor fault/deviation in physical and/or optical components will result in total loss of a number of connected users/systems.
- Latent or patent defects of optical system will lead to a reduction in performance.
- If critical applications are employed using an optical interconnect system, failure of it would result in catastrophic or critical events based on the applications.
- Components of the optical system should be highly reliable, as their repair and/or replacement would be cumbersome and would induce unwarranted downtime.

15.7   Typical Components of Optical Circuitry

Reliability could be defined as the successful operation of all functional components. The type and number of components required to establish the systemic transformation will determine the reliability of the functional block. Hence, in order to establish the reliability of a given circuitry of

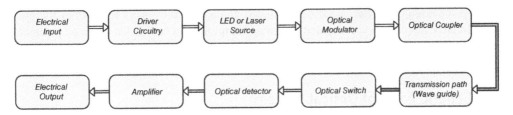

Figure 15.3 Typical optical block diagram.

functional block, the first step is to list all the components associated with it. The significance of the reliability of individual components varies based on the configuration within which the component is embodied. Lesser reliability in a serial connection causes greater deterioration in the system reliability than if the same part is part of a parallel connection. A typical list of components of an optical system is shown in Figure 15.3 (representative).

As most of the raw data are occurring in analog domain, an electrical signal is expected to be transmitted through the media. Electrical signals that are to be transmitted to some destination through an optical interconnect system must be converted to the optical domain for transmission. A brief description of each of the functional components is given below, for ready reference.

Electrical input is the raw data which is either sensorized or obtained through other physical behavioural changes. It is converted into an electrical signal. This is the signal or data which is required to be transferred through the media.

Driver circuitry processes the electrical signal and coverts it into bit stream with suitable encoding techniques. Data are ready to get transmitted if electrical, or undergo optical modulation, if optical.

Optical modulator converts the (often digitized and encoded) electrical signal into optical signal according to bit sequence in electrical signal. After the optical signal has been generated, it is fed into the optical path of transmission. Multiplexing techniques, typically WDM, are employed before feeding the signal to the optical path.

Optical couplers are structures that are used to inject the light into the optical system. The wavelength of the optical signal is changed in order to enable the receiver to selectively respond to the transmitted signal and hence receive only the intended signal.

Optical interconnects use guided wave and free space for signal transmission. Guided wave optics involves the use of waveguides to contain the optical signals within a board or package, or on a chip which consists of materials with a high index of refraction surrounded by a material with a lower refractive index. Free-space optics utilize diffractive optics and conventional lenses or microlens arrays to guide single or multiple parallel optical beams in free space.

Optical switches are devices that can selectively switch light signals that run through optical fibers or integrated optical circuits from one circuit to another. They are used in optical routing networks to route the light travelling in waveguides to a different location. An optical switch is simply a switch which accepts a photonic signal at one of its ports and sends it out through another port based on the routing decision made. An optical switch has one or more input ports and two or more output ports that we usually call a $1 \times N$ or $N \times N$ optical switch.

The **receiver** side of the optical interconnect system is responsible for reconstruction of the electrical signal. Suitable decoding is required to be carried out at the detector.

An **optical detector** is the device for detecting the light pulses and converting them to photo current. An amplifier is finally used for amplifying the photo current and providing the digital signal in the form of conventional voltage/electrical output signal.

Based on the strength of the signal demodulated and as per the requirement from the end receiver circuitry, amplification is decided. A suitable amplifier can be decided based on parametric requirements.

15.8   Generic Types of Optical System

To ensure compliance of Optical System with RAMS requirements, it is very much essential to fully understand and comprehend the technologies of the components, possible failure modes, and expected failure rate. Those parameters are typically influenced by the usage profile and environment. Based on the technology, optical the path is often changed based on requirements and capabilities. With the improvements and advancements in technology, the optical path/system is often offered in a single package encompassing optical modulator and switch. While this option is highly miniaturized, the failure modes and effects associated in any of the tiny components fabricated cannot be ruled out. Hence it is highly essential to identify the internal building blocks of the device(s) and the possible failure modes which can creep into it.

In general, there are two kinds of optical switches: **O-E-O** (optical-electrical-optical) switch and **O-O-O** (optical-optical-optical) switch, also known as all-optical switch. An OEO switch requires the analog light signal first to be converted to a digital form, then to be processed and routed before being converted back to an analog light signal. An all-optical fiber-optic switching device maintains the signal as light from input to output. Traditional switches that connect optical fiber lines are electro-optic.

Functional and component block diagram within each of the types of the switches are to be critically examined and possible failure modes are to considered well in advance. The failure rate and hence the reliability of each of these types of switches varies dramatically.

15.8.1   Factors Influencing Reliability in Optical Networks

Although the optical signal is immune to electromagnetic interference such as electrical noise and lightning, the physical disturbance to any of the constituent parts of the optical switch and hence the system degrades the reliability significantly. The factors that can generally affect Reliability in an optical network are *operating temperature, humidity, chemical environment, radiation loads, mechanical stress, vibration levels,* and *components operational conditions: optical power, current, voltage, frequency, etc.*

15.8.2   Initial Insight of Failures

Optical system components are to be precision-controlled so as to ensure that the optical parameters are within acceptable limits. The functional components shall ensure that the expected optical wavelength is unaffected either by way of attenuation, distortion, or loss.

For the purposes of carrying out RAMS study, complete working mechanism of the optical components within the system shall be understood along with significant details of interfaces and interconnects. Following details are required further to carry out the reliability tasks.

- Define functional blocks.
- Determine interrelationship among the connected components.

- Identify criticality of each of the components in achieving a given functionality.
- Trace effects.
- Assign severity.
- Trace for detection mechanism.
- Locate mitigating provisions.

Effective RAMS studies/analyses/feedback will only be possible with mutual understanding of the processes, failures, and failure modes by both the RAMS Team and the Design Team. Analyses carried out without involvement of either of the teams will not yield fruitful results in terms of optimization and compliance. Following section is aimed at ensuring compliance of design to RAMS requirements.

15.9   Ensuring RAMS for the Optical System

During various phases of the product life cycle, in this case the optical system, there are number of Reliability, Availability, Maintainability, and System Safety analyses to be carried out which will add value to the development of the product. Activities are categorized under the acronym **R A M S**. The applicability of each and every activity differs, depending on the phase of the life cycle. The list given below is elaborate but not exhaustive.

15.9.1   Reliability – An Essential Insight

Reliability is conventionally defined as the probability of achieving intended functions, for a duration of time, within the operating environmental conditions. Functional capability along with required reliability requirements are to be captured and "designed in" at the conceptual stages of the project/product development, failing which an enormous amount of the life cycle cost will be spent on maintenance and availability enhancements. Probabilistically, it is the minimum amount of time during which there are statistically no failures. Statistically MTBF is used to define life of those components, which is defined as the time by which approximately 33% of the population of components would survive.

Due to ignorance of the failure process and insight into the failure phenomena resulting from the following factors, reliability is treated using both probability and random processes:

- physical
- chemical
- mechanical
- electro mechanical.

In general, failure events (malfunction/broken/open/short/change in value. . .) are treated as a probability and occurrences (time/cycle/km. . . to failure) are treated as a random variable.

Random variables are assumed to take values in accordance with some probability distributions. There are two major types of probability distributions. Commonly used probability distributions with which to depict the occurrence of failures are listed below. Details of the each of the distributions can be found in any statistical handbook.

- Discrete distributions: *binomial* and *Poisson*.
- Continuous distributions: *exponential*, *normal*, *lognormal*, *Weibull*, *beta*, and *gamma*.

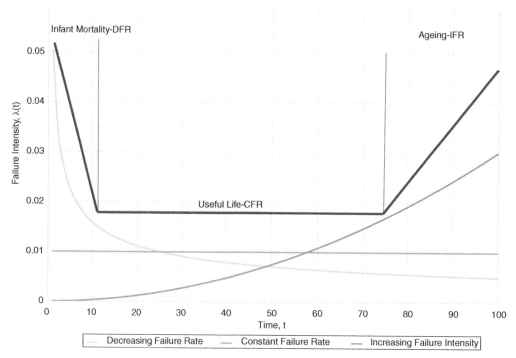

Figure 15.4 Bath tub curve (failure Intensity vs time).

Those distributions are representative of failure phenomena over time. Components may exhibit any/many of the following trend(s) in failure rate:

- Increasing Failure Rate (IFR)
- Decreasing Failure Rate (DFR)
- Constant Failure Rate (CFR).

A typical bathtub curve (Figure 15.4) represents various stages of the component and applicable failure rates.

Typically three types of configurations are possible for optical interconnect systems or optical networks, namely, a) series; b) parallel; and c) complex

15.9.1.1 Typical Reliability Configurations

Series: An arrangement of functional components in series, which will fulfil the end objective(s), is referred to as a reliability configuration or Reliability Block Diagram. Any or many of the components of the serial link would result in non-availability of the intended function. Greater the number of physically connected or logically and serially interconnected components, lesser would be the reliability. Mathematically, the reliability of "n" serially connected components is

$$R_s\left(t\right) = \prod_{i=1}^{N} R_i \tag{15.1}$$

Reliability is higher if the number of serially connected components is as low as possible. Adding functionally non-essential components as part of the serial link would adversely affect the reliability of the system. **Parallel:** Multiple components are added in parallel in order to ensure that the required functionality is achieved even under the conditions of failure of any/ many of the success paths. A success path is a list of serially connected non-failed components which ensures that the end functionality of the system is achieved. There could be many such success paths in a parallel system redundancy. The optimal number of such success paths is decided by the user requirements, with criticality of end system functionality. It should be very clearly understood that multiple parallel paths would require additional resources such as volume, power, weight, space, cost, maintenance efforts, spare parts etc. It is proven and well established that reliability enhancement by virtue of multiple parallel paths will be linear initially and tapers to an exponential pattern.

$$R_p(t) = 1 - \prod_{i=1}^{n} (1 - R_i) \tag{15.2}$$

Complex: There are often requirements wherein simple series and/or parallel reliability configuration would result in either under-design or over-design. Complex configurations, which are a mix of interconnections, will utilize lesser resources and are optimized for given target requirements. The inherent reliability of each of the components also plays a major role in determining the requirements of additional redundant components. For the purposes of optical networks, the level of redundancy is determined by the criticality of applications in which the system is employed. Where

$R_p(t)$: Reliability of parallel configuration

$R_s(t)$: Reliability of serial configuration

R_i: Reliability of component i

n: No. of components, given by i

DfR: Design for Reliability

The design and configuration for reliability is obtained as given in Figure 15.5.

15.9.1.2   Reliability Metrics

Various metrics of reliability are to be considered while drafting the reliability requirements for an optical network. Reliability requirements are often ignored at the initial stages of conceptualization. Design corrections during the initial stages would be cost-effective and easily achievable, rather at the later stage of the project or product development. In the case of non-specification of the reliability requirements by the customers, it is often considered to be advisable to conceive reliability requirements which are either better than those of competitors or captured using Quality Function Deployment (QFD) methods.

The following are the parameters which are significant to specify:

- Reliability
- MTBF
- Confidence Level.

Reliability:

Reliability of the optical system shall be explicitly specified by specifying the all the relevant factors. Typically, it is better explained and specified if "R" at "t" hours with confidence level "CL" are

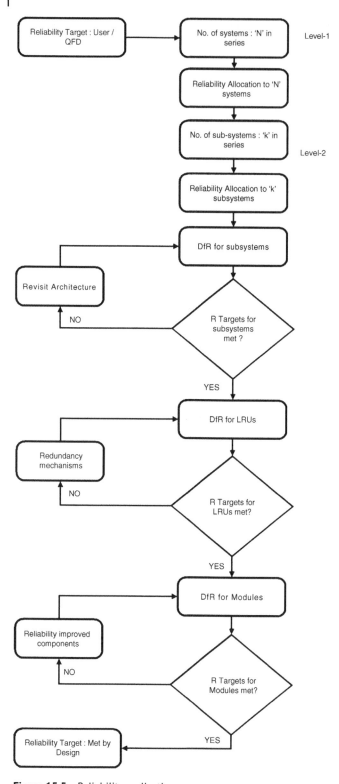

Figure 15.5 Reliability realization process.

given. Missing of any of the parameters would consequently lead to misinterpretations and justifications. Clear definition of reliability would ease demonstration of the parameters without ambiguity and comply with contractual obligations.

MTBF:

Although reliability is a useful measure for defining the target, it is more customary to specify the Mean Time Between Failure (MTBF) for a repairable component and Mean Time to Failure (MTTF) in the case of non-repairable components. Another component called Mean Time Between Critical Failures (MTBCF) is also defined, in cases where minor/acceptable failures are tolerated within the performance.

FFOP:

Failure-free operating period is the continuous period of operation in which a major unacceptable system failure is eliminated. However, there shall be minor degradations/deviations, which are still within the acceptable limits of operations. Faults are permitted, but failures are not allowed to occur during this FFOP. The major difference between a failure rate based approach and the FFOP is that the conventional scenario accepts no faults, whereas the latter can tolerate errors/faults. But there are no failures in both cases.

MFOP:

Instead of using the more common terms such as MTBM/MTTF/MTBCF, another way of defining the mean time with "acceptable level of degradation" is the Maintenance Free Operating Period (MFOP). There is an increasing trend towards using this key term, especially in the aerospace industry, where fault-tolerant systems are embodied. If MFOP is specified as one of the reliability metrics, it is to be noted that minimal maintenance can be performed during the MFOP, called the Maintenance Recovery Period (MRP). During the MRP only essential checks and inspections are carried out, such as CLAIR (Clean, Lubricate, Analyse, Inspect & Repair). If detailed checks or repair are required, those checks shall be performed during detailed investigations. The period of duration of extensive checks shall be outside the purview of MFOP (Figure 15.6).

Repair or replacement actions are usually performed when the diagnosis points out that the decision on repair versus replacement from the economical point of view. For a capital item, the

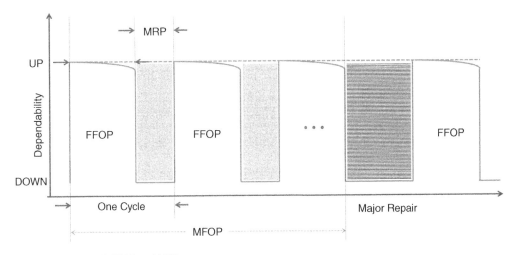

Figure 15.6 MFOP, FFOP, and MRP.

repair scheme is often chosen as the optimal solution, rather than replacement. MTTR is the terminology often used to specify how quickly a failed component can be restored back into operational state, when maintenance actions are performed in accordance with the prescribed procedures and processes. There are very well established and documented maintenance schema available which reduce the downtime of the system.

Confidence Level & Limits: Confidence Level is a useful measure to define the expected range of reliability parameters. It is a statistical measure which defines the confidence of the analyst on the statistic. Higher the Confidence Level, wider is the acceptable region and hence the more likely that the statistic will deviate from the true value. A lower Confidence Level ensures that the value estimated is much closer to the true value. Confidence Level is bound by two limits on either side of the statistical distribution, called the confidence limits. There are three types of confidence limits specified in statistical terms:

- upper single-sided confidence limit
- lower single-sided confidence limit
- two-sided confidence limit.

While any of the reliability parameters are estimated from the field trials, Confidence Level aims to locate the true statistic. From the data, any of the limits mentioned above shall be utilized in approximating the estimated value to the true value of the statistic/parameter. Any estimated parameter will be suspected for its correctness and accuracy, due to the variability involved in data collection, sample selection, model selection, appropriateness of model, assumptions made, analysis parameters chosen, and interpretation of results. It is quite impossible to nullify all the variability, of which some are subjective and some are objective. In statistics, a confidence interval is a type of interval estimate of a population parameter and is used to indicate the reliability (probability of probability) of an estimate, whereas in the case of Reliability and MTBF, the confidence bounds are used to provide the possible acceptable variations in the estimated parameter. The variations could either be single-sided (tailed) or two-sided (tailed). Figures 15.7 & 15.8 refer to these respectively, where x refers to the LCL-Lower Confidence Limit and y refers to the UCL-Upper Confidence Limit. To be more precise, confidence bounds provide information about the outcome of a trial, if an experiment is repeated again and again. How frequently the observed interval contains the population parameter is determined by the confidence (significance) level or confidence coefficient. If confidence intervals are constructed across many separate data analyses of repeated and possibly different experiments, the proportion of such intervals that contain the true value of the parameter will match the Confidence Level. The significance level of the confidence interval would indicate the probability that the confidence range captures this true population parameter given a distribution of samples.

Certain factors may affect the confidence interval, including size of sample, level of confidence, and population variability. A larger sample size normally will lead to a better estimate of the population parameter. In order to ensure that reliability is imbibed in the optical system during design, following activities are to be carried out.

- Reliability Apportionment
- Hardware Reliability Prediction
- Software Reliability Prediction
- Derating Analysis
- Stress-Strength interference Analysis
- Failure Mode Effects and Criticality Analysis
- Failure Mode Effects Test Cases

Figure 15.7 One-sided confidence bounds.

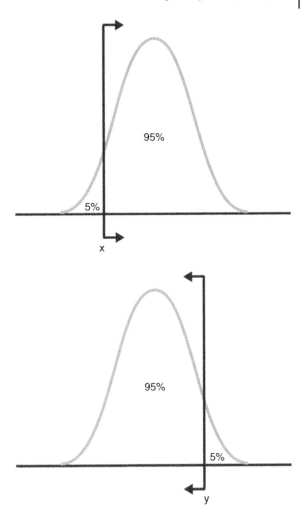

Figure 15.8 Two-sided confidence bounds.

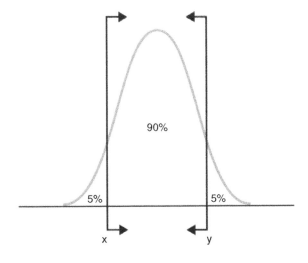

- Reliability Estimation
- Reliability Assessment
- Human Error Analysis
- Reliability Growth Analysis
- Software Reliability Assessment
- Design-Cost Trade-off Studies.

Brief descriptions of the activities listed above are presented in the next subsections.

15.9.1.3 Reliability Apportionment

The starting point of the reliability programme, after capturing the user requirements, is the Reliability Apportionment. This exercise would result in assigning of second level reliability. This level could further be assigned to third level and so on, until the desired indenture level is reached for realization. There are methods available to apportion the top-level reliability to the next-level reliability. In general, in the aircraft industries, the aircraft-level reliability requirement for the stated duration is further assigned/apportioned to the next system level reliability. The design objective of each system should also consider the reliability goal specified in addition to functional requirements to be realized. Reliability Apportionment should be followed by Reliability Prediction during the preliminary design stage of system/subsystems/items, to ensure that the chosen design would satisfy the reliability requirements apportioned. To achieve DfR (Design for Reliability), following apportionment methodology shall be adopted. Following flow chart represents a typical apportionment exercise which is to be carried out initially. Weibull distribution can be effectively utilized to represent IFR, DFR, & CFR. Refer to Figure 15.9. For detailed allocation methodology [8] could be referred.

15.9.1.4 Hardware Reliability Prediction

Very often the Reliability Prediction and estimation are misunderstood, if not erroneously interpreted, and, on most occasions, interchangeably used. Reliability Prediction is based on the "Handbook data" approach on most occasions, where the vendor-tested data are not available, or the specific part of the design is not finalized. The failure rate data/MTBF required for the Reliability Prediction would be called from any of the following sources of data. The order of priority and the correctness to the real value, with reduced number of assumptions is as follows: *Field Tested*, *Vendor Tested*, *Handbook*.

It is impractical for any procuring agency to field test each and every component of the design, owing to the program priorities and resources required. Hence, it is desirable to use the vendor-provided data, if available. Not meeting this requirement, the data for reliability prediction could be tailored from handbooks. MIL-HDBK-756B draws guidelines for carrying out reliability prediction. Following are typical methods used for the purpose:

- Parts Count
- Parts Stress
- Similar Item
- Similar Circuit
- Active Element Group.

Early prediction of reliability of a proposed design would be helpful in

- evaluating the design proposed in terms of R&M
- designing cost trade-off studies

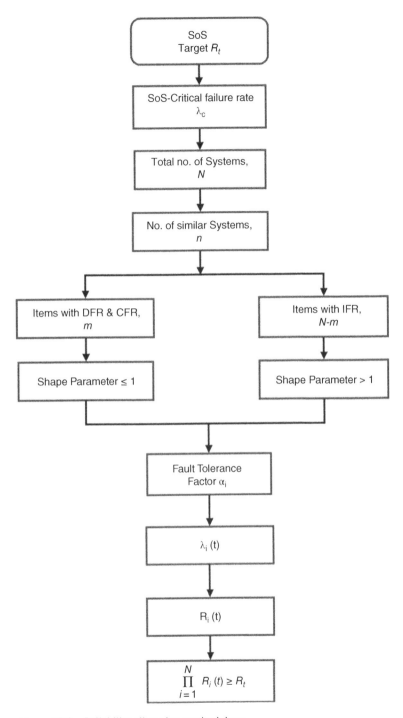

Figure 15.9 Reliability allocation methodology.

- meeting customer requirements/apportioned requirements of reliability
- comparing various design options in terms of reliability
- feasibility studies.

15.9.1.5 Software Reliability Prediction

There are well established tools and techniques available to predict, estimate, assess, and verify hardware reliability. However, the same is not applicable in the case of the software domain, for many reasons. Moreover, the definition of Software Reliability is the probability of failure-free software operation for a specified period of time in a specified environment. Although Software Reliability is defined as a probabilistic function, and with the notion of time, it is different from traditional Hardware Reliability. Software Reliability is not a direct function of time.

As most of the present systems are being run in conjunction with software, Software Reliability is also an important factor affecting system reliability. It is very difficult to identify a system, which may be simple or complex, without being controlled by software. However, software reliability differs from hardware reliability in that it reflects design perfection, rather than manufacturing perfection. The high complexity of software is the major contributing factor to Software Reliability problems. There are many models to choose from for software reliability prediction. Careful selection of an appropriate model that can best suit the case is essential. However, measurement in software is still in its infancy, meaning that the models have excessive assumptions and limitations.

Software Reliability is an important attribute of software quality, together with functionality, usability, performance, serviceability, capability, installability, maintainability, and documentation. It is difficult to achieve higher Software Reliability, because the complexity of the software tends to be high. Developers tend to include more and more complexity into the software layers, with the rapid growth of system size and functionality requirement, by upgrading the software every now and then. Although complexities associated with optical system is expected to be much less when compared to many other complex applications, the role of software failures cannot be ignored. There are many SRGMs (Software Reliability Growth Models) published in the literature, which are applicable for a given set of assumptions and boundary conditions and shall be appropriately selected and utilized for quantifying software reliability. A survey of software reliability growth models could be found in [9].

15.9.1.6 Derating Analysis

Derating is defined as the mechanism of operating the component/module/assembly below its rated or design limit, in order to have a wider tolerance limit, in the case of excursions, and to increase the life of the item. Main objective of derating is to reduce the stress levels applied on components, herein this case those components of optical circuitry. Derating analysis is carried out during the design stage of the item and accordingly the reliability of the items would improve. The stresses such as voltage, temperature, current, power, duty cycle, load, frequency of operation etc. could be derated. Derating is one of the methods to improve reliability.

$$Derating = 1 - \frac{\text{Actual parametric value}}{\text{Rated parametric value}} \tag{15.3}$$

From the perspective of electrical components, detailed methodology has been drafted in [10].

15.9.1.7 Stress-Strength Interference Analysis

For the case of mechanical components, where the failure phenomenon is influenced by mechanical and thermal load/stress applied on it and the inherent characteristics of the material, reliability is determined by the stress vs strength interference. The underlying fact for the same is attributed to failure mechanisms of mechanical components such as fatigue, leakage, wear, thermal shock, creep, impact, corrosion, erosion, lubrication, elastic deformation, radiation damage, de-lamination, buckling etc., which depend upon the characteristics of the components chosen. These parameters could be described by probability distributions. When the strength of a

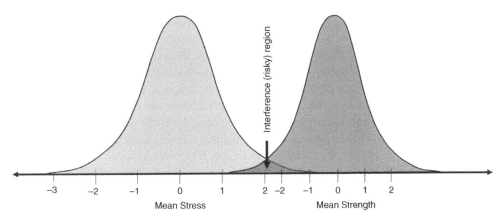

Figure 15.10 Stress strength interference.

material/component is less than the applied stress, the material fails. If the mean values of stress and strength are wider apart and variations about the means are less, then the probability of failure could be reduced to a minimum. A plot using Standardized Normal for both stress and strength indicating the region of interference is given in Figure 15.10.

15.9.1.8 Reliability Estimation

Estimating the reliability of an item later during the design phase provides more insight than Reliability Prediction carried out during the conceptual or PDR stage. The former assumes logical and functional modeling, whereas the latter ignores all the features of the design. This means that Reliability prediction provides the approximate (too conservative to consider) value of the Reliability or Failure rate of a system, by assuming the "All Series-Basic Reliability Modeling" concept, ignoring the functional importance of each and every item involved in the design. With the Reliability estimation, the MTBF of a system/item can be discovered, leading to a comparison with the apportioned values of Reliability metrics. The data for Reliability Estimation could be "field tested" or "vendor specified" or "handbook tailored", based on the appropriateness and availability of data.

In the process of reliability modeling, the logical and functional relationship with the system failure are given due importance. Understanding system functionality is a prerequisite for the RBD modeling. Complexity and inter-relationship among the components are indicated and modeled carefully for a meaningful RBD-based estimation. This would be a better and more accurate approach for estimating system reliability, compared to Reliability Prediction.

15.9.1.9 Failure Mode Effects and Criticality Analysis

FMECA is a design validation tool which, when carried out in the initial stages of design, will provide valuable information to the designers as regards identifying the potential critical failure modes and their effects on the overall system functionality and the mitigation methods possible. The potential failure modes could be identified for each component or function. Accordingly the FMECA could have functional failure modes or Engineering failure modes as its starting point. There are two major types of FMECA:

- Design FMECA
  - Functional FMECA
  - Engineering FMECA
- Process FMECA

The outcome of FMEA could be expressed in terms of Risk Priority Number (RPN), which will rank the components based on the Severity, Criticality, and Detection Provisions. Those components with higher RPN, which is above the limit of acceptance from the risk point of view, could be considered for improvement/change in design, fault monitoring, fault tolerance, or redundancy.

If carried out by the team, which consists of members from design, RAMS, and manufacturing, FMEA will help in developing a highly reliable product. This exercise will ensure that the potential failure modes will not appear in the field, and, if they do appear, there will not be any major safety/mission consequences, as a result of detection and mitigation features having been incorporated. This analysis is a proactive analysis for validating the design. If carried out systematically, FMECA provides valuable information such as potential failures, root causes of identified failures, critical components, safety/Reliability of critical components, detection provisions, mitigation plans, and (non-)acceptability of the failure modes in the design.

Criticality Analysis is a part of FMECA. This analysis is carried out to find out potentially critical components with respect to the loss of functionality. Severity and Probability of failure are the two governing factors for criticality analysis. The criticality number could be obtained from the CA, *Item Criticality Number* and *Mode Criticality Number*. These numbers are to be plotted on a 2D chart with severity and probability of failure on the X & Y axes, respectively (see Figure 15.11). Those items/modes which is/are posing danger to the system functionality with more severity (classification-I:catastrophic) and with more probability of occurrence (> 0.7) to be addressed with due concern.

15.9.1.10 Failure Mode Effects Test Cases

FMET is a method which reveals inherent design weaknesses first predicted by the FMECA process. By exposing a design to a single or combined set of environmental/input conditions, a distribution of multiple failure modes could be obtained.

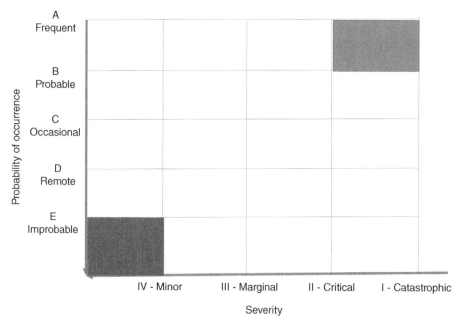

Figure 15.11 Criticality matrix.

FMET could be useful to:

- identify real failure modes and their root causes
- increase product quality by exposing failure modes that may not be detected in traditional testing
- reduce the development time by identifying failure modes at design verification
- reduce testing time and costs
- reduce number of required prototypes for testing
- increase reliability growth
- reduce warranty claims
- improve design maturity.

15.9.1.11 Reliability Assessment/Demonstration

Assessment is a technique by which identification, quantification, and ranking could be carried out on any system under study. From the perspective of RAMS, assessment indicates the establishment of realistic/achieved objectives with actual data pertaining to it. A quantitative assessment of reliability usually employs mathematical modeling, applicable results of tests conducted, failure data, estimated reliability values, and non-statistical engineering estimates. Testing is a prerequisite for any reliability assessment. The assessment could be carried out for hardware or software or both. Accordingly, the failure data need to be segregated and categorized.

Reliability assessment could be carried out in a sequential manner split into phases according to programme priority and attained system maturity. In every phase, achieved reliability would be established and system design improvements may be initiated to eliminate the failures observed, if applicable, and/or improvements/features could be added at the subsequent stages.

15.9.1.12 Human Error Analysis

It is a known fact that to err is human. It has also been proved in many fatal accidents that the slip and/or mistake in human action is the cause in most cases. Owing to the very nature of the severity of aircraft accidents, it is very much essential to assess probability of human error and thus human reliability. The assumption of human reliability as '1' is to be ignored on the above grounds. There are methods established to assess the human reliability for nuclear applications. The approach for safety-critical applications could be established. The methodology in assessing human error probability for flight control applications is to be devised and assessed. This activity may be carried out if prerequisites are met and available.

There are many databases compiled to find out the approximate value of human reliability (HEP – Human Error Probability). In simpler terms, it is the ratio of the number of errors to total opportunities for making the error.

$$HEP = \frac{\text{No. of errors}}{\text{Total number of opportunities}} \tag{15.4}$$

Human error probabilities are being studied in most domains, due to the effects of the outcome. Sun et al. [11] discusses quantification of HEP in railway dispatching tasks. Usually, HEP is given significance if the outcome of the slip/mistake results in loss of life or irrecoverable damages or huge economic damages [12].

15.9.1.13 Reliability Growth Analysis

Reliability growth techniques enable management to plan, evaluate, and control the reliability of a system during its development stage. Any product developed with the reliability goal set forth would possibly not meet the goal at the first instance of development due to the introduction of errors, deficiencies in design & development, and manufacturing flaws [13]. It is a common practice to produce engineering items/prototypes to demonstrate the design and correct the design/manufacturing errors/flaws to meet the objectives in terms of functionality and reliability. With the above assumptions, reliability achieved during the initial phase of product development would certainly be lesser than the desired goal. Hence, further corrections/modifications would be introduced to enhance/improve the reliability or to achieve the set goal.

If the areas of improvements are identified clearly and modifications are implemented correctly, then there could be a possible improvement/growth in reliability, when compared with the initial design. An analysis called Reliability Growth Analysis [RGA] would eventually capture and estimate the growth established by way of design improvements. Testing is carried out on the prototypes, to assess the achieved reliability and to identify the gap between the goal and the achieved reliability. Further to this, the cause for the lower reliability and the areas of improvement could be identified through failure analysis. FRACAS would be integrated with RGA results to improve system design and improvements.

RGA would call for planned and unplanned upgradings of system design. Well defined tracking of requirements and FRACAS plan would achieve a better RGA process.

15.9.1.14 Life Data Analysis

Life Data Analysis is an effective method in characterizing the life of a product. The unit of life could be expressed in terms of hours, cycles, meters, or any other metric which represents the life of an item. Data from testing and field usage could be combined appropriately to calculate life or estimate remaining life of an item.

Using LDA, it is possible to predict the life of all products in the population by fitting a statistical distribution (model) to life data from a representative sample of units. Parameterized distribution for the data set can then be used to estimate important life characteristics of the product, such as reliability or probability of failure at a specific time, the mean life, and the failure rate. Selecting a lifetime distribution that will fit the data is an important task in LDA. Life is to be modelled with the selected distribution and the parameters estimated from the distribution depict various parameters of life of the product.

15.9.1.15 Physics of Failure

Physics of Failure (PoF) is a technique that leverages the knowledge and understanding of the processes and mechanisms that induce failure to predict reliability and improve product performance. This approach to reliability assessment is probabilistic in nature based on modelling and simulation that relies on understanding the physical processes contributing to the onset of the critical failures. Evaluating materials, structures, and the technologies are within the scope of PoF methodology. Identification and elimination of susceptibility to potential failure mechanisms in order to prevent operational failures, drives the PoF. It focuses on characterizing the life-cycle usage and environmental stress load profiles of an application and understanding the cause and effect physical processes and mechanisms they produce which causes degradation and failure in materials and components. PoF modelling facilitates reliability and durability assessment of design alternatives and trade-offs. Failures could be due to over-stress, wear out, and variations.

The PoF approach to reliability aims to address the following aspects:

- designing system reliability
- establishing IVHM [Intelligent Vehicle Health Management]
- elimination of failures prior to testing and usage
- increasing use level reliability
- improving diagnostic and prognostic techniques and processes
- decreasing operation and support costs
- eliminating potential failure modes
- analyzing the root cause of the failures
- formulating Condition-Based Maintenance (CBM).

15.9.1.16   Design-Cost Trade-off

Designing any product involves cost, which often has a trade-off with other performance parameters. A trade-off is a scenario by which improvement in one aspect pulls down the other aspect and hence a proper balancing act is needed to achieve the optimally desirable point. It often implies a decision to be made with full comprehension of both the advantages and disadvantages of a particular choice. In the context of RAMS, every aspect of RAMS – Reliability, Availability, Maintainability, and System Safety – would either directly or indirectly be related to cost and is always directly proportional. For this reason, various design considerations are to be studied for, from the cost perspective, and an optimal value of RAMS parameters could be achieved with acceptable reduction to contain the cost within the budget. If the cost is not the criterion for selection of various alternate design methodologies (particularly applicable in military products, where RAMS are given most importance), then the design with the best RAMS combination could be selected for realization. In order to solve these problems in a simpler way, functional requirements and RAMS requirements are to be kept separate initially, and the former is solved first, within the given budget and further improvements in respect of RAMS. The latter RAMS parameters are to be further studied for improvements.

15.9.2   Availability Measures of Optical Networks

Availability of an item is determined by both Reliability and Maintainability. Analyses based on simulation of statistical behaviour of the systems under consideration are essentially carried out for the purposes of establishing the Availability of any system. Typical availability analyses include:

- availability assessment
- reliability centred maintenance
- competing failure modes
- warranty analysis
- trend analysis.

15.9.2.1   Availability Assessment

Assessment of availability plays a vital role in establishing the "On-demand Reliability". The inherent availability of any item is dictated primarily by its inherent reliability and maintainability factors. Availability of simple configurations could be assessed using conventional methods. Markov analysis is one of the methodologies usually used to ascertain the availability of a system which contains *n* items with *m* states. The total number of possible combinations is $m \times n$. While estimating availability using Markov analysis, the failure rate and repair rates are required to be fed in, because the transition from a healthy state to an unhealthy state is determined by the failure rate and the reverse is deterrmined by the repair rate. For each state, the rates of transition into and out of the particular state are used to write the differential equations as a function of failure rate and repair rate. Those differential equations formed by all the intermediate and final states can be solved to obtain the time-dependant probability of being in a given state of the *m* states.

Markov Analysis is one of the means of analyzing the reliability and availability of systems whose components exhibit strong dependencies between their states. Other analyses for the availability assessment generally assume component independence. Typical cases of dependency are given below, for example:

- components in cold or warm standby
- *k* out of *m* : [Good/Failed] redundancy
- common maintenance personnel
- common spares with a limited inventory.

There are usually two variables involved in Markov Analysis, the Time and the State of the system. The time parameter could be either in continuous domain or discrete domain. The state may be in any of acceptable operable condition or in inoperable condition. When the state is unacceptably inoperable, maintenance action would get initiated. This will enhance availability of the system, where redundancy is involved and there are failures. The major drawback of Markov methods is that Markov diagrams for large systems are generally exceedingly large and complicated and difficult to construct and analyze. However, Markov models may be used to analyze smaller systems with strong dependencies requiring accurate evaluation.

The state transition diagram of Markov model identifies all the discrete states of the system and the possible transitions between those states. In a Markov process the transition frequencies between states depend only on the current state probabilities and the constant transition rates between states. In this way the Markov model does not need to know about the history of how the state probabilities have evolved in time in order to calculate future state probabilities. Although a true Markovian process would only consider constant transition rates, computer programs allow time-varying transition rates. These time-varying rates must be defined with respect to absolute time or phase time.

15.9.2.2 Reliability-Centered Maintenance

Reliability-Centred Maintenance provides a structured framework for analysing the functions and potential failures for physical assets in order to develop a maintenance plan that will provide an acceptable limit of operability, with an acceptable limit of risk, in an efficient and cost-effective manner. RCM ensures higher availability, with the available reliability designed in. This is an analytical process to determine the optimum failure maintenance strategies based on data/information to ensure safety and cost-effectiveness. The maintenance strategies could be adopted and modified dynamically based on the behavior of the system. The main goal of RCM may not necessarily be avoiding the failure from occurring, but rather be avoiding the consequence of the functional failure. Task Evaluation helps in determining the appropriate proactive tasks for the functional failure and determines the best interval in which the Repair/Replacement tasks have to be performed. With the implementation of RCM, the availability, downtime, maintainability, cost-effectiveness and efforts can be improved.

The dynamic RCM process involves systematic execution and evaluation of system design and life through FMECA and FTA. The maintenance plan gets modified according to the present state of the system.

15.9.2.3 Competing Failure Modes

This analysis could be carried out with the failure time data available for all possible failure modes of a product. There could be a greater number of failure modes for every component in a system which can result in failure of the system. The product could fail due to occurrence of any one of the failure modes, but predominantly due to a single failure mode, for any non-repairable product. From the perspective of reliability analysis, the failure modes compete to cause the failure of the product. To carry out this analysis, all the potential failure modes of the product are known a priori, in addition to the failure rate and failure mode ratio. This can be represented in a reliability block diagram as a series system in which a block represents each failure mode. Pulido [14] details life data analysis using the competing failure modes.

Competing failure modes analysis segregates data pertaining to each failure mode and then combines the results to provide an overall reliability model for the product. The first step in analyzing data sets with more than one competing failure mode is to perform a separate analysis for each failure mode. In the analysis for each failure mode, the failure times for the mode being analyzed are considered to be failures and the failure times for all other modes are considered to be suspensions. These are suspension times because the units would have continued to operate for some unknown amount of time if they had not been removed from the test when they failed due to another mode.

A major outcome of the analysis to to eliminate, if not mitigate/tolerate, the competing failure modes of the product, in order to improve availability.

15.9.2.4 Warranty Analysis

Reliability is the most important dictating factor in the case of warranty. The cost involved in maintaining the warranty (by way of repair or replacement) depends on the failure rate of the product. The cost incurred for maintaining the warranty depends on the type of warranty policy which is adopted. In determining the value of a warranty, a Cost Benefit Analysis is used to measure the life cycle costs of the system with and without the warranty, to determine whether the warranty will be cost-beneficial to the producer. Following are some of the essential factors to be considered when developing a warranty policy:

- cost of the warranty
- cost of warranty administration
- compatibility with total program efforts
- cost of overlap with contractor support
- intangible savings
- reliability
- maintainability
- supportability
- availability
- life-cycle costs
- technology
- size of the warranted population
- likelihood of achieving performance requirements
- warranty period of performance
- inventory cost.

The type of warranty, by taking into account of all the above factors, could be arrived at, based on the optimal value of warranty cost to the company and benefit to the customers. Detailed studies are required for the establishment of the warranty of items, ranging from failure rate, repair rate, repair cost, logistics, repair resources to inventory etc.

15.9.2.5 Trend Analysis

RAMS analyses help to reduce the LCC of the product/system. Based on the quantitative outcomes of the analyses, the performance of a product could be evaluated and its life cycle cost could be predicted. Trends in reliability could indicate the growth or deterioration, based on the improvements carried out on the product/system. From the detailed study of the trends, the LCC could be forecasted and the life of the product could be established, before its failure. Recommendations in respect of maintenance policies and management decisions are also inherited from the trend analysis, for effective management of project. The data source for the trend analyses may be from warranty, test, process, or field data. Trend analyses play a vital role in imparting an active reliability-centered maintenance (RCM) system and enabling the prognostic nature of an Integrated Vehicle Health Management (IVHM) system. The following are typical parameters of interest from trend analyses, from a RAMS perspective: *fault propagation, types of failure modes, competent failure mode, life consumed, remaining life, expected time to failure, achieved reliability.*

15.9.3 Maintainability Aspects of Optical Networks

Similarly to the concept and analytical activities related to Reliability, Maintainability also plays a significant role in dictating the dependability of the product. The following is a list of essential activities related to maintainability which would be carried out during the life cycle of the product:

- maintainability apportionment
- maintainability prediction
- maintainability estimation
- maintainability assessment
- maintainability demonstration
- maintenance strategy (plan/philosophy)
- spare parts optimization
- failure reporting and corrective action system.

15.9.3.1 Maintainability Apportionment

Similarly to the concept of Reliability, the Maintainability of the LRUs/Systems/Subsystems should be dealt with from the initial stages of design. Maintainability Apportionment is the starting point of the maintainability analyses, given that the stated requirements of the overall maintainability are known. Both Reliability and Maintainability concepts are to be built into the system design from the design stage itself, to ensure better Availability of the system. The apportioned value of Maintainability could be expressed in MTTR, which consists of defect diagnosis, rectification, and retest, assuming that everything else required is immediately available. The other metric of maintainability, the MMH/FH, could also be apportioned.

Maintainability Apportionment is needed for:

- providing designers and manufacturers of each part of the system with maintainability requirements
- providing R&M figures for comparison with assessments made during design and development
- enabling trade-offs to be studied at an early stage.

15.9.3.2 Maintainability Assessment

Assessment is more important for those cases where the parameters are demonstrable. As in the case of Reliability, Maintainability can also be assessed statistically by modeling. To assess maintainability statistically, a list of maintainability factors for assessment should be devised. The outcome of the maintainability assessment is establishing the maintenance measure (MM), with weightings assigned to the maintainability factors. Items with higher MM should be attended to for further improvement in maintainability or better design. Setting up a threshold for MM can be done based on customer requirements and the expertise of the RAMS team with the maintainability factors.

15.9.3.3 Maintainability Demonstration

Maintainability Demonstration is a methodology to verify whether the stated and designed-in maintainability requirements have been achieved. Maintainability Demonstration should be part of the Maintenance Test Plan, which should clearly bring out the phases of plan, starting with Maintainability Verification through Demonstration to Evaluation of stated Maintainability. This environment shall be representative of the working conditions, viz., tools, support equipment, spares, facilities, and technical publications, that would be required during operational service use.

15.9.3.4 Maintainability Estimation/Evaluation

The impact of the actual operational, maintenance, and support environment on the maintainability parameters of the system will be evaluated during this stage of Maintainability Plan. Correction of deficiencies, if any, could be also part of the estimation. Evaluation is usually carried out in an integrated manner, simulating the actual environment. This exercise should be follow on after maintainability prediction.

Similar to the case of RBD, the logical activities of maintainability are modeled in the Maintainability Verification phase during the design stage, commencing with initial design and continuing through hardware development from components to the configuration item. The maintainability model would be developed at this stage to verify the claim on Maintainability. At the least, some minimal maintainability verification test could also be carried out, if required to support the design from the Maintainability point of view.

15.9.3.5 Maintainability Prediction

It would be a better approach in realizing a product to have both Reliability and Maintainability features built in. The relevance of Maintainability early during the design stage is justified in order to achieve a higher level of accessibility, testability, and availability. Hence, the Maintainability should be predicted very early in the design stage. The numerical value predicted would suggest the acceptability of the design as regards maintainability and serviceability. Design without the concept of Maintainability designed in would eventually result in exorbitant increase in life cycle cost in general, and maintenance cost and efforts in particular.

15.9.3.6 Maintenance Strategy [Plan/Philosophy]

There are methods in carrying out maintenance activities that fall under either proactive maintenance or reactive maintenance. If the maintenance activities are based on calendar time or operating time, then this should be clearly highlighted in the maintenance philosophy or manual. The maintenance strategy could be arrived at based on numerical computations or experience, in cases where the necessasry data are not available. The requirements for the maintenance tasks, such as maintenance facility, maintenance personnel (skilled and skill level), tools, test equipment, spares, supporting items, and logistics, should also to be outlined clearly.

15.9.3.7 Spare Parts Optimization

In order to support the maintenance actions, as identified in the maintenance philosophy, it is required to maintain a minimum level of inventory. This will affect the life cycle cost including warranty associated with the item. Hence the spare parts inventory should be optimized by considering all the other applicable factors of LCC. Standardization is one of the simpler means to achieve spare parts optimization. As in the case of warranty, spare parts optimization is also dictated by Reliability and Maintainability. In addition to this, interchangeability helps in reducing the inventory. These are design factors which should be conceptualized during the initial stages of product development.

15.9.3.8 Failure Reporting and Corrective Action System

Sustaining the RAMS objectives will be possible only with continued monitoring and improvement throughout the life cycle. Irrespective of the maturity level of the system, failures may happen randomly due to various factors associated with it and the randomness involved. FRACAS is one of the closed-loop systems, which essentially have a tracking and implementation mechanism for the failures encountered. In FRACAS, failures related to hardware and software are formally reported without losing evidence. Analyses are performed to the extent possible to understand and identify the failure cause and positive corrective actions are identified, implemented, and verified to prevent further recurrence of the failure. This system would play a major role in combining data in respect of failures where the system is deployed to various territories.

- The risk involved in the "Probability of Loss of Control of Commercial Aircraft" should lower to the order of 1×10^{-9} per hour.
- The fatalities arising out of nuclear industry in terms of core damage frequency should be lower to the order of 1×10^{-05} per reactor year.
- The "Probability of Loss of Control of Fighter Aircraft" should be lower to the order of 1×10^{-7} per hour.

Hence, in order to satisfy the safety goals, the risk assessment should be carried out by including software logics, interlocks, redundancy management, and failures.

15.9.4 Optical Networks for Safety-Critical Applications

Risk assessment is the major safety analysis process (Figures 15.12 and 15.13), which is essentially to be carried out to identify and mitigate/control safety:

- Common Cause Analysis
- Common Mode Analysis
- Zonal Safety Analysis
- Hazard Risk And Operability Studies
- Software Risk Assessment
- Fault Tree Analysis
- Functional Hazard Analysis

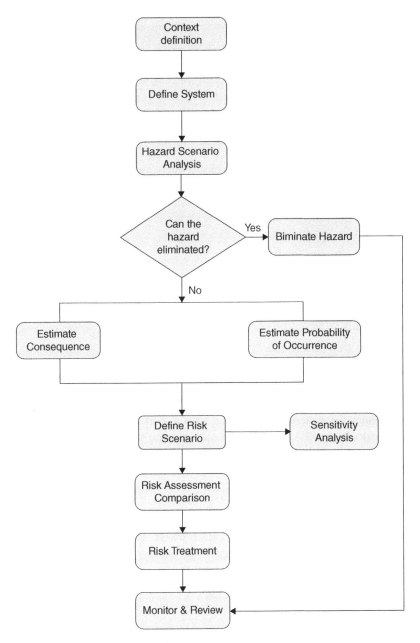

Figure 15.12 Risk-based decision process.

- Warning Time Analyses
- Software Risk Assessment
- Event Tree Analysis.

15.9.4.1 Common Cause Analysis

Common Cause Analyses are gaining momentum in RAMS techniques due to the very nature of the severe effects of them on all the systems. Independent failures have a low probability of

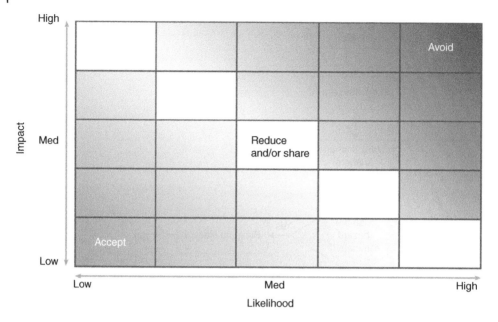

Figure 15.13 Risk acceptance criteria.

occurrence compared with dependent failures. Common cause failure modes result in the loss of independence, which dramatically increases probability of failure. Zonal Safety Analysis (ZSA) is one of the most important system safety analytical methods, used as part of CCA. ZSA combined with two other methods – Particular Risks Analysis (PRA) and Common Mode Analysis (CMA) – forms CCA. In summary, CCA consists of the following:

- ZSA
- PRA
- CMA.

The common cause could either be internal or external. CCA is used to find and eliminate/mitigate common causes for multiple failures, usually external to the system/LRU under consideration. Intersystem effects are of major concern for the external CCA. Examples of common causes which are external are:

- lightning
- EMI/EMC
- over voltage
- under voltage
- grounding.

There are cases where the common causes are resident within, or applicable to, the intra LRU. Those effects are also to be studied for, due to its criticality. The following are some of the typical cases of internal CCA:

- lower tolerance limit
- upper tolerance limit
- worst case tolerance effects.

From the aspect of common power supply and grounding scheme to the systems on aircraft, the common causes, viz., over voltage, under voltage, and grounding, though applicable intra-system, could be covered under external CCA.

15.9.4.2 Common Mode Analysis

The basic assumption of failure independence in the safety analysis is not valid due to the system design and implementation. One of the most important modes of failure, and one which can severely degrade actual safety, is a common mode failure. This type of failure involves the simultaneous outage of two or more components due to a common cause. Common Mode Analysis (CMA) provides evidence that the failures assumed to be independent are truly independent. In reality, this analysis is extremely complex due to the large number of common mode failures that may be related to the different common mode types such as design, operation, manufacturing, installation, and others.

Common Mode Analysis, which is the subset of CCA, is performed to verify that the events are truly independent. The effects of design, manufacturing, maintenance errors, and failures of system components which defeat the failsafe design should be analyzed as part of CMA. Consideration should be given to the independence of functions and their respective monitors.

The following are some examples of Common Mode Faults:

- design deficiency
- functional deficiency
- maintenance error
- situation-related stress (e.g., abnormal flight conditions or abnormal system configurations)
- installation error
- requirements error
- normal environmental factors (e.g., temperature, vibration, humidity, etc.)
- external source faults.

15.9.4.3 Fault Tree Analysis

Fault tree analysis [FTA] is a deductive failure analysis which focuses on one particular undesired event and provides a method for determining causes of this event. In other words, a fault tree analysis is a "top-down" (vertical) system evaluation procedure in which a qualitative model for a particular undesired event is formed and then evaluated. The analyst begins with an undesired top-level hazard event (Failure Approach) and systematically determines all credible single faults and failure combinations of the system functional blocks at the next lower level, which could cause this event. The analysis proceeds down through successively more detailed lower levels of the design, until a primary event is uncovered or until the top-level hazard event requirement has been satisfied. A primary event is defined as an event which, for one reason or another, has not been further developed. That is, the event need not be broken down to a finer level of detail in order to show that the system under analysis complies with applicable safety requirements. A primary event may be internal or external to the system under analysis and can be attributed to hardware failures/errors, software, or human errors.

Graphical representation of FTA is triangular tree in shape and takes its name for the branching that it displays. It is the format which makes this analysis a visibility tool for both engineering and the regulatory agencies. It is concerned with ensuring that design safety aspects are identified and controlled.

A Binary Decision Diagram (BDD), also known as function graph/directed acyclic graph (DAG), is one of the techniques used to find the unreliability/unavailability of an event. The sequential propagation/failure occurrence of all possible events are listed in BDD. The flow of binary logic is employed, whereas the onset of any event associated with the occurrence of the other events is developed like a interlinked tree structure. The BDD approach will be very useful in developing phased mission reliability analyses, owing to the fact that the system configuration, component behaviour, success criteria, and time varies from phase to phase. It will be simpler to analyze the system if the same is modelled into a logical reliability graph. The outcome of FTA could also be obtained by using the BDD technique, apart from the Exact method and the Esary–Proschan method. A BDD could be employed where sensitivity analysis, minimum cut sets, minimum path sets, and unreliability need to be computed.

15.9.4.4 Functional Hazard Analysis

More often, an accident occurs as the result of a sequence of causes. A hazard is defined as a condition, event, or circumstance that could lead to or contribute to an unplanned or undesirable event. A hazard analysis will consider system state and failures or malfunctions. While in some cases risk can be eliminated, in most cases a certain degree of risk must be accepted, by considering the cost, effect, and resources. The risk could be quantified by establishing the severity (consequence) and the probability of occurrence.

FHA is a tool to identify and evaluate the hazards by rigorous examination and evaluation of the system and subsystem configuration and functionalities, including software. The series of system functionalities are broken down to sub-levels of functionalities and then hazards are identified. FHA is an inductive approach, similar to FTA. The main focus of FHA is on the functions of the systems/subsystems. FHA is more productive and supportive for design if carried out during the very early phases. System hazards are identified by evaluating the safety impact of a function failing to operate, operating incorrectly, or operating at the wrong time. FHA may be carried out for the single system, a subsystem, or integrated systems. A basic understanding of the system concepts, functionalities, and experience with the similar systems is essential to generate the list of potential hazards. Once the functional hazards are identified, then further analysis of the hazards is required based on the severity and probability of occurrence assumed. The following are the factors involved in FHA:

- identification of functional hazards
- identification of safety-critical functions
- prediction of hazard casual factors such as failures, design errors, human errors
- risk assessment, for those highly risky hazards
- safety provisions to mitigate the hazards.

15.9.4.5 Hazard and Operability Studies

Hazard and Operability Studies is a methodology to identify potential hazards in a system and the issues in terms of operability, where the operation is conceived but not catered for in the design. This technique is usually well suited for process-related domains, in examining the system built and the risk associated with it, in case of deviations from the design intent. The outcome of the studies will be utilized for risk management. HAZOP is carried for those health and safety-related systems and subsystems, in order to assess the safety features built in to the design, in case of inputs/process that are not within the design scope of the item.

The HAZOP procedure involves taking a full description of a process and systematically evaluating/questioning every part of it, to establish how deviations from the design intent can arise. Once identified, an assessment is made as to whether such deviations and their consequences can have a negative effect upon the safe and efficient operation of the system. If considered necessary, suggestions for the actions to be taken to remedy the situation are also presented. Risk Management i.e., Risk Assessment, Control, Review and Communication also considered part of the HAZOP. HAZOP could typically be categorized as follows: *Process HAZOP, Procedure HAZOP, Human HAZOP, Software HAZOP.*

15.9.4.6 Zonal Safety Analysis

Zonal safety analysis is carried to ensure that every zone of the aircraft is safe and free from possible hazards. The objective of the analysis is to provide confidence that the installation of the equipments meets the safety requirements in terms of installation and interference. The effects of failures of equipment should be considered with respect to their impact on other systems and structures falling within their physical sphere of influence. When an effect which may affect safety is identified, it is to be highlighted and installation/design of equipment is revisited appropriately. Various zones are to be studied carefully, by clearly observing the systems in that zone and constructive/destructive interference among the systems. The identified interference, with an initiator turned on, may lead to catastrophic consequences. Event Tree Analysis can also be used to study the end effects of the adverse events identified.

15.9.4.7 Particular Risk Assessment

During the design and development of new system, it is essential to carry out Particular Risk Assessment, which relates to threats to the aircraft/system/subsystem from the outside environment (bird strike, lightning, hail) and threats to the systems from events originating in other systems. PRA is carried out as part of System Safety Analyses, called the CCA. Accordingly, the PRA could either be external or internal to the aircraft. These assessments are carried out to ensure the robustness of the design to survive the potential threats identified. Formation and collection of all possible potential threats to the system under consideration are required to evaluate each of them. The results should be collated to have a single point of reference: the ability to survive all known external threats. If PRAs have been accomplished on previous programs, they can be used as a starting point for the new assessment, then the systems are reevaluated against the new design and differences created by new design features need to be added to the list. Following are the typical risks: *fire, high energy devices, leaking fluids, hail, ice, snow, bird strike, tread separation from tire, wheel rim release, lightning, high intensity radiated fields, failing shafts, etc.*

Each of the identified risks should be studied appropriately with respect to the design under consideration, and the simultaneous, or cascading, effect(s) of each risk, if any, are documented.

15.9.4.8 Software Risk Assessment

Software risk assessment plays a vital role in getting certification of safety-critical installation/equipment/setup, as most of them are controlled by processor-based application. In addition to this, these days COTS components are also being used in the Military/Automotive industry due to non-availability of MIL-grade components in the market. Hence, the risk associated is manifest multi-fold along with the software ported.

Software is being used in every application, from day-to-day commercial applications to communications to safety-critical nuclear applications to aerospace applications. As the safety of humans and the environment is considered to be the most important consideration, there are requirements for those control applications to meet the stipulated safety constraints/norms. In order to meet the safety norms, the risk associated with software-based application should be acceptably low-level. To demonstrate this, risk assessment of software needs to be carried out in quantifiable terms.

15.9.4.9 Event Tree Analysis

An event tree is a graphical, inductive, analytical representation of the Boolean logic model that identifies and quantifies possible outcomes following an initiating event, by taking into account whether the safety barriers are functioning or not. Forward logic is employed in event tree analysis and it is constructed through an inductive approach, unlike fault trees, which use a deductive approach. FTAs are constructed by defining top events and then useing backward logic to define causes. However, ETA and FTA are closely linked. FTAs are used to quantify the failure probability/unavailability of system events that are part of the event tree. The logical processes employed in FTA and ETA is the same.

The main objective of ETA is to identify, design, and avoid procedural weaknesses, and probabilities of the various outcomes from an anticipated accidental event. The secondary aim is to identify the improvements that could be possible in the protection systems and safety barriers, to reduce the probability of occurrence of the end event. The initiating event is usually the first significant deviation that may lead to unwanted consequences and that could be due to equipment failure, human error, or process upset. Consequences of a single occurrence of the initiating event are studied in relation to the other barriers of safety.

15.10 Process Control in Optical Components

It is known that, due to the sensitivity of an optical signal to any imperfections or impurities, careful monitoring and control of the process parameters are crucial. In this aspect, critical parameters of optical components are to be measured and controlled. Statistical Process Control techniques shall be employed to ensure that deviations are very well within the acceptable limits.

For this purpose, critical physical parameters or components shall be drawn from FMECA. Critical failure modes identified as part of the study sheets of FMECA need careful attention. Variabilities are to be minimized to the extent technologically possible. Employment of SPC in optical system is essential to achieve/reap projected benefits.

15.11 Hardware – Software Interactions (HSI) in Optical Networks

Irrespective of the fact that the modulation and multiplexing, and then the demodulation and demultiplexing, are predetermined and selected based on the applications, the need for control and management of the characteristics of communication media on-the-fly is still required. This would be possible with increasing usage of software or firmware components as part of the network or switch elements. Usage of software/firmware enables flexibility in switching and routing of data. This would also result in miniaturization, and optimization of resources, volume, power, and weight. All of this results in easier configuration and management of hardware resources without any downtime.

In order to achieve configuration management and control on-the-fly, there shall be enough provision to effectively utilize the bandwidth and efficiency of the optical network, while optimally allocating the resources. At the same time, it is imperative to ensure that the RAMS aspects are also adhered to. Based on the area of application and the criticality of the functionality, the interactions among the hardware and software require significant consideration and characterisation.

Enough studies have been carried out to quantify the percentage of failures triggered by pure software and it is found that 35% of errors are converted into faults and eventually to failures which are attributable to errors in pure software elements. Of those remaining, the portion of failures due to hardware-software interactions are expected to be greater in recent times, due to the functional dependency on software/firmware. From field experience it is expected that the pure hardware-related faults are minimal. The underlying reason for the smaller number of faults which are related to hardware is that hardware components are often designed and tested for varying environmental applications. Inherent faults are getting screened and infant mortality and design-related flaws are eliminated by the process of qualification testing. After qualification testing every deliverable component is screened prior to shipping to customers.

In most the cases, the application and other specific software components are loaded into hardware platform and then a hardware software integration test is carried out. Involvement of more and more sophisticated firmware/software requires certification norms to be fully complied with. In the case of the aerospace domain the Federal Aviation Authorities (FAA) and Federal Aviation Regulation (FAR) dictate the procedures and processes to be adopted to ensure that functionality is achieved without compromising safety. In the case of other domains, SAE ARP-761, and 4754 elaborate detailed requirements of acceptance of software-intensive items.

15.12 Typical RAMS Realisation Plan for an Optical System

There are numerous activities relating to RAMS, which are to be carried out during the early stages of design. Those activities during various phases of the product/project life cycle are enumerated in the preceding sections. Prior to the phase-wise classification of RAMS activities, the hierarchical levels of activities are defined below.

From the aspect of realization of an optical system, all RAMS activities could broadly be categorized under two major headings: *system-level RAMS activities* and *item-level RAMS activities*.

15.12.1 System-level RAMS Activities

The system-level RAMS activities are highlighted in Figure 15.14. Upon completion of system levels RAMS, further RAMS aspect can be applied to lower indenture levels, as depicted in Figure 15.15.

Taking an optical interconnect or network or switch as a system, further indenture levels can be derived using a deductive top-down approach. Those activities/analyses listed and briefed in the previous sections would be given the utmost importance and the outcome of each of the analyses has to be fed back to the design team for design modifications/improvisation/mitigation/tolerance as the case may be. Following the flow chart in Figures 15.14 and 15.5 give an insight into the typical activities of product development, indicating the relevant RAMS tasks by the side at the system level. It is to be noted that system realization will fall into and overlap with the item development stage, and accordingly the RAMS activities would require detailed design consolidation.

There will be a large number of reviews carried out to ensure that the design proposed meets all the requirements required by the user explicitly as well as any other implicit requirements.

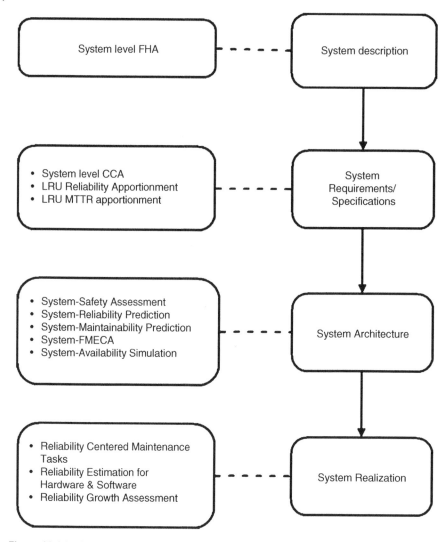

Figure 15.14 System-level RAMS activities.

15.12.2 Item-level RAMS Activities

Having ensured that the architectural concept addresses all the observations brought during the system level, compliance of those attributes at each item level will also be ensured. In order to carry out these tasks, clarity on the items is assumed to be available only when the system description is complete and precise in nature. Activities could be initiated on an item by item basis if the item-level design details are finalized. More detailed design analyses, simulation, test, evaluation, verification, and validation are required at the end component or item level. Applicable RAMS tasks at the unit/item/component level are enumerated in the form of a flow chart for easy visualisation. Any miss or delay in the execution of any/many of the tasks may delay execution of the project towards the completion in terms of certification or handing over to customer.

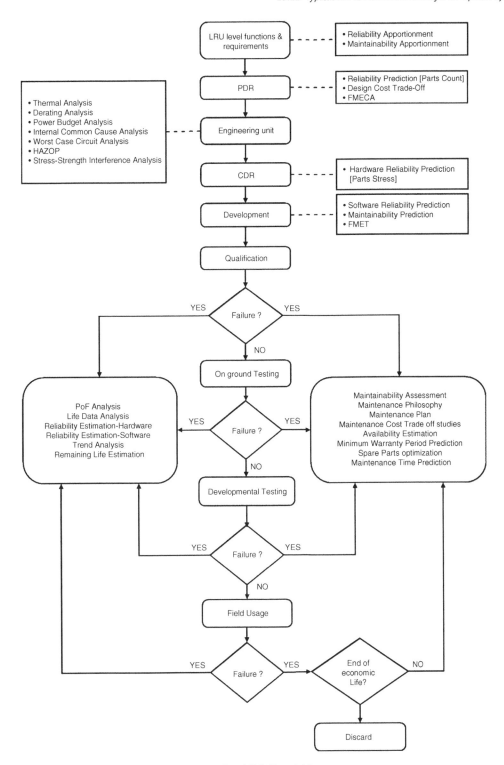

Figure 15.15 LRU/Item/Unit/Component-level RAMS activities.

15.13   Trade-off Factors of Optical Networks

In terms of functional and bandwidth capabilities, an optical network-based system outweighs a conventional electrical and electronics-based system. Added to this, the voluminous amount of data handling is easily handled by the optical system. Careful attention must be given to optical system, as the physical medium used for communication does not tolerate impurities.

Hence the usage environment plays major role in determining the efficacy of optical systems for their intended purpose. Although the need for maintenance does not arise as compared to other conventional communication media, once it has occurred, the downtime is relatively higher and requires specialized tools, techniques, and experienced personnel. The following points need to be addressed:

- controlled environment/fault-tolerant design/insulation
- qualification of the optical system to the usage profile
- validation of error-free software/firmware
- proficient maintenance philosophy/technicians/tools.

If all the RAMS attributes are addressed adequately, an optical system will be far superior in terms of performance and life cycle cost.

15.14   Some Open Problems in RAMS-Optical System

From the perspective of RAMS within the optical domain, the following aspects still need to be addressed:

- reliability modelling of optical switches
- identification of functional failure modes
- contribution of each of failure modes – Failure Mode Ratio
- reliability modelling of hardware software interactions
- database for basic failure rate of optical switch and other components
- failsafe design, if used in safety-critical applications
- maintainability enhancement of optical system using advanced tools and techniques
- prognostics of optical system.

15.15   Conclusion

The RAMS activities are essential in establishing the dependability of the system, from conceptualization through design and development to phase-out. Requirement capturing is a vital task in any project. Unless the requirements are stated clearly, realization is impossible, meaning that the time, cost, and efforts invested are wasted in realizing a product that will not be as per the expectations of the customers. Capabilities of optical network based system are realizable, only if RAMS features are built-in. In addition to the functionalities, the RAMS parameters are also to be specified during conceptualization, failing which the item developed may be *down for maintenance, functionally fit but not user-friendly, difficult to access, harder to troubleshoot, difficult to repair and maintain,* or have *increased life cycle cost.* These activities are to be carried out depending upon applicability at various stages and based on the program/project directives. At every stage of the RAMS activities, compliance to user requirements needs to be verified. Sequential integration of design changes into the RAMS analyses is essential to ascertain the RAMS parameters achieved during the evolution of products. Guideline documents as applicable are to be referred to in carrying out every analytical activity identified. The procedural and systematic execution of the activities as per the applicable guidelines/handbooks are to be adhered to, to meet the industrial standards. Timeliness and proper execution of the listed RAMS activities will be of greater impact and usefulness in realizing a dependable system, by ensuring that design feedback emanating from RAMS activities are provided on time with a broad objective of minimizing the life cycle cost and improving/maximizing system dependability. The advantages of optical system would be sustainable, if and only if RAMS attributes are addressed and incorporated as part of the design.

Bibliography

1 W.A. Hansen, B.N. Edson, and P.C. Larter. Reliability, availability, and maintainability expert system (RAMES). In *Annual Reliability and Maintainability Symposium 1992 Proceedings*, pages 478–482, 1992. doi: 10.1109/ARMS.1992.187868.

2 L. Du, Z. Wang, H.-Z. Huang, C. Lu, and Q. Miao. Life cycle cost analysis for design optimization under uncertainty. In *2009 8th International Conference on Reliability, Maintainability and Safety*, pages 54–57, 2009. doi: 10.1109/ICRMS.2009.5270241.

3 C. Manso, R. Munoz, N. Yoshikane, R. Casellas, R. Vilalta, R. Martinez, T. Tsuritani, and I. Morita. TAPI-enabled SDN control for partially disaggregated multi-domain (OLS) and multi-layer (WDM over SDM) optical networks [Invited]. *IEEE/OSA Journal of Optical Communications and Networking*, 13(1):A21–A33, 2021. doi: 10.1364/JOCN.402187.

4 P. Wiatr, J. Chen, P. Monti, and L. Wosinska. Energy efficiency versus reliability performance in optical backbone networks [invited]. *IEEE/OSA Journal of Optical Communications and Networking*, 7(3):A482–A491, 2015. doi: 10.1364/JOCN.7.00A482.

5 S.B. Olejnik and B. Al-Diri. Reliability vs. total quality cost – part selection criteria based on field data, combined optimal customer and business solution. In *2011 IEEE International Conference on Quality and Reliability*, pages 408–412, 2011. doi: 10.1109/ICQR.2011.6031751.

6 N.E. Strifas, P. Panayotatos, and A. Christou. Optical interconnect reliability. *Optical Engineering*, 37(8):2416–2418, 1998. doi: 10.1117/1.602023.

7 N.E. Strifas, C. Pusarla, and A. Christou. Reliability of commercial optical interconnect systems. In R.T. Chen and P.S. Guilfoyle, editors, *Optoelectronic Interconnects and Packaging IV*, volume 3005, pages 54–57. International Society for Optics and Photonics, SPIE, 1997. doi: 10.1117/12.271111.

8 A.G. Varuvel and P.X. Pruno. Reliability allocation technique for complex system of systems. *International Journal of Reliability and Safety*, 13(1–2):61–82, 2018.

9 M.I.M. Saidi, M.A. Isa, D.N.A. Jawawi, and L.F. Ong. A survey of software reliability growth model selection methods for improving reliability prediction accuracy. In *2015 9th Malaysian Software Engineering Conference (MySEC)*, pages 200–205, 2015. doi: 10.1109/MySEC.2015.7475221.

10 G.G.K. Morris. Streamlining the electronic component derating process: An electrical engineering perspective. In *2018 Annual Reliability and Maintainability Symposium (RAMS)*, pages 1–7, 2018. doi: 10.1109/RAM. 2018.8463046.

11 Y. Sun, Q. Zhang, Z. Yuan, Y. Gao, and S. Ding. Quantitative analysis of human error probability in high-speed railway dispatching tasks. *IEEE Access*, 8:56253–56266, 2020. doi: 10.1109/ACCESS.2020.2981763.

12 J. Jin, K. Li, L. Yuan, and G. Wei. A human error quantitative analysis of the railway ATO system based upon improved CREAM. In *2019 IEEE Intelligent Transportation Systems Conference (ITSC)*, pages 3200–3205, 2019. doi: 10.1109/ITSC.2019.8917181.

13 I. Gunawan and N.S. Fard. Reliability evaluation of optical multistage interconnection networks. In *2007 ICTON Mediterranean Winter Conference*, pages 1–4, 2007. doi: 10.1109/ICTONMW.2007.4446915.

14 J. Pulido. Life data analysis using the competing failure modes technique. In *2015 Annual Reliability and Maintainability Symposium (RAMS)*, pages 1–6, 2015. doi: 10.1109/RAMS.2015.7105142.

16

Protection, Restoration, and Improvement

Arighna Basak[1] and Angsuman Sarkar[2]

[1] *Department of Electronics and Communication Engineering, Brainware University, Kolkata, West Bengal, India*
[2] *Department of Electronics and Communication Engineering, Kalyani Government Engineering College, Kalyani, West Bengal, India*

16.1   Introduction

Thanks to the convergence of information and communications technology and the rapid growth of fiber-optic communication systems, today's telecommunications systems can provide end users with fast, high-quality services. The type of service has expanded from voice-only to include a extensive range of multimedia options. As the quantity of corporate customers involved continues to grow, even a minor interruption in service can have a serious impact. Therefore, how to avoid service interruption and keep service loss low becomes an important issue when it is unavoidable; in other words, survivability must be built into the architecture of the telecommunications network. The capability of the network to deliver uninterrupted services in the event of failure is called survivability. Due to its benefits in ability, dependability, cost, and scalability, optical fiber has become the main means of transmission in telecommunications networks. The incredible high capacity of optical fiber (several megabits per second) is one of its most attractive features. The only medium that can provide high-bandwidth services at a reasonable cost is optical fiber. Optical fiber is often used in today's backbone networks. However, so far, simply a minor part of the complete potential of the connected fiber has been apprehended. The main cause of this inefficiency is a massive mismatch between peak electronic processing and source rate, as well as fiber capacity [1].

The capability of a network to maintain a endless level of quality of service (QoS) in the face of varied failure scenarios is measured by network survivability. It is a QoS assurance library provided by network providers to communication operators. It will be difficult to ensure the QoS needs in the case of a failure if additional spare resources cannot be appropriately prepared. In the occasion of a failure, these solutions use a collection of tools to prepare ahead and use backup methods to increase QoS guarantees. Network failures are caused by network element defects, inadequate/incorrect maintenance procedures, software problems, hurricanes, earthquakes, lightning, floods and traffic accidents, and disruptions (resentful

Optical Switching: Device Technology and Applications in Networks, First Edition. Edited by Dalia Nandi, Sandip Nandi, Angsuman Sarkar, and Chandan Kumar Sarkar.
© 2022 John Wiley & Sons, Inc. Published 2022 by John Wiley & Sons, Inc.

employees, external influences, hackers) [2]. The majority of these failure scenarios are difficult to forecast and dismiss. Survivability mechanisms can be included in the network design phase to reduce the impact of a certain set of failure situations. Network recovery and survivability network architecture are the two major characteristics of traditional network survivability solutions [3].

A survivable system must be able to deliver important services in the case of a failure, assault, or accidents [4]. The minimum level of such quality characteristics associated with critical services should be specified. For example, if the defense system's performance deteriorates to the opinion that the target's range surpasses the system's ability to launch, the defense system's missiles will become ineffective. Survivability is typically defined as the ability to balance a variety of quality qualities, including safety, consistency, modifiability, presentation, obtainability, and affordability. Recent advances in optical network, particularly WDM, have enabled next-generation systems to track at rates of many terabits per second [5]. Single or supplementary optical fiber lines are connected to optical switching nodes in a wavelength-driven optical network. Because each optical fiber has a very significant capacity, a failure of (node or link) in this sort of network can result in a large amount of data being lost [6]. The optical WDM network's capacity to respond gracefully to such breakdowns is referred to as the optical network's survivability [7]. The optical network is a high-capacity network that practices optical technology. They can exchange gigabytes of data in one second and therefore provide more bandwidth because they can run at very high data speeds. Therefore, any failure may result in data loss, resulting in economic loss and loss of confidential information. Human errors, such as the excavator cutting the fiber-optic cable or the operative dragging the erroneous connector or spinning off the wrong switch, are the most collective reasons of failure. Fiber interruption is the most prevalent cause of link failure; it is the most common failure to occur. The loss of active components in network equipment (such as sources, receivers, or processors) is becoming the most common failure mode. In most cases, network equipment is set up with redundant controllers. Furthermore, the failure of the controller has no influence on the flow, but only on network management knowledge. Another situation is the failure of a node. Due to catastrophic disasters such as fire, flood, or earthquake, the entire central office is often destroyed. Switch or router problems owing to software catastrophes are another cause of node failure. Therefore, survival is essential to recover from these failures. Survivability refers to the network's capability to improve from failures [1].

Recovery (reactive) or protection (active) techniques can be used to ensure the survival of optical systems. Only after a failure in a recovery-based program will an alternate path be developed. Each incoming connection request in a protection-based system is allocated a primary route and a backup route where the link is not bridged at the time of setup. A protection plan, which predetermines and reserves backup bandwidth in the event of single/multiple link/node failures, is a common strategy for constructing a survivable network. The fundamental challenge of flexible optical networks is to design methods to regulate main and backup paths [7], thus reducing resource consumption and increasing the number of networks. As part of the protection strategy, backup resources should be retained before failure. This includes setting aside some network resources for failures during connection setup or network strategy, and only activating these resources after the faults have occurred. Although the protection strategy has a faster recovery time, it wastes resources until there are no more failures. Use this method to protect in contrast to link and path faults. Therefore, protection is the most important way to ensure survival. These protection technologies must provide some redundant network capacity and use this capacity to automatically divert traffic around the fault.

16.2 Objectives of Protection and Restoration

Telecommunications companies have a challenging responsibility to confirm that their networks can endure to operate smooth in the event of fiber interruption, equipment failure, and human error. Contemporary optical networks combine a redundancy layer and a durability layer to achieve extremely high reliability. Using modern protection and recovery techniques can quickly identify network failures, and can dynamically redirect operations around the affected location.

Advanced recovery algorithms built into optical network routers or switches can usually protect modern networks. Until recently, the optical layer's network recovery options were limited compared to the more advanced methods used on the top layer. As the requirements for bandwidth and high-speed operation increase, network capabilities and wavelength speeds also increase. As network functions increase, the cost of using routers or optical network layers to provide recovery also increases, because all services must be terminated at each optical network router or switch at each node location. The cost of transponders and optical interfaces that terminate hundreds of gigabit (or even TB) of network capacity at the WDM and router layer in each location represents a high cost. On the other hand, the optical network provides relatively inefficient and efficient transmission through metropolitan area, regional, and long-distance. Optical networks have only recently had complex recovery procedures found in optical network routers and switches. However, new optical technologies increasingly include more advanced optical layer protection and restoration. These solutions combine the advantages of operating in the optical layer with strong network resilience.

16.3 Current Fault Protection and Restoration Techniques

In traditional networks, a wide range of protection systems are used and it is important to understand the concepts of work routes and protection routes. The working channel carries traffic when everything is in usual process; the protection path delivers a backup path to avoid problems. Because the occupied path and the protection path often employ distinct steering systems, both paths will not be lost if one fails. The protection method is applicable to various network configurations.

Some people focus on point-to-point connections. In optical networks, ring topologies are very popular. The simplest topology is a ring, which provides an alternative method to avoid failure. Many protection systems in the optical layer are designed to be used with true mesh topologies. These methods are designed to work in the event of a physical failure. It is generally believed that a single failure rather than a double failure is most likely. Multiple failures can be contemplated, although the likelihood of multiple failures is considerably decreased with proper design. One or more connections at the client layer will fail due to a physical failure. A single component failure, such as a transceiver failure, can create a single link failure. Cutting an optical fiber that contains various wavelengths will result in many link failures at the client layer. A shared risk link group consists of links that are disconnected at the same time due to a single failure event (SRLG). A single switch or router failure can potentially trigger SRLG because all links connecting to the switch or router would fail [8].

It is possible to have dedicated or shared protection. All operation links have their individual dedicated network capacity, which can be redirected under dedicated protection in the event of a failure. We employ shared protection to take advantage of the circumstance that not all of the network's active connections are terminated at the same time (for example, if they are in dissimilar portions of the network). Various occupied networks can share bandwidth protection as a result of careful design. This cuts down on the amount of network bandwidth needed for security. Additional advantage of shared protection is that the protection bandwidth can be used to transmit low-priority traffic under normal circumstances. This low-priority data will be rejected on failure when bandwidth is required to secure the connection [9, 10].

Reversible or irreversible protection systems are available. In the event that one of the schemes fails, traffic will be diverted from the functioning route to the protection route. The traffic remains on the protective path in the unrecoverable scheme until it is manually switched to the original functional path, which is normally done through the network controlling scheme. Traffic will immediately return from the protection route to the work route in the recovery plan once the work route has been chosen. The recovery permits the network to restore to its prior condition once the failure has recovered. Shared protection systems are virtually always recoverable, but dedicated protection systems are either recoverable or unrecoverable. Because several functional connections share the same protection bandwidth, it should be released as soon as feasible once the original fault has been fixed so that it can be recycled to protect other networks in the event of a later failure.

Even more perplexing, protection switching can be either unidirectional or bidirectional. This is not to be confused with one-way or two-way fiber transmission. In unidirectional protection networks, traffic in all directions are routed independently. Only one type of traffic is transferred to the protection fiber when a single fiber is cut, while the other direction stays on the original functioning fiber. Both directions are switched to the protective fiber in bidirectional switching. When using bidirectional transmission, the exchange is almost always bidirectional by default, because when the fiber is cut, traffic will be lost in both directions (if there is a device failure, the two directions may not be lost, instead staple fiber). Because it is simple to switch the function service to the protection path at the receiving end without the need for a signaling system between the receiver and the host. Unidirectional protection switching is commonly employed in combination with a dedicated protection scheme. When the service is transmitted simultaneously on the one-way dedicated protection function and the protected route, the receiver at the end of the route will actively select the best signal from the two incoming signals. However, if a two-way switch is mandatory, the receiver must notify the transmitter that the disconnect has occurred. This requires the use of an automatic protection switching signal system (APS).

If the node's receiver detects that the fiber is broken, it turns off the working fiber's transmitter and switches to the protected fiber to broadcast traffic. When the running fiber loses signal, the receivers of other nodes will detect it and transmit the traffic to the protection fiber. The actual APS protocols castoff in SONET and optical networks are much more difficult than those described here because they must deal with various situations.

In a two-way communication system, both the source and the destination can identify fiber cuts, where information is communicated in both directions through a single fiber. Although the fiber cut does not require the use of the APS protocol, one-way failures and other maintenance functions will require the use of the APS protocol. In the event of a shared protection scheme, the APS protocol is also compulsory to entree the shared protection bandwidth in a synchronous manner. Because bidirectional protection switching is easier to administer and handle in more complicated networks, most shared protection systems use it instead of unidirectional switching [11].

16.3.1 Link Protection

During the connection establishment, the alternate path and wavelength are mapped about every link on the main route. When a link breaks, any connections that go via it are diverted around it, and the connection's source and destination nodes remain unaffected [12, 13] (Figure 16.1).

16.3.2 Path Protection

Each connection's source and destination nodes allocate active and standby routes from one end to the other during the connection setup procedure. When a link breaks, backup resources are used, and the source and destination nodes for every connection that travels over the misplaced link are notified of the loss. The entire light path, from source to destination, is protected by these mechanisms. When a failure occurs, it is not essential to locate the issue; rather, traffic is routed to a backup path where the link and node do not cross [12, 13]. A detailed explanation of the road safety system is shown in Figure 16.2

16.3.2.1 Current Fault Protection Techniques

Contemporary optical network protection approaches are classified as indicated in Figure 16.3 by Guido Maier et al. [9], and are fleetingly addressed as follows.

16.3.2.2 Path Protection in Mesh Network

In a mesh network, optical path protection technologies include 1+1 (repetition), 1:1 (dedicated), and 1:N (shared). The traffic from the source to the destination is simultaneously transmitted on two different optical fibers (usually through discontinuous routes). The 1+1 protection mode is adopted. The destination end is only one of the two received lines, assuming unidirectional protection switching. If the fiber is changed, the destination

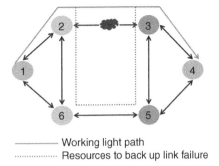

——— Working light path
·············· Resources to back up link failure

Figure 16.1 Link protection.

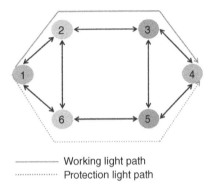

——— Working light path
·············· Protection light path

Figure 16.2 Path protection.

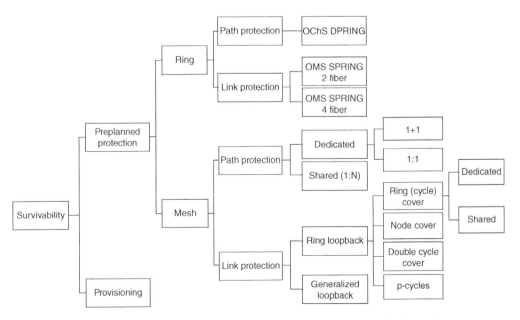

Figure 16.3 Classification of different fault protection techniques. Source: [9]/Springer Nature.

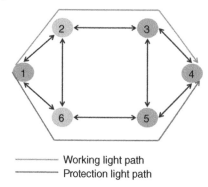

————— Working light path
————— Protection light path

Figure 16.4 1+1 dedicated protection.

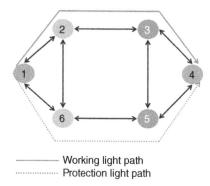

————— Working light path
·············· Protection light path

Figure 16.5 1:1 dedicated protection.

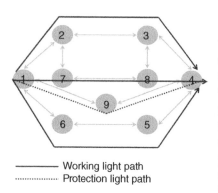

————— Working light path
·············· Protection light path

Figure 16.6 1:N shared protection.

immediately changes to another fiber and receives data as usual. This category of security is fast and does not require a communication mechanism between the two parties. There are still two threads from source to destination, 1:1 protection. On the other hand, traffic is sent through individual fiber at a period, just like a fiber in operation. If the fiber is expurgated, both the source and target will change to another protection fiber. As mentioned above, the motioning between the source and the destination needs to use the APS protocol. Due to the additional communication overhead, 1:1 protection cannot recover traffic as fast as 1+1 unidirectional protection. However, associated with 1+1 protection, it has two main advantages. The first is to protect the fiber from being used in normal operation. Therefore, it can be used to send lower priority data [12–18]. Another benefit is that the 1:1 shield can be extended to allow multiple working fibers to segment a single shield fiber. N occupied fibers share a single shield fiber in a more general 1:N shield arrangement. This configuration can withstand the loss of a single functional fiber. The APS protocol must verify that in the situation of multiple faults, only the traffic on single of the unsuccessful fibers is transferred to the protection fiber [18] (Figures 16.4–16.6).

16.3.2.3 Path Protection in Ring Networks

OChDPRing technology is useful to a ring that practices two optical fibers to transmit signals in reverse directions. Path protection is achieved by spending two optical fibers to create two reverse-broadcasting optical channels around the ring [9]. The source portion continuously communicates on the occupied connection and the protection connection. This method is effective in point-to-point, ring, and grid designs. This technology is also named OChDPRing (OCh dedicated protection ring) in the context of the ring (Figure 16.7). Also, subsequently protection bandwidth is not communal across multiple networks, and this strategy wastes bandwidth. However, it remains one of the most basic protection mechanisms, and several vendors have incorporated it into optical add/drop multiplexers and crossovers [17].

16.3.2.4 OMS Link Protection-OMS-SPRing (Optical Multiplex Section-Shared Protection Ring)

OChSPRing (shared protection ring) works at the optical channel layer, not the optical multiplex section layer. The shortest path along the ring is used to create a working optical path. When the working optical path fails, use an interval switch or a ring switch to restore it. Non-overlapping optical channels in the ring can represent similar protection wavelengths. This spatial multiplexing makes OChSPRing more efficient than OChDPRing in processing distributed traffic. OChSPRing works in a similar way, but the fiber now corresponds to the wavelength and the connection to the optical path. OMSSPRing can be split into two- or four-fiber mode depending on the physical configuration. Each fiber pair in a four-fiber system has a second pair assigned to backup traffic. In a dual fiber system, half of the WDM channels carry working traffic from a fiber pair in one direction on each fiber, while the other half is used as a backup resource to protect the other connection [9].

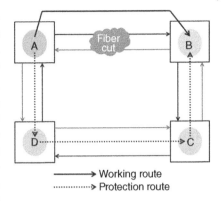

Figure 16.7 OCh-DPRing.

16.3.2.5 Ring Loopback

The network is separated into multiple groups of fibers, each of which is handled as a particular ring. Every ring is provided through an OMS protection mechanism after disintegration [3]. Ring loopback is supplementary divided into three categories: Ring Cover, Double Cycle Cover, and P-Cycles.

Ring Cover: The mesh network is represented graphically, with each node acting as a vertex and every link acting as an edge. The system is separated into smaller rings (rings) so that each edge is covered by at least one of them. Alternative protection fiber parallel to the cycle defends the entire cycle. In this technique, an edge can be protected by more than one period. Every cycle is a four-fiber ring based on fiber protection. Consequently, an edge with two loops needs 8 fibers and an edge with n loops requires 4n fibers. The main purpose of ring coverage technology is to trace a collection of rings that cover all networks and then use these rings to defend the network from faults. Redundancy reduction is the goal of this strategy, because certain network traffics in the ring coverage can be used in multiple rings, resulting in higher network redundancy [19, 20].

Double Cycle Cover: Dual-cycle coverage is a four-fiber ring coverage method. The premise is that every network link has a couple of neutralization protection and working fibers [21]. Directed graphics are used to represent mesh networks (graphs). There are a duo of unidirectional working fibers (bidirectional working connection) and a pair of unidirectional backing fibers (bidirectional shielded connection) at the edge of each graph. Therefore, every edge is perfectly shielded by two rings. The dual-sheath cycle strategy maintains 100% redundancy by providing a protective fiber for each working fiber (such as in a SONET ring). The solution was originally to solve the redundancy problem added by the ring coverage scheme [22] (Figure 16.8).

Figure 16.8 Double cycle ring covers technology.

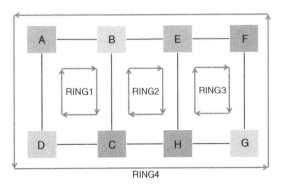

P-Cycles: The p-cycle technology relies on the ability of the ring to cover not only its own connection, but similarly any probable string loop linking two non-adjacent ring nodes. The cycle reduces the amount of redundancy required to defend the mesh network from link failures. The connection cycle and the node cycle are two categories of cycles. The p period of the link protects all its channels, although the p period of the node protects all its connections. One of the important benefits of loops is that they can save key factors, and they are widely regarded as the most effective capacity minimization protection structure. P-cycles can protect cycle links. In addition, p-cycles can protect crossover links [17] (Figure 16.9).

Generalized Loopback Technique: Ring-based techniques might also involve a technique known as the generalized loopback technique. Though it is not strictly speaking one of the mesh-based ring protection systems, its usage of a loopback process to convert the signal from functioning to redundant capacity is comparable to the APS operation in rings.

This method of avoiding the utilization of rings was explored by Medard et al. [23]. Instead, a secondary digraph is supported up by a primary digraph. After a fault, the stream passed by the primary digraph is transmission onto the secondary digraph via the failed connection.

Restoration: Restoration is a reactive technique to failure recovery in which network resources are not reserved ahead of time. When a network fails, a exploration is conducted to locate accessible network resources so that network functionality can be restored utilizing these resources [24].

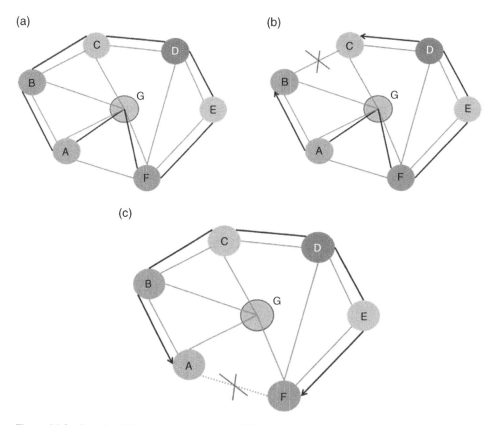

Figure 16.9 P-cycle: (a) Navigating nodes A-G. (b) Protection of a fault on cycle. (c) Protection of a fault on straddling path.

Restoration can also be used in the occurrence of a link or path failure. Depending on the sort of rerouting, restoration can be characterized as link, sub-path, or path-based.

- **Link Restoration:** In the happening of a failure, the failed link's end nodes participate in a distributed mechanism that dynamically finds an alternative path about the link for every dynamic wavelength that passes through it. When a fresh wavelength channel path is exposed about a unsuccessful link, the failed link's end nodes modify their OXCs (Optical Interconnects) to redirect that channel to the fresh route. The connection is terminated if no alternate route and corresponding wavelength can be discovered for an interrupted connection.
- **Path Restoration:** On an endways basis, the source and destination nodes of every connection autonomously determine a backup way, which can be on multiple wavelengths channel. When a fresh route and wavelength network for a linking is identified, network essentials like OXCs are altered and the connection is swapped to the new path. If there is no novel route, the connection is terminated.
- **Sub-Path Restoration:** When a linking fails, the upper node identifies the failure and identifies a alternative path for every episodic connection from itself to the associated destination node. Subsequent OXCs are adjusted accordingly and the link changes to the novel path after fruitful detection of resources for the novel backup route. If there aren't enough resources available, the connection is terminated.

16.3.2.6 Current Restoration Techniques

Restoration approaches are classified by Ilyas and Muftah [25] based on the functionality of the cross-connects. Recovery solutions can be static, where the network matrix is known ahead of time, or dynamic, where connections are made as and when they come with no prior information of the appearance time, depending on the traffic difficulties. Depending on the category of network restoration method, it can be integrated, which is used in reduced networks with a organized traffic matrix, or scattered, which is used in enormous networks where connections attain in a circulated fashion and every node executes its own routing and provisioning calculation. Restoration techniques are also characterized by their major performance metric, such as restoration speed, blockage likelihood, link consumption, scalability, or restorability. Figure 16.10 shows a brief summary of the restoration processes.

- **Proactive Restoration:** Alternate routes are determined accordingly in this method. The connection is simply redirected to the previously computed path after the problem occurs.
- **Reactive Restoration:** After the fault has occurred, alternate pathways are designed. After a failure, classic reactive methods flood packages into the network to search for obtainable capability and create a new path.
- **Link-Based Scheme:** Demand should be rerouted around the broken link.
- **Path-Based Scheme:** As an alternate path, a completely different path is used. Dedicated alternate schemes backup the standby route for the specific demand.
- **Backup Multiplexing:** This authorizes additional backup pathways on one or more lines to share the same wavelength. This technique is created on the single failure hypothesis, which circumstances that only a link or node would flop at any one moment. Under the single fault assumption, two fault disjoint backup paths can share the same backup wavelength because only one of the allotment standby tracks will be triggered, resulting in no clash.
- **Primary Multiplexing:** A primary path on a wavelength can be put up as a backup for one or more needs. As a result, the backup becomes unusable during the primary demand and returns to facility where the primary request has accepted.

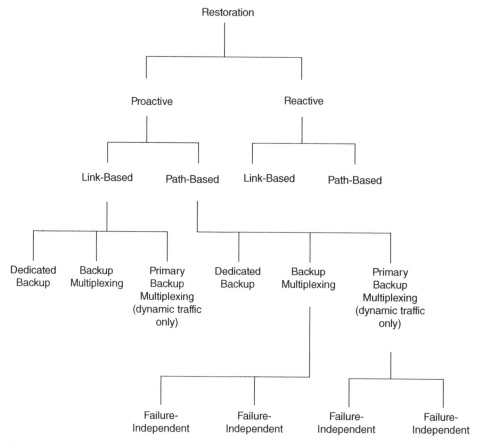

Figure 16.10 Classification of different restoration techniques. Source: [25]/CRC Press.

- **Failure-Dependent Scheme:** This is a subsection of path-based arrangements in which the alternate path is resolute by the specific failure. Therefore, for each probable failure on the path, a backup path is provided.
- **Failure-Independent Scheme:** These systems look for a link and node detached track to use as a backup. This path can thus be used irrespective of which connection has failed.

16.4 Energy Efficiency of Optical Switching Technology

Optical layer recovery redirects and restores the data around the network interference without the network loss of dedicated 1+1 optical protection, which is 50%. Operators benefit from automatic service restoration by improving network efficiency and capacity utilization, thereby reducing costs and gaining more space to access shared revenue-generating services. When a failure happens, the system automatically recognizes, determines a novel restoration path, and redirects any affected wavelengths to the new path without manual intervention. In fact, the network can control itself and repair itself. The access protection function and the ability to redirect wavelengths in real time are critical to improving network efficiency. Increase network capacity, improve network usage, and reduce overall costs. A single shared recovery channel in a real-world mesh network can support up to four or five working paths, requiring as little as 15–20% of the total capacity to protect the network. Operators can also combine 1+1 protection for critical business services with restored wavelengths in the optical layer on the same network for optimal performance and efficiency.

16.5   Signal Quality Monitoring Techniques

Due to the speedy expansion of WDM technology in the past era, the bandwidth of optical fiber systems has increased to more than 50 Tb/s. Since each optical fiber carries such a large quantity of traffics, even a passing interruption may cause catastrophic values to all areas of society. Therefore, network service providers must be very careful to avoid network failures. Previously, SONET/SOH layer bit interleaved parity (BIP) bytes were commonly used for this purpose. On the other hand, WDM networks are quickly moving from one point to another to continuously reconfigure the all-optical network. In other words, monitoring the BIP bytes before the signal reaches the terminal is no longer possible. Contemporary active WDM systems, on the other indicator, are more vulnerable to optical layer faults such as ASE noise, fiber nonlinearity, residual scattering, polarization mode scattering, and other polarization-based possessions, as well as crosstalk from multipath interference, due to their higher transparency and extreme range. Therefore, the ability to directly monitor the restrictions that distress the performance of the network in the optical layer is vital for the correct process and managing of these active WDM networks [26]. The primary goal of the optical monitoring and evaluation technology is to reduce downtime and increase network accessibility. It can similarly support to optimize the use of network resources by dropping the system boundary essential for error-free performance.

In light wave systems, optical layer controlling technology has been used to detect fiber extensively and connection failures and deliver feedback indications for adaptive compensators. Owing to the extensive use of ROADM, this technology is now commonly used in today's WDM networks to verify signal eminence like channel strength and wavelength. Rather, optical performance analysis is now widely seen as a significant technology driver for the evolution to the succeeding generation of flexible all-optical networks. The Optical Performance Monitor (OPM) on the market today is primarily a miniature version of a spectrum analyzer that uses a tunable bandpass filter or deflection grating. Therefore, these OPMs can identify the optical power and wavelength of every WDM channel and evaluation optical signal-to-noise ratio (OSNR) by linearly introducing the ASE level left from the signal.

Pilot tone is another monitoring technique [27–30]. For example, pilot tones have been used to display various optical features of the WDM broadcast, such as wavelength, channel power, and OSNR. These characteristics can be monitored using pilot-based techniques instead of expensive demultiplexer filters. Therefore, this technique has the potential to be extremely profitable. Furthermore, because the pilot tone must monitor its consistent optical signal everyplace in the network, it is very much appropriate for use in active WDM networks [27]. Therefore, observing the tone frequency of every WDM signal can be used to control its optical path [27, 30]. However, the cross-gain modulation and the enthused Raman effect of EDFA will ominously reduce the performance of this method [27]. Pilot tone, RF spectrum analysis, clock amplitude capacity, eye diagram investigation, and DOP investigation belong to the different checking methods discussed for chromatic scattering and polarization mode scattering [31–34]. As mentioned above, several OPM methods have been projected to display various metrics on active WDM networks. But it is almost incredible to observe every parameter at every point in the network. Several strategies have been recommended to observe multiple constraints instantaneously to overcome this challenge [35–37]. The approach based on the amalgamation of asynchronous sampling and pattern appreciation seems particularly interesting [35]. In addition, some efforts have been made to observe the performance by determining only characteristics that are susceptible to changes like ASE noise and MPI crosstalk [37].

16.6 Challenges and Recent Research Trends

Network survivability is projected to confirm that network operators can remain to deliver services in the occurrence of a disaster. According to research and news reports, network disappointments caused by physical outbreaks and natural calamities have a substantial effect on optical networks. Such network faults can cause part of the network to stop working, causing service disruption and increasing congestion on the respite of the network. Consequently, fault-tolerant and disaster-tolerant optical networks have concerned the devotion of the research community and have developed a major topic in research over the last ten years. To deal with network component failures, some research on protection and recovery methods has been completed. This research reviews previous research to critically examine issues such as protection, recovery, cascading failures, disaster-related failures, etc. In [38–42], various survival technologies for optical networks based on WDM have been considered. References [38] and [43] discuss various restoration methods for mesh and ring optical networks. Reference [44] even describes a set of modular techniques for handling normal and protection routes. To reduce network congestion at an affordable price, protection and recovery are provided on a link-based or route-based basis [45–48]. The reserve capacity is reserved in the protection scheme and remains free till a fault happens. As an outcome, the effectiveness of network exploitation is low, but the recovery speediness is assured.

In addition, according to reference [48], only in the design of the backup network, the cumulative effect of failures can be reduced by up to 40%, indicating that the protection is not fully effective. When a node or link fails, recovery refers to the willingness to redistribute blocked traffic through a dynamically discovered route. Resources used for recovery will not reserve unused network capacity, but will accommodate it. Although recovery time and overload may be longer, it is more effective than a protection strategy in terms of capacity. To use these two systems, a hybrid strategy (combination of protection and recovery) can be implemented [49]. The recovery plan is pre-built in the hybrid plan, but the possessions are assigned in actual time. Failures of fiber-optic connections and optical crossover (OXC) can cause the channel to collapse, resulting in a large quantity of data loss. Ramamurthy et al. [50] studied protection switching and recovery time of dispersed protocols for path and link recovery. Ramamurthy et al. investigated how to adjust capacity usage and vulnerability to a large number of connection failures. Reference [51] proposed various types of network protection solutions, including SONETSDH, Multiprotocol Label Switching (MPLS), Internet Protocol (IP), and so on. In reference [49], if the protection approach is used in a capacity effective way, it will permit quicker recovery from failures. On the other hand, path protection decreases capability because every path is enclosed by a separate backup path [40, 52]. Dedicated protection expands service continuity, QoS, and ASR connection, however does not allow optimization of resource exploitation [53]. In WDM optical networks, a shared backup path protection arrangement can be used to elucidate this problem, however, the recovery time of dedicated protection is shorter than that of shared protection [40, 52, 54, 55].

It is further recommended to separate the main route from its vulnerable parts (sub-routes). Associated with path protection, sub-path protection delivers speedy scalability and recapture at the expense of resource usage. For each link formed according to the requested connection, link protection can be realized through a backup path with disjoint links [56–58]. Based on preliminary research, most research emphases on a single node or single link failure at a specific time. Single-link failures are easier to manage than multilink failures, and are likely to be the root cause of dual-link loss. The multiple failure model is used to evaluate the impact of two link failures for different protection strategies, and in their white paper, Kim et al. provide a recovery strategy for damaged and congested shared paths [59]. In examining trade-offs the cost of capacity and survivability is also calculated. Choi et al. [56] studied double link failure retrieval using a failure model, in which two links can flop at any time. It is claimed that the use of the proposed technology can achieve 100% recovery from dual link failures by progressively cumulative the backup capacity. For simultaneous double bond failures, a mixture of protection and recovery methods is projected to re-establish the linking with the lowest existing unexploited capacity [60]. For an optical network with two link failures, a mixture of protection and recovery strategies is recommended [61], and the additional link failure happens during the recovery period of the first unsuccessful link. Consequently, shorter and longer path distances are used for the main path and the backup path. The backup capacity is reserved by the protection element of the hybrid equipment so that the affected demand can be reinstated through the pre-planned backup path, while the restoration element constantly looks for another route to restore the remaining affected demand. Many cyclo-based strategies have been recommended to protect against single connection failures, but Feng et al. [62] developed a cycle-based integer linear programming (ILP) model was developed to reduce two link failures using fixed load recovery. In [62], two additional active flow algorithms are also designed and tested for different traffic loads and capacities, namely Shortest Path Pair Protection (SPPP) and Short Full Path Protection (SFPP). Based on simulation data, SFPP is more effective than SPPP in cumulative traffic capacity, but is less likely to hang when there is heavy traffic.

When a failure occurs, a backup path is provided according to the updated link state information of the nodes and links in the recovery scheme [39, 40, 63–66]. It occupies more resources than protection, but the recovery time is longer than protection, because once a failure occurs, this technology will dynamically identify the backup path. There are three types of recovery: link-based recovery, (ii) route-based recovery, and (iii) segment-based recovery [39]. The recovery link takes the least time to recover from a failure, while the recovery path takes the longest. Providing a more survivable path for network connections can recover the consistency of network connections, which is crucial for several critical network facilities (banking, big data transmission, etc.). To achieve effective recovery in the WDM optical network [67] an HCA was proposed.

This method is only used to handle single link failures and relies on adapted resource distribution to reserve wavelengths in progress during connection formation for the backup optical path. An active recovery strategy is proposed in [68], which increases the use of network resources while reducing the possibility of congestion. In [69], Chen et al. established the Dynamic Load Balancing Shared Path Protection (DLBSPP) process, which uses a flow-conscious recovery apparatus to compute disjoint backup paths of links for multilink failure traffic. Jara et al. [70] proposed a wavelength measurement and routing method, called the cheapest path-based fault-tolerant method, to study the simultaneous occurrence of k link faults in a dynamic WDM optical network (FTBCP). FTBCP calculates the main route and backup route offline with the maximum traffic load to minimize the longest recovery time, and then distributes these calculated routes as necessary to reduce failures k $\geq$1.

Blocking with cut-off frequency of fiber and heavy traffic loss possibly, as traffic flooding and network congestion becomes a serious problem, network survivability converts an significant issue in network strategy and real-time process. It is important for network workers to develop actual strategies to improve from network and node connection failures. Most WDM network survivability research focuses on single link failure recovery. In a real network, many failures (that is, occurring almost simultaneously) are also possible, and appropriate recovery strategies can be implemented.

16.7 Conclusion

Optical networks are being installed all over the world to provide large-scale connections and data broadcast. Network constituent failures, especially fiber-optic outages, are common, and disaster-related failures are infrequent; but they were forced to change the network. Therefore, due to the huge data traffic on the optical path in the optical network, it is essential to develop appropriate protection and recovery technologies to avoid any loss of data or revenue. One can use the existing optical network protection and recovery solution classification. In the future, real-world networks are conceived as a combination of multiple elastic systems. A hybrid strategy will be selected based on service differentiation, optical network type, speed, efficiency, recovery time, and cost trade-offs. At the network interface, the optical network topology provides the intelligence necessary for efficient routing and rapid recovery from faults. The work of optical networks must be expanded, as well as the intelligence necessary to properly manage network problems and provide protection and recovery. Furthermore, it is important for network operators to develop solutions that not only solve the aforementioned network difficulties, but also solve problems such as cascading failures and network congestion. By optimizing capacity, a multi-method framework can provide greater protection and survivability from disasters. This will support fault-tolerant supervision to improve the disaster recovery capabilities of the network. In addition, new optical devices like DWDM multiplexers, ADM (add/drop multiplexers), and OXC allow the use of smart all-optical cores, where data packets can be sent over the network without leaving the optical domain.

Bibliography

1 S. Subramaniam and D. Zhou. Survivability in optical networks. *IEEE Network*, 14(6):16–23, 2000.

2 J.C. McDonald. Public network integrity/spl mdash/avoiding a crisis in trust. *IEEE Journal on Selected Areas in Communications*, 12(1):5–12, 1994.

3 L. Nederlof, K. Struyve, C. O'Shea, H. Misser, Y. Du, and B. Tamayo. End-to-end survivable broadband networks. *IEEE Communications Magazine*, 33(9):63–70, 1995.

4 R.C. Linger, N.R. Mead, and H.F. Lipson. Requirement Definition for Survivable Network Systems. *Requirements Engineering, Proceedings, Third International Conference*: 14–23, 1998.

5 A.K. Todimala and B. Ramamurthy. A dynamic partitioning protection routing technique in WDM networks. *Cluster Computing*, 7(3):259–269, 2004.

6 C.V. Saradhi, L.K. Wei, and M. Gurusamy. Provisioning fault-tolerant scheduled lightpath demands in WDM mesh networks. *IEEE Xplore*, 2004. doi: 10.1109/BROADNETS.2004.70.

7 R. Shenai and K. Sivalingam. Hybrid survivability approaches for optical WDM mesh networks. *Journal of Lightwave Technology*, 23(10): 3046, 2005.

8 S. Baroni, J.O. Eaves, M. Kumar, M.A. Qureshi, A. Rodriguez-Moral, and D. Sugerman. Analysis and design of backbone architecture alternatives for IP optical networking. *IEEE Journal on Selected Areas in Communications*, 18(10):1980–1994, 2000. doi: 10.1109/49.887918.

9 G. Maier, A. Pattavina, S. De Patre, and M. Martinelli. Optical network survivability: protection techniques in the WDM layer. *Photonic Network Communications*, 4(3):251–269, 2002. doi: 10.1023/A:1016047527226.

10 R. Batchellor. Optical layer protection: Benefits and implementation. In *Proceedings of National Fiber Optic Engineers Conference*, 1998.

11 B.T. Doshi, S. Dravida, P. Harshavardhana, O. Hauser, and Y. Wang. Optical network design and restoration. *Bell Labs Technical Journal*, 4(1):58–84, 1999. doi: 10.1002/bltj.2147.

12 S. Neeraj Mohan. Network protection and restoration in optical networks: a comprehensive study. *International Journal of Research in Engineering and Technology (IJRET)*, 2(1):50–54, 2013.

13 H. Saini and A.K. Garg. Protection and restoration schemes in optical networks: a comprehensive survey. *International Journal of Microwaves Applications*, 2(1):5–11, 2013.

14 G. Ellinas, A.G. Hailemariam, and T.E. Stern. Protection cycles in mesh WDM networks. *IEEE Journal on Selected Areas in Communications*, 18(10):1924–1937, 2000. doi: 10.1109/49.887913.

15 S. Ramamurthy and B. Mukherjee. Survivable WDM mesh networks. Part I – Protection, *IEEE Xplore*, 2:744–751, 1999. doi: 10.1109/INFCOM.1999.751461.

16 S. Ramamurthy and B. Mukherjee. Survivable WDM mesh networks. II. Restoration. *IEEE Xplore*, 3:2023–2030, 1999. doi: 10.1109/ICC.1999.765615.

17 A. Askarian, S. Subramaniam, and M. Brandt-Pearce. Evaluation of link protection schemes in physically impaired optical networks. *IEEE Xplore*, 1–5, 2009. doi: 10.1109/ICC.2009.5199008.

18 A.E. Kamal. OPN04-3: 1+N Protection in mesh networks using network coding over p-cycles. *IEEE Xplore*, 1–6, 2006. doi: 10.1109/GLOCOM.2006.378.

19 M. Medard, R.A. Barry, S.G. Finn, W. He, and S.S. Lumetta. Generalized loop-back recovery in optical mesh networks. *IEEE/ACM Transactions on Networking*, 10(1):153–164, 2002. doi: 10.1109/90.986592.

20 M. Keshtgary, F.A. Al-Zahrani, and A.P. Jayasumana. Network survivability performance evaluation with applications in WDM networks with wavelength conversion. *IEEE Xplore*, 11:15–27, 2004. doi: 10.1007/s11107-006-5320-4.

21 R.J. Ellison, R.C. Linger, T. Longstaff, and N.R. Mead. Survivable network system analysis: a case study. *IEEE Software*, 16(4):70–77, 1999.

22 G. Ellinas, A.G. Hailemariam, and T.E. Stern. Protection cycles in mesh WDM networks. *IEEE Journal on Selected Areas in Communications*, 18(10):1924–1937, 2000. doi: 10.1109/49.887913.

23 M. Medard, R.A. Barry, S.G. Finn, W. He, and S.S. Lumetta. Generalized loop-back recovery in optical mesh networks. *IEEE/ACM Transactions on Networking*, 10(1):153–164, 2002. doi: 10.1109/90.986592.

24 S. Ramamurthy, L. Sahasrabuddhe, and B. Mukherjee. Survivable WDM mesh networks. *Journal of Lightwave Technology*, 21(4):870–883, 2003. doi: 10.1109/JLT.2002.806338.

25 M. Ilyas and H.T. Muftah. *The Handbook of Optical Communication Networks*, Series Editor Richard C. Dorf University of California, Davis, CRC Press, 2003.

26 Y.C. Chung. Performance monitoring in optical networks (Tutorial). APOC 2003, 2003.

27 H. Ji, K. Park, J. Lee, H. Chung, E. Son, K. Han, S. Jun, and Y. Chung. Optical performance monitoring techniques based on pilot tones for WDM network applications. *Journal of Optical Networking*, 3(7):510–533, 2004.

28 G. Rossi, T.E. Dimmick, and D.J. Blumenthal. Optical performance monitoring in reconfigurable WDM optical networks using subcarrier multiplexing. *Journal of Lightwave Technology*, 18(12):1639–1648, 2000. doi: 10.1109/50.908673.

29 H.C. Ji, P.K.J. Park, H. Kim, J.H. Lee, and Y.C. Chung. A novel frequency-offset monitoring technique for direct-detection DPSK systems. *IEEE Photonics Technology Letters*, 18(8):950–952, 2006. doi: 10.1109/LPT.2006.872325.

30 K.J. Park, C.J. Youn, J.H. Lee, and Y.C. Chung. Optical path, wavelength and power monitoring technique using frequency-modulated pilot tones. *OFC 2004*, 2:3, 2004.

31 K.J. Park, C.J. Youn, J.H. Lee, and Y.C. Chung. Performance comparisons of chromatic dispersion-monitoring techniques using pilot tones. *IEEE Photonics Technology Letters*, 15(6):873–875, 2003. doi: 10.1109/LPT.2003.811337.

32 G.-W. Lu, M.-H. Cheung, L.-K. Chen, and C.-K. Chan. Simultaneous PMD and OSNR monitoring by enhanced RF spectral dip analysis assisted with a local large-DGD element. *IEEE Photonics Technology Letters*, 17(12):2790–2792, 2005. doi: 10.1109/LPT.2005.859162.

33 F. Buchali, S. Lanne, J.-P. Thiery, W. Baumert, and H. Bulow. Fast eye monitor for 10 Gbit/s and its application for optical PMD compensation. 2001 OSA Technical Digest Series (Optical Society of America, 2001), 2001.

34 M. Petersson, H. Sunnerud, M. Karlsson, and B.-E. Olsson. Performance monitoring in optical networks using Stokes parameters. *IEEE Photonics Technology Letters*, 16(2):686–688, 2004. doi: 10.1109/LPT.2003.822244.

35 T.B. Anderson, S.D. Dods, K. Clarke, J. Bedo, and A. Kowalczyk. Multi-Impairment Monitoring for Photonic Networks. *33rd European Conference and Exhibition of Optical Communication*, 1–4, 2007. doi: 10.1049/ic:20070143.

36 B. Kozicki, A. Maruta, and K. Kitayama. Experimental demonstration of optical performance monitoring for RZ-DPSK signals using delay-tap sampling method. *Optics Express*, 16(6):3566–3576, 2008.

37 H.Y. Choi, S.B. Jun, S.K. Shin, and Y.C. Chung. Simultaneous monitoring technique for ASE and MPI noises in distributed Raman amplified systems. *Optics Express*, 15:8660–8666, 2007.

38 S. Subramaniam and Dongyun Zhou. Survivability in optical networks. *IEEE Network*, 14(6):16–23, 2000. doi: 10.1109/65.885566.

39 J. Zhang and B. Mukheriee. A review of fault management in WDM mesh networks: basic concepts and research challenges. *IEEE Network*, 18(2):41–48, 2004. doi: 10.1109/MNET.2004.1276610.

40 B. Mukherjee. *Optical WDM Networks*. New York, Springer Science & Business Media, 2006.

41 S. Rani, A.K. Sharma, and P. Singh. Survivability strategy with congestion control in WDM optical networks. *International Symposium on High Capacity Optical Networks and Enabling Technologies*, 1–4, 2007. doi: 10.1109/HONET.2007.4600270.

42 Z. Zhang, Z. Li, and Y. He. *Network Capacity Analysis for Survivable WDM Optical Networks. Advances in Intelligent and Soft Computing*. Berlin, Heidelberg, Springer, 2012. doi: 10.1007/978-3-642-27334-6_33.

43 S. Sengupta and R. Ramamurthy. From network design to dynamic provisioning and restoration in optical cross-connect mesh networks: an architectural and algorithmic overview. *IEEE Network*, 15(4):46–54, 2001. doi: 10.1109/65.941836.

44 R. Gupta, E. Chi, and J. Walrand. Different algorithms for normal and protection paths. *Fourth International Workshop on Design of Reliable Communication Networks, (DRCN 2003)*, 189–196, 2003. doi: 10.1109/DRCN.2003.1275356.

45 S. Ramamurthy and B. Mukherjee. Survivable WDM mesh networks. Part I – Protection, *IEEE Xplore*, 2:744–751, 1999. doi: 10.1109/INFCOM.1999.751461.

46 H. Saini and A. Garg. Protection and restoration schemes in optical networks: a comprehensive survey. *International Journal of Microwaves Applications*, 2:5–11, 2013.

47 B.C. Chatterje, N. Sarma, P.P. Sahu, and E. Oki. Literature survey. In *Routing and Wavelength Assignment for WDM-Based Optical Networks*. Cham, Switzerland, Springer, 17–34, 2017.

48 P. Gill, N. Jain, and N. Nagappan. Understanding network failures in data centers. *ACM SIGCOMM Computer Communication Review*, 41(4):350–361, 2011.

49 S. Neeraj Mohan. Network protection and restoration in optical networks: a comprehensive study. *International Journal of Research in Engineering and Technology*, 2:50–54, 2013.

50 S. Ramamurthy, L. Sahasrabuddhe, and B. Mukherjee. Survivable WDM mesh networks. *Journal of Lightwave Technology*, 21(4):870–883, 2003.

51 J.-P. Vasseur, M. Pickavet, and P. Demeester. *Network Recovery: Protection and Restoration of Optical, SONET-SDH, IP, and MPLS*. Elsevier: San Francisco, CA, USA, 2004.

52 G. Maier, A. Pattavina, S. De Patre, and M. Martinelli. Optical network survivability: Protection techniques in the wdm layer. *Photonic Network Communications*, 4:251–269, 2002.

53 S.N.F. Binti Halida, S. Idrus, M. Farabi, and N. Zulkifli. Dedicated protection scheme for optical networks survivability. In *Proceedings of the 2011 4th International Conference on Modeling, Simulation and Applied Optimization (ICMSAO)*, Kuala Lumpur, Malaysia, 1–5, 2011.

54 H. Alshaer. Dynamic connection provisioning with shared protection in IP/WDM networks. *International Journal of Communication Systems*, 27:2832–2850, 2014.

55 C. Ou, H. Zang, N.K. Singha, K. Zhu, L.H. Sahasrabuddhe, R.A. MacDonald, and B. Mukherjee. Subpath protection for scalability and fast recovery in optical WDM mesh networks. *IEEE Journal on Selected Areas in Communications*, 22:1859–1875, 2004.

56 H. Choi, S. Subramaniam, and H.-A. Choi. On double-link failure recovery in wdm optical networks. In *Proceedings of the Twenty-First Annual Joint Conference of the IEEE Computer and Communications Societies INFOCOM*, New York, NY, USA, 808–816, 2002.

57 S. Ramasubramanian and A. Chandak. Dual-link failure resiliency through backup link mutual exclusion. *IEEE/ACM Transactions on Networking*, 16:157–169, 2008.

58 Y. Guo, F. Kuipers, and P. Van Mieghem. Link-disjoint paths for reliable qos routing. *International Journal of Communication Systems*, 16:779–798, 2003.

59 S.-I. Kim and S. Lumetta. Multiple Failure Survivability in WDM Mesh Networks; Report No. UILU-ENG-06-2205; Coordinated Science Laboratory: Urbana, IL, USA, 2006.

60 M. Sivakumar, C. Maciocco, M. Mishra, and K.M. Sivalingam. A hybrid protection-restoration mechanism for enhancing dual-failure restorability in optical mesh-restorable networks. In *Proceedings of the OptiComm 2003: Optical Networking and Communications*, Dallas, TX, USA, 37–48, 2003.

61 L. Ruan and T. Feng. A hybrid protection/restoration scheme for two-link failure in wdm mesh networks. In *Proceedings of the Global Telecommunications Conference (GLOBECOM 2010)*, Miami, FL, USA, 1–5, 2010.

62 T. Feng, L. Long, A.E. Kamal, and L. Ruan. Two-link failure protection in WDM mesh networks with p-cycles. *Computer Network*, 54:3068–3080, 2010.

63 E. Bouillet. Path Routing in Mesh Optical Networks. Chichester, John Wiley & Sons, 2007.

64 B. Jaumard, M. Bui, B. Mukherjee, and C.S. Vadrevu. IP restoration vs. optical protection: Which one has the least bandwidth requirements? *Optical Switching and Networking*, 10:261–273, 2013.

65 Y. Zhao, X. Li, H. Li, X. Wang, J. Zhang, and S. Huang. Multi-link faults localization and restoration based on fuzzy fault set for dynamic optical networks. *Optics Express*, 21: 1496–1511, 2013.

66 A. Kadohata, T. Tanaka, W. Imajuku, F. Inuzuka, and A. Watanabe. Rapid restoration sequence of fiber links and communication paths from catastrophic failures. *IEICE Transactions on Fundamentals of Electronics, Communications and Computer Sciences*, 99:1510–1517, 2016.

67 D.S. Yadav, S. Rana, and S. Prakash. Hybrid connection algorithm: A strategy for efficient restoration in wdm optical networks. *Optical Fiber Technology*, 16:90–99, 2010.

68 S. Rani, A.K. Sharma, and P. Singh. Restoration approach in wdm optical networks. *Optik*, 118:25–28, 2007.

69 B. Chen, J. Zhang, Y. Zhao, C. Lv, W. Zhang, S. Huang, X. Zhang, and W. Gu. Multi-link failure restoration with dynamic load balancing in spectrum-elastic optical path networks. *Optical Fiber Technology*, 18:21–28, 2012.

70 N. Jara, G. Rubino, and R. Vallejos. Alternate paths for multiple fault tolerance on dynamic WDM optical networks. In *Proceedings of the 2017 IEEE 18th International Conference on High Performance Switching and Routing (HPSR)*, Campinas, Brazil, 1–6, 2017.

17

Optical Switching for High-Performance Computing

Rajendra Prasath, Bheemappa Halavar, and Odelu Vanga

Indian Institute of Information Technology (IIIT) Sri City, Chittoor, Andhra Pradesh, India

17.1   Introduction

Data centers are growing with the ever-increasing quantity of data used across various applications and services in the world wide web (WWW). It has been observed that the size and the complexity of the data are also increasing with the ever-increasing demand of High Performance Computing (HPC) applications and services. In order to meet the growing need of the data centers, it has become essential to transfer a high volume of data across multiple servers (services/applications) very fast. Several applications in modern scenarios need to transfer a large amount of data, consuming a high bandwidth. This in turn pushes the service providers and data centers to provide sufficient bandwidth so as to provide users with faster access to various services/application programming interfaces. For many applications in recent days, it is important to ensure low network latency to achieve high throughput and scalability.

Optical networks are considered as a suitable paradigm to handle bandwidth and low latency issues in various data centers and service providers. The inherent "unlimited" nature of high-capacity data transfers of optical fiber links allows transmission of data over hundreds of wavelength channels at very high data transfer rates. Additionally, optical switching provides low latency compared to traditional electronic-based switching. It has been shown that an optical packet switching may take a latency of 25 nanoseconds whereas an equivalent electronic-based switching may require 90 nanoseconds latency [1]. Optical switching technologies are considered a suitable candidate due to inherent Wavelength Division Multiplexing (WDM) to improve bandwidth and latency-related issues. Further, an optical interconnect provides low loss of data, low power consumption, and a high capacity of data transfer for point-to-point connections.

Optical switching helps to reduce large volumes of wiring of optical fibers without any pitfall in the connectivity-related issues. The waveguide bandwidth capacity of optical networks leads to the separation of signals using space-domain switching and wavelength-based routing.

Optical Switching: Device Technology and Applications in Networks, First Edition. Edited by Dalia Nandi, Sandip Nandi, Angsuman Sarkar, and Chandan Kumar Sarkar.
© 2022 John Wiley & Sons, Inc. Published 2022 by John Wiley & Sons, Inc.

The configuration speed of the underlying paths is determined by the efficiency of the switches. The following aspects are essential in classifying the class of switches on optical networks:

1) **Provisioning and Protection:** Optical add and drop multiplexers that are reconfigurable in backbone networks are used to make or terminate connections in a fraction of time units (in milliseconds).
2) **Packet Switching:** Changes in huge data traffic may allow a statistically measurable multiplexed gain whenever connections are made or terminated in nanoseconds.
3) **Bit-level Processing:** All optical techniques are used to perform various operations in picoseconds and offer significant functionalities for efficient signal processing, minimizing the energy requirements.

Switching technologies enable electronically configured routing in nanoseconds. The elements used for enabling cross-point switches for the transfer of high-radix elements with promising broadcast functionality and enhanced seemless connectivity. In the next section, we will focus on exploring the basic aspects of optical switching.

17.2 Optical Switching

Optical routing has become popular in the past few decades and the need for advanced switching technologies has risen with the growing increase in bandwidth, faster data transfer with high throughput, and low latency [2, 3]. The primary purpose of the traditional switches is to route data transmission in point-to-point communication networks. Optical switching enables transmission of data from one source node to a target node without any overhead being incurred for converting the signals into electrons. This saves the time that needs to be used for signal conversion and takes full advantage of all-optical networks. Optical switches hardly require transmission by itself, rather it is a direct approach to send the signal on its path. In the early 2000s, data transmissions sent over fiber-optic mediums were routed through electronic switches. This is not a straight-forward data transmission and the signals were required to be converted into electrons and then back into photons again. The conversion resulted in slowing the speed of data transmission and reducing the effect/performance of optical communications. Optical switching is a promising approach to minimize the cost of data transfer, to improve the data transmission rates, and to help enhance the quality of signals [4].

17.2.1 Basics of Optical Switching

Switching is an important activity in any interconnection network. In traditional interconnection networks, the date transfer rate and bandwidth are very limited. Optical networks/technologies have grown up in quick succession to enable fast data transfers with huge bandwidths. This is due to the medium of communication – optical fiber – which works on total refractive index that reflects light back to the core in specific patterns. Depending on the mode of data transmission, these patterns may vary. The reflection of light signals allows the fiber-optic cables to bend around corners without the loss of integrity of the light-based signal. Optical fiber is the fundamental medium of transmission in optical networks and several tasks including packet switching, message signalling and processing are handled electronically. Optical switches were developed to achieve the conversion of optical to electrical signals. An optical switch is defined as follows: it is a switch that enables signals either in optical fibers or in integrated optical circuits to be switched from one source to another. An optical switch may physically shift an optical fiber to drive one or more alternative fibers. There are several types of optical switching. We will explore these switching types in the next section.

17.2.2 Types of Optical Switching

An optical switch is basically a switch that accepts a photonic signal and sends it out based on the routing decision made. Optical switches are of two kinds: Optical-Electronic-Optical (O-E-O) switches and Optical-Optical-Optical (O-O-O) switches. The latter is known as an all-optical switch. An O-E-O switch requires the analogue light signal that need to be first converted into a digital form, then to be processed and routed before being converted back to an analogue light signal. O-O-O switching is achieved through photonic means only. Readers may refer to Figure 17.1.

17.2.2.1 Optical Packet Switching

Data packets can be multiplexed together, making use of the capacity efficiently and increasing the flexibility of optical packet switching [5]. Efficient uses of optical buffers and to achieve enhanced switch throughput, the wavelength dimension is also incorporated inside the optical packet switch. Traditional packet switches analyze the headers and determine the destination to forward the packets. Optical packet-switching technologies enable quick allocation of Wavelength Division Multiplexing channels on-demand on a microsecond time scale. Increasing the transmission bit rate to match increasing bandwidth demand could be achieved incrementally with no big impact on switching nodes. Optical packet switching also provides high-speed data transfer rate, transparency, and configurability.

Networks that are based on optical packet switching can be divided into two types: *slotted (synchronous)* and *unslotted (asynchronous)*. In the former type, all packets are of same size and placed together with the header inside a fixed time slot. In the later type, packets received at the input ports must be aligned in phase with a local clock reference. Fixed-length packets get segmented at the network edge and these segments may reassemble at the other edge. This may not be suitable for high-speed networks. So it is good to consider asynchronous data transfer with variable length packets. Packets arriving at an asynchronous network enter the switch without being aligned. Also, it is not possible to predict the behavior of packets in an unslotted network.

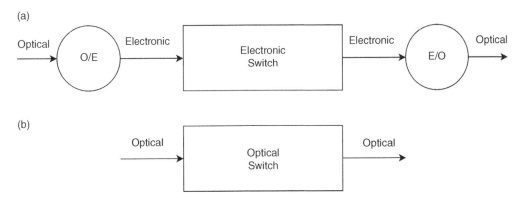

Figure 17.1 Optical switching: (a) classical O-E-O switch (here O/E stands for Optical to Electronic Conversion and E/O refers to Electronic to Optical Conversion; (b) all optical switch.

This could possibly result in packet contention, thereby affecting the network throughput and increasing the packet loss ratio. The use of the wavelength domain for contention resolution could sort this out [6, 7].

17.2.2.2 Circuit Switching

This class of routing is a popular communications mechanism in which a path is established with dedicated bandwidth before the the the actual communication takes place. This is a kind of on-demand path establishment strategy that is followed on request generation. After the communication gets across, the reserved path and the associated bandwidth is freed immediately. A well known example of this category of switching is traditional telephone network calls. Circuit switching requires a dedicated path for relatively lesser duration. This type of routing is different from circuit-based routing in which the path established may possibly stay for longer duration. Circuit-based routing is a kind of semi-permanent communication infrastructure that is recently used in optical networks and high bandwidth requiring networks like Virtual Private Networks (VPN) [8].

17.3 Communication vs Computation

One of the primary trade-offs in optical switching is the following [9, 10]: Does a specific multiplexing close the gap between data transmission and data generation/consumption speeds? In order to use the high bandwidth of all-optical networks in various interconnection networks/data centers, the speed mismatch between various parts of the underlying communication system has to be sorted out. One such imminent drawback is the speed mismatch between the fast transfer of data in the optical networks and slow network control that is executed in the electronic domain. Another important point to be considered is the high transmission rate of optical networks and comparatively very slow generation and consumption of data at sending and receiving nodes. So the following question arises: *Does there exist a significant mismatch between communication and computation of data at the sending and receiving nodes?* This problem has been addressed by Yuan et al. [9, 11] with Time Division Multiplexing techniques.

17.4 Path Reservation Algorithms

In order to achieve the fullest use of optical networks, it is necessary to make use of circuit switching like strategy for transmitting the optical signals in high-performance computing systems. This may not require operations like buffering, encoding/decoding, and routing at the intermediate nodes during the message transfer due to the fact that these operations depend on optical-electronic or electronic-to-optical converters.

All optical networks require a connection to be established that spans over multiple links and uses the same link on all links so as to avoid the conversion of wavelength or interchange of a time slot at intermediate nodes [10]. The control strategies for the assignment of the wavelength or time slot in multiplexed networks are not scalable in huge networks. Thus it is very essential to design and develop path reservation protocols for all-optical networks. We consider two such path reservation algorithms, namely *forward reservation protocols* and *backward reservation protocols*. In the former, the source node, on receipt of a request, begins the allocation of virtual channels along the links towards the destination node. In the latter, a probe packet is sent through the network, collecting the virtual channel usage information before starting to assign the actual allocation. These protocols are derived from the generalizations of the control protocols in non-multiplexed circuit-switching networks.

Many all-optical networks assume the assignment of virtual channels and a smaller number consider online control mechanisms to find these assignments. Qiao and Melhem [12] proposed a distributed control algorithm for establishing connections in multistage networks. The performances of dynamic path reservation protocols are compared while considering the signaling overheads. These protocols belong to the forward reservation category. Backward reservation schemes for multiplexed networks are slightly more complex than the forward reservation protocols.

The path from a source node to a destination node has to be ensured by a path reservation algorithm before the connection is utilized.

1) **Forward vs. Backward:** The distributed path reservation protocols require locking mechanism to ensure the exclusive usage of a virtual channel. The virtual channels are set by a control message traveling from a source node to a destination node in forward reservation protocols. A separate probe and control messages travel through the links; one control message is used to probe the path and another control message is used to lock the available virtual channels.

2) **Dropping vs. Holding:** The dropping approach works as follows: once the establishment of a connection is not progressing, then it releases the virtual channels locked on the path that is partially established and informs the source node that the reservation has failed. In holding approach, the virtual channels on the partially established path kept in the locked state until the time-out occurs.

3) **Aggressive vs. Conservative:** The protocol under aggressive reservation attempts to establish a connection by locking as many virtual channels as possible during the process of reservation and only one of the locked channels is used for communication and others are released. The protocol under conservative reservation, locks only one virtual channel during the process of reservation.

17.5 High-Performance Optical Switching and Routing

Large-scale scientific challenges are driven by the need for computation, which has led to a rise in the speed of HPC systems [13]. There are plenty of challenges in High performance computing that could potentially depend on efficient optical switching.

17.5.1 HPC Interconnection Challenges

The current high-performance computing systems are used in various big data applications which need to operate on several terabytes of data. In HPC, large amounts of data are processed at high speed using multiple computing devices and storage. The amount of data processing and communication is increasing significantly in order to achieve the required computing power. In HPC systems, as the computing requirements increase, the number of processing and storage devices will also increase. On the other hand, overall power consumption of these electronic devices should be minimized [13].

The overall performance of an HPC system depends on factors such as type of programming model used, power consumption management, and memory management mechanism. Among these, the important factor that affects performance is the rate at which communication occurs between the processing devices. The medium which provides the communication between the processing and the memory elements is referred to as the interconnection network. The traditional communication framework fails to reach the speed of microprocessors. The performance of interconnect networks depends on the propagation speed though the transmission medium [14].

To avoid interconnection network bottlenecks, the performance of interconnection networks must satisfy the requirements of the application. As a result, interconnection networks are one of the major criteria that must be considered while developing HPC systems. Topology, routing, and switches are basic components of an interconnection network. The performance of interconnections depends on the network design and switching elements [15].

17.5.2 Challenges in the Design of Optical Interconnection Network

Switches communicate the information between various elements. Signal conversion is used in traditional electrical switches to exchange data. The latency and power overhead of signal conversion have an impact on the overall system performance. HPC-based applications need high-speed ultra switching and link capabilities [16].

1) I/O bandwidth and power are the limiting factors for expanding radius in the ES.
2) Nanophotonics technology is used to solve the bandwidth problem.
3) Fiber waveguide with Dense Wavelength Division Multiplexing (DWDM) in the chip rather than the electric data pins [17].

The use of optics will enable speed upto 50 times higher than the regular electrical communication. We have different classes of optical interconnects in data centers and HPC such as on-board, board-to-board, and rack-to-rack. These classes are based on the hierarchical communication level. Optical sources, optical amplifiers, modulators (which convert data from electrical to optical), and photodetectors are the primary components of an optical connection. The challenges in building optical interconnects include choosing the right components based on power and performance limitations, as well as other typical component-related issues such signal attenuation, losses, errors, packaging temperature dependency, and so on citecha7.

17.6 Optical Switching Schemes for HPC Applications

The use of dense wavelength multiplexing allows for large data rates in optical switching. Different optical switching methods based on hybrid, fast OS, arrayed waveguide routers have been presented to fulfil the criteria such as low-latency, bandwidth, and power.During information switching, contention may occur, and it is resolved by using the appropriate routing methods, wavelength path, and buffer scheme.

Crossbar switches, matrix vector multipliers (MVM), tree-structured splutters, and combiners are different types of switching elements. The types of optical switches are depicted in Figure 17.2. Any one of the switching elements can be used to create larger switches (switch fabric). N-column switching elements make up an N-stage switching fabric. The blocking and nonblock natures of switching fabric are determined by the behaviour of the connection assignment. The Clos network is an example of a three-stage switch configuration.

Optical switching architecture is mainly classified as optical packet switching and optical circuit switching. Optical switching provides lower communication delay and high bandwidth. HPC requires high bandwidth communication with low latency infrastructure. Features of optical

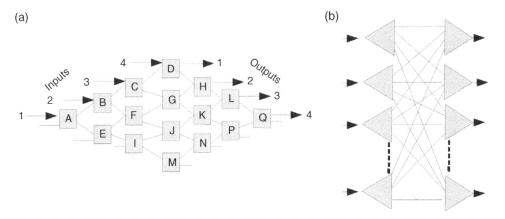

Figure 17.2 Optical switches: (a) 4 × 4 crossbar switch; (b) tree architecture.

switching architecture can overcome the latency bandwidth bottleneck in HPC. Appropriate optical switching architecture is decided based on the type of traffic generated by the application. As the application traffic is unpredictable and it has wide variation, is bursty, and requires more bandwidth (video streaming application). Optical packet switching can be used for the application which has large variation and bursty data. An application which requires high bandwidth and fixed communication delay with no loss of packets can employ the optical circuit switching. On the other hand, a combination of these two switches can also be used as a hybrid approach. Lightness [18] and LIONS [19], are hybrid optical switch architectures [20]. Various optical interconnection network architectures have been proposed by many researchers, such as SYMNET [21] high-speed optical interconnect for scalable shared-memory multiprocessor [22]. An pptical interconnect consists of the optical switch and the link connecting to it.

17.6.1 Routing Scheme (Avoid Packet Loss, Contention, etc.)

The general issues we encounter while using optical packet switching are packet loss, packet reordering, cost, signal loss, and an effective buffering scheme for storing of information. The fiber delay lines can hold the packets for some amount of time by delaying the signal. The overhead of the fiber delay line is that it can not hold the packets longer and the signal quality degrades. Sometimes due to overload traffic the optical packet switching may face issues of contention, which lead to packet loss. The contention can be resolved by different methods such as packet transmission on different wavelength channels, packet delaying, and diverting of packets.

17.6.1.1 Buffering Schemes

In general the buffering is applied at Input, Output and shared. There are different schemes to buffer packets in optical switching, such as delay lines, deflection routing, and buffered deflection routing. There two main categories in optical delay-line buffering scheme: feed-forward delay and feedback delay. Each of these can be implemented in single-stage and multistage switches. Broadcast and select is a type of single-stage forward switch [23]. Multistage and feed-forward cascaded approaches have been proposed, such as COD [24], SDL [25], and single buffer deflection routing switches. To provide a larger optical buffer research has proposed a combination of many small single stages to provide a higher optical buffer [26].

17.7   Security Issues in Optical Switching

There are several security threats reported in an optical switching infrastructure, since optical networks support a huge transmission rate and switching techniques are quite flexible. A variety of modern infrastructures support all optical networks and this has been commercially viable for deploying small to medium to large applications in the internet era, supporting activities in various domains including academics, e-commerce, military needs, healthcare services, and many other professions that evidence a steep increase for recent years. In such scenarios, it becomes essential to provide accessibility to these networks in a secure manner with certain promises as to reliable service delivery. In addition to modern end-to-end encryption, it is also necessary to incorporate additional security metrics that could be necessary to provide enhanced security features in all-optical networks.

The primary security issue is the *physical security* of data transmitted over optical networks. Physical security is supposed to ensure that the quality of service and minimum guaranteed security of the data are provided. In this case, users may be given information about the availability of such measures or delivery failures due to pitfalls in the privacy and service quality. There is another type of security, namely *semantic security*, in which the attacker may have access to the data stream, but the underlying semantic security protects the meaning and integrity of data.

The following two types of attacks are common on physical security: *service disruption* (SD) and *tapping*. The former presents access to the communication and the later provides access to unauthorized personnel for eavesdropping or network traffic analysis [27]. Service disruption attacks may be prevalent through three important components: optical fibers, optical amplifiers and wavelength selective switches (WSSs).

17.7.1   Network Vulnerabilities

Optical networks are evolved as heterogeneous and flexible networks with a wide spectrum of applications. As the data exchanged across the network grows every day, the optical networks become vulnerable to several types of attacks that include eavesdropping [28] and tapping as mentioned above. These types of disruption and security breach may result in a huge data loss or an income loss. Therefore, it becomes essential to be aware of various vulnerabilities, security issues, and attacks faced by modern optical networks.

17.7.1.1   Eavesdropping

The exposure of optical networks to unauthorized accesses poses a huge threat to the security of AON. Given modern network infrastructures, eavesdropping may occur on every network layer from the application layer to the physical layer, and these attacks are mainly targeting government websites, and financial and various commercial and service sectors [29]. Two kinds of eavesdropping attaches are given below: *attacks with direct access* and *attack via key access*. The former attack is achieved by accessing the optical channel that is not encrypted and the later attack is due to key access in an optical network that is encrypted [30, 31].

(1) **Channel Access-based Eavesdropping:** A common method of realizing eavesdropping attacks is directly accessing the optical channel via fiber tapping, i.e., removing the fiber cladding and bending the fiber to cause the signal to leak out of the core and onto the photo detector, capturing the information [31, 32]. Tapping devices which can be clipped onto the fiber and cause micro-bends to leak signals and deliver them into the hands of the eavesdropper are easily accessible on the market. Furthermore, existing tapping devices cause losses below 1 dB

and can go undetected by commonly used network management systems (NMSs). To detect such intrusions, the NMS needs to be enhanced with intrusion detection alarms triggered by insertion loss changes on fiber connections. Obviously, such detections require an active monitoring system running across the network. Another possible way of accessing the channel is via monitoring ports, which are typically present at different network components, such as amplifiers, wavelength selective switches (WSSs), or (de)multiplexers. The optical signal is mirrored by an optical splitter to allow connection of monitoring devices without traffic interruption. By obtaining onsite access, an attacker could use these ports to listen to the carried traffic.

(2) **Key Access-based Eavesdropping:** In order to protect the carried data from eavesdropping, encryptions methods are used, implemented in optical transponders. Such encryption cards are commercially available from most vendors. An example solution by Alcatel Lucent [33] relies on encryption of the data packets using encryption keys which are transferred over the NMS isolated from the data payload. Typically, encryption keys are managed by the end user. However, key management software is installed on the user side, which can serve as another point of attack reaching the operator NMS system.

17.7.2 Jamming Attacks (or Types of Attacks)

In this subsection, we have described a few jamming attacks that are probable in optical networks.

(1) **High-power Jamming Attacks:** It is realized that there could be a possibility of inserting an optical signal of high power over the permitted/legitimate signal strength on a legitimate wavelength in optical networks. Such high-intensity signals could damage already co-propagated signals by other users on common optical fibers/switches/amplifiers. This may be possible in the absence of any wavelength blocking functionality in the underlying optical networks. It is also observed that the jamming signals may also affect legitimate signals communicated at the same wavelength [34]. This may be due to the increase in in-band crosstalk. Signals that are propagated through a physical channel with the jamming signals may suffer from out-of-bound effects in optical fibers and amplifiers [35]. Jamming signals may introduce out-of-bound effects on crosstalk by spilling over to adjacent channels. In some specific fiber amplifiers, gain competition is caused by a jamming signal out of the specified range in which a legitimate signal may decay by a stronger jamming signal and thereby boost the attacking signal.

(2) **Alien Wavelength Attacks:** Network operators used to apply alien wavelengths for various reasons including network upgrades and better transmission of huge data over the existing network infrastructure. In the absence of the alien wavelength support, connections tend to get terminated or regenerated by a node in the network. But alien wavelength could traverse across various domains without the use of optical to electronic and electronic to optical conversions. This type of alien wavelength is used to upgrade legacy systems with high gigabit transponders. These solutions are very well used in modern application deployments. Depending on the use and maintenance of alien wavelengths, they could create potential threats to the network security. If the nodes in a network are in a broadcast and select mode of the configuration, alien wavelengths are induced in the network unfiltered. In such settings, these wavelengths are explored to identify various attack methods and thus providing a huge challenge for network operators.

(3) **Signal Insertion Attacks:** An intermediate and cost-effective method for network upgrading from legacy 10 Gigabit lightpaths to high capacity is the use of Mixed Line Rate (MLR) networks. These networks allow various modulation formats over the existing network infrastructure. A primary security threat of MLR networks emerges from nonlinear effects signals and legacy neighbouring channels. Amplitude - modulated on-off keyed (OOK)

channels – strongly degrades the quality of the higher bit rate, and cross-phase modulation (XPM) affects phase-modulated channels. In polarization multiplexed channels, cross-polarization modulation (XPolM) affects optical data transmission. Even though it is possible to space different channels in 50 GHz spacing, an extra OSNR penalty is imposed for the high bandwidth channels. The following factors define the severity of such penalties: *modulation format, channel launch power* and *guard bands*. In the deployed networks, it is hardly possible to alter the modulation format or launch powers, but leaving the guard bands between channels. One of the possible service degradation attacks in MLR networks may possibly happen by inserting an OOK channel without allowing for enough guard band. This leads to the deteriorate of the OSNR of legitimate signals by the attacking signal.

(4) **Signal Insertion on Monitoring Ports Attacks:** The components of all optical switches are bundled with external monitoring ports, which adds to specific security threats and vulnerabilities. These monitoring ports, in addition to enable eavesdropping, may also be utilized to insert signals into the optical networks and harm live traffic of the underlying network.

17.8 Optical Switching – Interesting Topics

There are plenty of open problems in the optical networking domain, especially in optical switching technologies. We have listed a few interesting topics that are worth investigating:

1) minimization of collisions
2) concurrency of packet transmissions
3) buffer utilization of NICs
4) low-latency switching capability
5) optical parallelism (to achieve high aggregation bandwidth).

17.9 Conclusion

In this chapter, we have addressed several components of optical switching in High Performance Computing architectures. We have described the electronic and optical switches with the limitations of electronic switches and the need for all-optical switches. Optical interconnect provides low power consumption with high capacity. This chapter covers various approaches including path provisioning and elasticity, security Issues in optical switching, and investigation and characterization of different HPC optical switching architectures. Finally, this chapter has highlighted some open challenges of optical switching in HPC, such as minimization of collisions in optical packet switching, low-latency switching capability, and effective channel modeling for high throughput using mathematical models and queuing principles.

Bibliography

1 W. Miao, J. Luo, S. Di Lucente, H. Dorren, and N. Calabretta. Novel flat datacenter network architecture based on scalable and flow-controlled optical switch system. *Optics Express*, 22(3):2465–2472, 2014. doi: 10.1364/OE.22.002465.

2 S. Goswami and S. Misra. *Network Routing: Fundamentals, Applications, and Emerging Technologies.* John Wiley & Sons Ltd., Chichester, 2017. doi: 10.1002/9781119114864.

3 K. Ramasamy and D. Medhi. *Network Routing: Algorithms, Protocols, and Architectures.* Morgan Kaufmann, 2017.

4 D.J. Bishop, C. Randy Giles, and S.R. Das. The rise of optical switching. *Scientific American,* 284(1):88–94, 2001.

5 D.K. Hunter and I. Andonovic. Approaches to optical internet packet switching. *IEEE Communications Magazine,* 38(9):116–122, 2000. doi: 10.1109/35.868150.

6 S.L. Danielsen, P.B. Hansen, and K.E. Stubkjaer. Wavelength conversion in optical packet switching. *Journal of Lightwave Technology,* 16(12):2095–2108, 1998. doi: 10.1109/50.736578.

7 G.I. Papadimitriou, C. Papazoglou, and A.S. Pomportsis. Optical switching: switch fabrics, techniques, and architectures. *Journal of Lightwave Technology,* 21(2):384–405, 2003. doi: 10.1109/JLT.2003.808766.

8 I. Chlamtac and A. Fumagalli. An optical switch architecture for Manhattan networks. *IEEE Journal on Selected Areas in Communications,* 11(4):550–559, 1993. doi: 10.1109/49.221202.

9 X. Yuan, R. Gupta, and R.G. Melhem. Does time-division multiplexing close the gap between memory and optical communication speeds? In *Proceedings of the Second International Workshop on Parallel Computer Routing and Communication,* PCRCW '97, page 261274, Berlin, Heidelberg, 1997. Springer-Verlag.

10 X. Yuan, R. Melhem, and R. Gupta. Distributed path reservation algorithms for multiplexed all-optical interconnection networks. *IEEE Transactions on Computers,* 48(12):1355–1363, 1999. doi: 10.1109/12. 817397.

11 X. Yuan. *Dynamic and Compiled Communication in Optical Time-Division-Multiplexed Point-to-point Networks.* PhD thesis, University of Pittsburgh, August 1998. URL https://www.cs.fsu.edu/~xyuan//paper/98dissertation.pdf.

12 C. Qiao and R. Melhem. Reconfiguration with time division multiplexed MINs for multiprocessor communications. *IEEE Transactions on Parallel and Distributed Systems,* 5(4):337–352, 1994. doi: 10.1109/71. 273043.

13 W. Wei, Q. Zeng, Y. Ouyang, and D. Lomone. High-performance hybrid-switching optical router for IP over WDM integration. *Photonic Network Communications,* 9(2):139–155, 2005. doi: 10.1007/ s11107-004-5583-6.

14 W.J. Dally and B.P. Towles. *Principles and Practices of Interconnection Networks.* Morgan Kaufmann Publishers Inc., San Francisco, CA, USA, 2004.

15 J. Escudero-Sahuquillo, E.G. Gran, P.J. Garcia, J. Flich, T. Skeie, O. Lysne, F.J. Quiles, and J. Duato. Efficient and cost-effective hybrid congestion control for HPC interconnection networks. *IEEE Transactions on Parallel and Distributed Systems,* 26(1):107–119, 2015. doi: 10.1109/TPDS. 2014.2307851.

16 F. Yan, C. Yuan, C. Li, and X. Deng. Fosquare: A novel optical HPC interconnect network architecture based on fast optical switches with distributed optical flow control. *Photonics,* 8(1), 2021. doi: 10.3390/photonics8010011.

17 N. Binkert, A. Davis, N.P. Jouppi, M. McLaren, N. Muralimanohar, R. Schreiber, and J.H. Ahn. The role of optics in future high radix switch design. In *2011 38th Annual International Symposium on Computer Architecture (ISCA),* pages 437–447, 2011.

18 J. Perell, S. Spadaro, S. Ricciardi, D. Careglio, S. Peng, R. Nejabati, G. Zervas, D. Simeonidou, A. Predieri, M. Biancani, H.J.S. Dorren, S. Di Lucente, J. Luo, N. Calabretta, G. Bernini, N. Ciulli, J.C. Sancho, S. Iordache, M. Farreras, Y. Becerra, C. Liou, I. Hussain, Y. Yin, L. Liu, and R. Proietti. All-optical packet/circuit switching-based data center network for enhanced scalability, latency, and throughput. *IEEE Network,* 27(6):14–22, 2013. doi: 10.1109/MNET.2013.6678922.

19 Y. Yin, R. Proietti, X. Ye, C.J. Nitta, V. Akella, and S.J.B. Yoo. Lions: An AWGT-based low-latency optical switch for high-performance computing and data centers. *IEEE Journal of Selected Topics in Quantum Electronics,* 19(2):3600409, 2013. doi: 10.1109/JSTQE.2012.2209174.

20 M. Renaud, F. Masetti, C. Guillemot, and B. Bostica. Network and system concepts for optical packet switching. *IEEE Communications Magazine*, 35(4):96–102, 1997. doi: 10.1109/35.570725.

21 A. Louri and A.K. Kodi. Symnet: an optical interconnection network for scalable high-performance symmetric multiprocessors. *Appl. Opt.*, 42(17):3407–3417, 2003. doi: 10.1364/AO.42.003407. http://ao.osa.org/abstract.cfm?URI=ao-42-17-3407.

22 A.K. Kodi and A. Louri. Design of a high-speed optical interconnect for scalable shared memory multiprocessors. In *Proceedings. 12th Annual IEEE Symposium on High Performance Interconnects*, pages 92–97, 2004. doi: 10.1109/CONECT.2004.1375210.

23 F. Masetti, M. Sotom, D. de Bouard, D. Chiaroni, P. Parmentier, F. Callegati, G. Corazza, C. Raffaelli, S.L. Danielsen, and K.E. Stubkjaer. Design and performance of a broadcast-and-select photonic packet switching architecture. In *Proceedings of European Conference on Optical Communication*, volume 3, pages 309–312, 1996.

24 R.L. Cruz and J.-T. Tsai. Cod: alternative architectures for high speed packet switching. *IEEE/ACM Transactions on Networking*, 4(1):11–21, 1996. doi: 10.1109/90.503758.

25 I. Chlamtac, A. Fumagalli, L.G. Kazovsky, P. Melman, W.H. Nelson, P. Poggiolini, M. Cerisola, A.N.M.M. Choudhury, T.K. Fong, R.T. Hofmeister, Chung-Li Lu, A. Mekkittikul, D.J.M. Sabido, Chang-Jin Suh, and E.W.M. Wong. Cord: contention resolution by delay lines. *IEEE Journal on Selected Areas in Communications*, 14(5):1014–1029, 1996. doi: 10.1109/49.510924.

26 D.K. Hunter, W.D. Cornwell, T.H. Gilfedder, A. Franzen, and I. Andonovic. Slob: a switch with large optical buffers for packet switching. *Journal of Lightwave Technology*, 16(10):1725–1736, 1998. doi: 10.1109/50.721059.

27 M. Medard, D. Marquis, R.A. Barry, and S.G. Finn. Security issues in all-optical networks. *IEEE Network*, 11(3):42–48, 1997. doi: 10.1109/65. 587049.

28 R. Rejeb, M.S. Leeson, and R.J. Green. Fault and attack management in all-optical networks. *IEEE Communications Magazine*, 44 (11):79–86, 2006. doi: 10.1109/MCOM.2006.248169.

29 S.K. Miller. Fiber optic networks vulnerable to attack. *Information Security Magazine*, 2006.

30 M. Medard, S.R. Chinn, and P. Saengudomlert. Attack detection in all-optical networks. In *OFC '98. Optical Fiber Communication Conference and Exhibit. Technical Digest. Conference Edition. 1998 OSA Technical Digest Series Vol.2 (IEEE Cat. No.98CH36177)*, pages 272–273, 1998. doi: 10.1109/OFC.1998.657391.

31 M. Furdek, N. Skorin-Kapov, S. Zsigmond, and L. Wosinska. Vulnerabilities and security issues in optical networks. In *2014 16th International Conference on Transparent Optical Networks (ICTON)*, pages 1–4, 2014. doi: 10.1109/ICTON.2014.6876451.

32 B. Everett. Tapping into fibre optic cables. *Network Security*, 2007(5):13–16, 2007. doi: 10.1016/S1353-4858(07)70036-3.

33 A. Lucent: 1830 photonic service switch. https://www.nokia.com/networks/products/1830-photonic-service-switch/. Last accessed Feb. 2021.

34 C. Mas, I. Tomkos, and O.K. Tonguz. Failure location algorithm for transparent optical networks. *IEEE Journal on Selected Areas in Communications*, 23(8):1508–1519, 2005. doi: 10.1109/JSAC.2005.852182.

35 Y. Peng, K. Long, Z. Sun, and S. Du. Propagation of all-optical crosstalk attack in transparent optical networks. *Optical Engineering*, 50(8):1–5, 2011. doi: 10.1117/1.3607412.

18

Software for Optical Network Modelling

Devlina Adhikari

Information and Communication Technology, School of Technology, Pandit Deendayal Energy University (PDEU), Raisan, Gandhinagar, Gujarat, India

18.1 Optical Networks

In optical networks, information is transmitted in the form of light from a source node to a destination node. Various advantages like enormous bandwidth, high-speed transmission, long-distance communication, etc. lead to optical networks being the most reliable and widely used for communication. Optical networks may be wired or wireless. In wired optical networks, light travels through optical fiber cables. Optical fiber-based communication is used for terrestrial and subsea networks. Wireless optical networks are known as free-space optical networks, in which the optical signal is transmitted in a vacuum. Free-space optical networks may also be implemented to establish connections between satellites and for short-distance communication purposes in terrestrial networks.

Commonly used optical networks are categorized based on three different areas:

Local Area Network (LAN): LAN is intended for users in a small area like an organization.
Metropolitan Area Network (MAN): MAN interconnects users in different cities.
Wide Area Network (WAN): WAN is used to provide long-distance communication like the connection between countries.

18.1.1 First Generation of Optical Networks

The first generation of optical networks includes synchronous optical networking (SONET) and synchronous digital hierarchy (SDH). These are basically transmission protocols used to transmit digital data streams synchronously through a medium, which is essentially optical fiber, using a coherent light source like a laser. SONET is developed in North America, and SDH is used in Europe and Japan. The basic SONET signal rate, Synchronous Transport Signal level 1 (STS-1), is 51.84 Mbps, which is lower than the basic SDH rate Synchronous Transport Module-1 (STM-1), 155 Mbps. The standard transmission rates for SONET/SDH are depicted in Table 18.1.

Optical Switching: Device Technology and Applications in Networks, First Edition. Edited by Dalia Nandi, Sandip Nandi, Angsuman Sarkar, and Chandan Kumar Sarkar.
© 2022 John Wiley & Sons, Inc. Published 2022 by John Wiley & Sons, Inc.

Table 18.1 SONET and SDH signal transmission rates.

SONET Signal	SDH Signal	Bit Rate (Mbps)
STS-1		51.84
STS-3	STM-1	155.52
STS-12	STM-4	622.08
STS-24		1244.16
STS-48	STM-16	2488.32
STS-192	STM-64	9953.28
STS-768	STM-256	39814.32

Source: [1]/IEEE.

18.1.2 Second Generation of Optical Networks

With the advent of wavelength division multiplexing (WDM), the second generation of optical networks started in the latter part of the twentieth century [2]. WDM networks were point to point at first, where network nodes are connected directly by physical fiber links. The next development was wavelength-routed optical networks (WRON). In WRON, a lightpath network topology or virtual topology is formed over the physical networks depending on the connection requests between source-destination pairs. Creating lightpath topology in WDM networks is achieved through Routing and Wavelength Assignment (RWA) [3]. The purpose of RWA algorithms is to find a suitable route between the source-destination pairs and allocate the available wavelength to establish a lightpath between them. If wavelength converters are not used, the same wavelength should be assigned in all the links of a route. This condition is known as the wavelength continuity constraint. The channel allocation process is fixed for WDM networks, i.e., a fixed 100 GHz wavelength channel is allocated to a connection irrespective of the requirement (Figure 18.1).

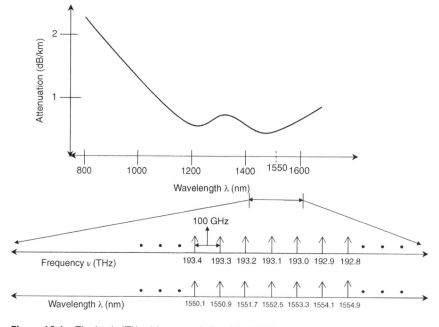

Figure 18.1 The basic ITU grid system defined for WDM systems. Source: [4]/Springer Nature.

Dense wavelength division multiplexing (DWDM) networks are an advanced version of conventional WDM networks, where denser channel spacing is used. The spacing is reduced to 50 GHz and even 25 GHz in the case of current DWDM applications. Advanced technologies, namely ultra-dense WDM, use 12.5 GHz channel spacing. The initial WDM network elements were simple multiplexers, demultiplexers, and optical add-drop multiplexers (OADMs). DWDM systems use erbium-doped fiber amplifiers (EDFAs), Raman amplifiers, and reconfigurable optical add-drop multiplexers (ROADMs) [5, 6]. EDFAs are used for amplification in the C-band (wavelength span of 1530 nm–1565 nm). Raman amplifiers are generally used for L-band wavelengths, i.e., 1565 nm–1625 nm.

18.1.2.1 Passive Optical Network

A passive optical network (PON) is a point-to-multipoint topology, where a single fiber serves the demand of multiple end-users. The name "passive" implies that the fiber bandwidth is divided between multiple endpoints using unpowered optical splitters, unlike the active networks, where dedicated connections are established for each user. The different types of PON technologies are Time Division Multiplexing PON (TDM-PON), Wavelength Division Multiplexing PON (WDM-PON), and Time and Wavelength-Division Multiplexing PON (TWDM-PON). The generalized structure of a passive optical network comprises three components [7, 8]:

- Optical line terminal (OLT): OLT is located at the central office of the service provider. It controls the upstream and downstream data flow in the optical distribution network (ODN). It also converts the incoming electrical signal to the required optical signal.
- Optical network terminal (ONT): ONT acts as a terminating point of the optical signal coming from the central office. It also converts the optical signal into an electrical signal.
- Optical distribution network (ODN): This connects the OLT and ONT.

18.1.2.2 Elastic Optical Network

The concept of an Elastic Optical Network (EON) was developed as a solution to the underutilization of optical spectrum caused in the case of traditional WDM networks [9]. A mini-grid system is proposed where the spacing between the ITU grids is 6.25 GHz [10]. EON provides flexibility in terms of spectrum allocation by allocating the exact amount of spectrum requested by users. EON is based on the Orthogonal Frequency Division Multiplexing (OFDM) technique [11]. The new and advanced bandwidth variable transceivers (BVTs) and bandwidth variable wavelength cross-connects (BV-WXCs) are required for flexible spectrum allocation in EONs (Fig. 18.2). The problem of resource allocation in EONs is termed Routing and Spectrum Allocation (RSA) [12]. The frequency slot unit (1 FSU= 12.5GHz) is defined, and the spectrum is allocated, in multiples of one FSU. The wavelength continuity constraint is converted to a spectrum continuity constraint, and a spectrum contiguity constraint is introduced [13]. RSA problems may be formulated by Integer Linear Programming (ILP) or Mixed Integer Linear Programming (MILP) models generally for small networks [14]. The software widely used by the researchers to solve the ILP/MILP is IBM ILOG CPLEX Optimization Studio, which uses the Simplex method for optimization [15]. For large networks, heuristic RSA algorithms are designed to establish lightpath connections [16]. The algorithms may be implemented in MATLAB or by using programming languages like C, C++, etc.

18.1.2.3 Cognitive Optical Network

Cognitive optical networks (COGNITIONs) is a new autonomous and user-controlled approach capable of reducing the complexity of optical networks [17]. COGNITION apprehends the current network state by introducing awareness on different planes like data, control, management, and service planes, so as to perform according to those network states [18]. The continuous process of "Observe–Orient–Plan–Decide–Act–Learn" makes the network adaptable to new challenges of the

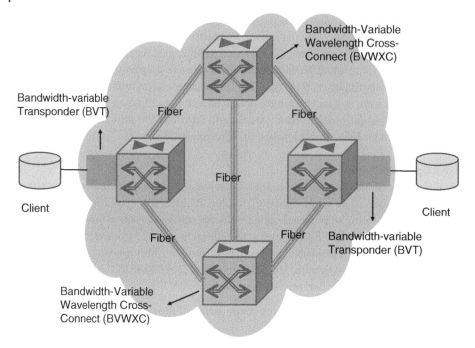

Figure 18.2 The architecture of an Elastic Optical Network.

network environment [19]. The key objective of designing optical networks considering the cognitive system is to develop a completely automated network that will not need any human interaction in any kind of situation [20]. The recently proposed Cognitive Zone-Based (CZB) spectrum assignment algorithm improves the fairness in elastic optical networks and outperforms the Spectrum Sharing First-Fit algorithm [21].

18.1.2.4 Optical Neural Network

Optical Neural Networks (ONNs) are optical implementations of artificial neural networks (ANNs) [22]. Optics has its own advantage that makes the ONNs computationally efficient compared to the electronically processed ANNs. The complete simulation technique for ONN was proposed in [23]. In addition, genetic algorithms (GA) and particle swarm optimization (PSO) may be applied to realize ONNs [24]. Various implementation techniques of ONNs have also been discussed [25]. The application area of ONNs includes pattern recognition, deep reinforcement learning (DRL), etc.

18.2 Simulation Tools for Planning of Optical Network

18.2.1 Network Simulators

18.2.1.1 NS-2

NS-2 is a widely used and readily available network simulator tool generally used for studying PONs [26]. Many researchers have adopted NS-2 for modelling and simulation of WDM networks [27, 28]. An extension of NS-2, namely Optical WDM network simulators (OWns), is a simulation tool that is used to study the different characteristics of WDM networks [29]. An NS2-based OBS simulation tool (nOBS) is also introduced in [30] to study the optical burst switching networks as NS-2 alone is unable to support all the components of an OBS network. The development and maintenance of NS-2 are no longer available.

Advantage: NS-2 is an open-source network simulator.

Limitations:
- Higher abstraction in the network layer and lower layers causes a gap between real world and simulation.
- Inadequate number of simulation tools.
- Model validation, and documentation is not sufficient.

18.2.1.2 NS-3

NS-3 is the latest version of network simulators that was initially released in 2008. Recently in January 2021, NS-3.33 was released [31]. C++ and Python are used to built NS-3. Photonic WDM Network Simulator (PWNS) is based on NS-3 and is extensively used to model core networks [32]. NS-3 is also utilized for the simulation of Next-Generation PON (XG-PON). The XG-PON module of NS-3 was developed to analyze the correct network behavior as much as possible [33]. The XG-PON module is further extended for the study of TWDM-PON [34].

Advantages:
- It is open-source.
- It provides a solution through the simulation that relates to the real-time implementation of the network.
- Efficient for the simulation of large networks.

Limitations:
- Although the realism is enhanced in NS-3 compared to NS-2, the simulation results are not always 100% true.
- All the functionality of a model cannot be correctly validated by simulation.
- Scalability limits, i.e., simulation time and required memory space are limited. A simulation cannot run for an indefinite time, and also, infinite memory space is not practically possible to achieve.

18.2.1.3 OMNeT++

Objective Modular Network Testbed in C++ (OMNeT++) is a network simulator implemented in C++. It was developed in 1992 at the Technical University of Budapest by Andras Varga [35]. OMNeT++ was initially designed for the performance evaluation of computer networks. It is compatible with Windows, Linux, and Mac operating systems. OMNet++ version 4.0/4.1 has been used to implement Ethernet Passive Optical Network (EPON) [36]. A Hybrid Optoelectronic Ring NETwork (HORNET) structure was built by using OMNet++ version 3.0 in Stanford University [37]. A simulation tool, namely Elastic Optical Network Simulation Framework for OMNeT++ (ElasticO++), was also designed for testing of resource allocation algorithms on EONs, considering various parameters and topologies [38]. Recently, an Omnet++ based simulator was developed for modelling of optical impairments in optical burst switched Dense Wavelength Division Multiplexed networks [39].

Advantage: It is freely available for non-commercial use, e.g., academic institutions.

Limitation: Inefficient for simulation of large networks.

18.2.1.4   OPNET

OPNET (Optimized Network Engineering Tools) is a versatile network simulator [40]. Various network topologies can be created and simulated using the built-in devices and protocols provided by IT Guru. OPNET Modeler can model and simulate Passive Optical Networks [41, 42]. OPNET is also utilized to analyze the performance of Optical Network-on-Chip (ONoC) [43, 44, 45].

Advantage: OPNET modeler may be used to get an overview of the system with many simulation examples.
Limitation: The pre-built models of the OPNET modeler cannot be customized.

More network simulators are available for simulation of the transport layer, like Scalable Simulation Framework Network (SSFNet), J-Sim, QualNet, MatPlanWDM, etc. [46].

18.2.2   Physical Layer Simulation

A few tools that are freely available for simulation of the physical layer are shown in Table 18.2.
Commercially available tools for physical layer simulation are displayed in Table 18.3.
Various physical layer attributes are included in the extended libraries of these software packages.

Table 18.2   Free tools for physical layer simulation.

Simulator	Developer/Distributor
FOCSS	TRLabs
LIGHTSIM	Softronix Software
SIMFOCS	Dr. Geckeler

Source: [47].

Table 18.3   Commercial tools for physical layer simulation.

Simulator	Developer/Distributor
ModeSYS	RSoft
OptiSYSTEM	Optiwave
OptSim	RSoft
PHOTOSS	P. I. Systemtechnik Jens Lenge
VPItransmissionMaker™	VPIphotonics

Source: [47].

18.3 New Technologies

18.3.1 Space Division Multiplexing (SDM)

Space division multiplexing or spatial division multiplexing (SDM) is a multiplexing scheme where the optical signal is transmitted through different spatial channels. Although the concept of SDM is not new in optical fiber communication yet, researchers have been focusing on SDM-based optical networks in the recent past only. The typical cause is the increasing need for bandwidth and the development of technologies capable of handling SDM channels. SDM-based EONs are studied to ensure efficient spectrum utilization [48, 49] and to improve the blocking performance [50]. Moreover, a trade-off between crosstalk and fragmentation is proposed in spectrally and spatially elastic optical networks (SS-EONs) based on the association rule mining technique [51]. CVX and Gurobi optimization tools are also used to study SDM-EONs with multi-core fibers [52].

18.3.2 Software-Defined Networking (SDN)

Software-Defined Networking (SDN) is an effective solution to address the increasing need for real-time traffic. SDN-enabled optical networks are one of the most recent research trends [53–55]. The service providers are adopting SDN technologies worldwide due to several advantages of this new technology. SDN offers dynamic traffic management, simplified administration due to software-managed devices, enhanced security, efficient utilization of resources, and reduced cost. The architecture of SDN is divided into three different layers, i.e., infrastructure, control, and application layer. The infrastructure layer is typically the data plane layer consisting of physical devices like routers and switches. This layer receives instructions from the control layer and forwards the packets accordingly. The control layer is the brain of SDN that provides centralized control over the entire network. The application layer is the management plane that instructs the controller to decide to offer various applications dynamically. ONOS (Open Network Operating System) simulation tool is used to study Hybrid SDN and PCE (Path Computation Element) Based Optical Networks [56]. A new simulation tool, namely Mininet-Optical, is used to estimate the real-time Quality of Transmission (QoT) through monitoring of SDN Control Plane [57]. An SDN-based emulation platform is also proposed to evaluate the performance of optical networks [58].

18.3.3 Artificial Intelligence/Machine Learning (AI/ML)

Artificial Intelligence/Machine Learning(AI/ML) is the key to achieving efficiency in optical networks in the modern era of automation. Incorporating intelligence into optical networks to expedite the operation and enhance the performance is now the latest topic in research. The development of AI/ML methods in optical networks has started and progressed rapidly [59, 60]. Various AI/ML aided modeling, and monitoring techniques have been analyzed for elastic optical networks in [61]. A machine learning Monte Carlo Tree Search (MCTS) optimization algorithm is also developed to improve the blocking performance and thereby reducing the cost for the inter-data center EONs [62].

Bibliography

1 C.A. Siller and M. Shafi. *SONET/SDH: A Sourcebook of Synchronous Networking*. Los Alamitos, CA, IEEE Press, 1996.

2 R. Ramaswami, K.N. Sivarajan, and G.H. Sasaki. *Optical Networks*. Morgan Kaufmann, Boston, third edition, 2010. doi: https://doi.org/10.1016/B978-0-12-374092-2.50018-7.

3 R. Ramaswami and K.N. Sivarajan. Design of logical topologies for wavelength-routed optical networks. *IEEE Journal on Selected Areas in Communications*, 14(5):840–851, 1996. doi: 10.1109/49.510907.

4 B. Mukherjee. *Optical WDM Networks*. Boston, MA, Springer, 2006. doi: 10.1007/0-387-29188-1.

5 D.H. Thomas and J.P. von der Weid. Impairments of EDFA dynamic gain-fluctuations in packet-switched WDM optical transmissions. *IEEE Photonics Technology Letters*, 17(5):1097–1099, 2005.

6 X. Jiang. Chapter 15 - optical performance monitoring in optical long-haul transmission systems. In Calvin C.K. Chan, editor, *Optical Performance Monitoring*, pages 423–446. Oxford, Academic Press, 2010. doi: 10.1016/B978-0-12-374950-5.00015-8.

7 H. Gupta, P. Gupta, P. Kumar, A.K. Gupta, and P.K. Mathur. Passive optical networks: Review and road ahead. In *TENCON 2018–2018 IEEE Region 10 Conference*, pages 919–924, 2018. doi: 10.1109/TENCON.2018.8650204.

8 M. Kumari, R. Sharma, and A. Sheetal. Passive optical network evolution to next generation passive optical network: A review. In *2018 6th Edition of International Conference on Wireless Networks Embedded Systems (WECON)*, pages 102–107, 2018. doi: 10.1109/WECON.2018. 8782066.

9 B. Chatterjee and E. Oki. *Elastic Optical Networks*. Boca Raton, CRC Press, 2020. doi: 10.1201/9780429465284.

10 I. Tomkos, S. Azodolmolky, J. Sol-Pareta, D. Careglio, and E. Palkopoulou. A tutorial on the flexible optical networking paradigm: State of the art, trends, and research challenges. *Proceedings of the IEEE*, 102(9):1317–1337, 2014. doi: 10.1109/JPROC.2014.2324652.

11 G. Zhang, M. De Leenheer, A. Morea, and B. Mukherjee. A survey on OFDM-based elastic core optical networking. *IEEE Communications Surveys Tutorials*, 15(1):65–87, 2013.

12 B.C. Chatterjee, N. Sarma, and E. Oki. Routing and spectrum allocation in elastic optical networks: A tutorial. *IEEE Communications Surveys Tutorials*, 17(3):1776–1800, 2015. doi: 10.1109/COMST.2015.2431731.

13 S. Talebi, F. Alam, I. Katib, M. Khamis, R. Salama, and G.N. Rouskas. Spectrum management techniques for elastic optical networks: A survey. *Optical Switching and Networking*, 13:34–48, 2014. doi: 10.1016/j.osn.2014.02.003.

14 D. Adhikari, R. Datta, and D. Datta. Design methodologies for survivable elastic optical networks with guardband-constrained spectral allocation. In *2018 3rd International Conference on Microwave and Photonics (ICMAP)*, pages 1–6, February 2018. doi: 10.1109/ICMAP.2018.8354464.

15 I. IBM D.P. IBM ILOG CPLEX Optimization Studio, 2014.

16 D. Adhikari, D. Datta, and R. Datta. Impact of BER in fragmentation-aware routing and spectrum assignment in elastic optical networks. *Computer Networks*, 172:107167, 2020. doi: 10.1016/j.comnet.2020.107167.

17 R.W. Thomas, L.A. DaSilva, and A.B. MacKenzie. Cognitive networks. In *First IEEE International Symposium on New Frontiers in Dynamic Spectrum Access Networks, 2005. DySPAN 2005.*, pages 352–360, 2005. doi: 10.1109/DYSPAN.2005.1542652.

18 G.S. Zervas and D. Simeonidou. Cognitive optical networks: Need, requirements and architecture. In *2010 12th International Conference on Transparent Optical Networks*, pages 1–4, 2010. doi: 10.1109/ICTON. 2010.5549176.

19 A.Y. Grebeshkov. Cognitive optical networks: architectures and techniques. In Vladimir A. Andreev, Anton V. Bourdine, Vladimir A. Burdin, Oleg G. Morozov, and Albert H. Sultanov, editors, *Optical Technologies for Telecommunications 2016*, volume 10342, pages 40–47. International Society for Optics and Photonics, SPIE, 2017. doi: 10.1117/ 12.2270294.

20 R.W. Thomas, D.H. Friend, L.A. DaSilva, and A.B. MacKenzie. *Cognitive Networks*, pages 17–41. Springer Netherlands, Dordrecht, 2007. ISBN 978-1-4020-5542-3. doi: 10.1007/978-1-4020-5542-3_2.

21 R.S. Tessinari, D. Colle, and A.S. Garcia. Cognitive zone-based spectrum assignment algorithm for elastic optical networks. In *2018 International Conference on Optical Network Design and Modeling (ONDM)*, pages 112–117, May 2018. doi: 10.23919/ONDM.2018.8396116.

22 X. Sui, Q. Wu, J. Liu, Q. Chen, and G. Gu. A review of optical neural networks. *IEEE Access*, 8:70773–70783, 2020. doi: 10.1109/ACCESS.2020.2987333.

23 W.B. Marvin, D.S. Phatak, and W.P. Burleson. Full simulation of optical neural networks. In S.T. Kowel, W.J. Miceli, J.L. Horner, B. Javidi, S.T. Kowel, and W.J. Miceli, editors, *Photonics for Processors, Neural Networks, and Memories*, volume 2026, pages 497–508. International Society for Optics and Photonics, SPIE, 1993. doi: 10.1117/12.163599.

24 T. Zhang, J. Wang, Y. Dan, Y. Lanqiu, J. Dai, X. Han, X. Sun, and K. Xu. Efficient training and design of photonic neural network through neuroevolution. *Optics Express*, 27(26):37150–37163, 2019. doi: 10.1364/OE.27.037150.

25 R. Xu, P. Lv, F. Xu, and Y. Shi. A survey of approaches for implementing optical neural networks. *Optics and Laser Technology*, 136:106787, 2021. doi: 10.1016/j.optlastec.2020.106787.

26 NS-2 official website. http://www.isi.edu/nsnam/ns/.

27 P. Sakthivel and P. Krishna Sankar. Multi-path routing and wavelength assignment (RWA) algorithm for WDM based optical networks. *International Journal of Engineering Trends and Technology (IJETT)*, 10:323–328, 2014. doi: 10.14445/22315381/IJETT-V10P262.

28 S. Pandey and V. Tyagi. Performance analysis of wired and wireless network using NS2 simulator. *International Journal of Computer Applications*, 72:38–44, 2013.

29 B. Wen, N. Bhide, R. Shenai, and K. Sivalingam. Optical wavelength division multiplexing (WDM) network simulator (OWNS): Architecture and performance studies. 2001.

30 G. Gurel, O. Alparslan, and E. Karasan. nobs: an ns2 based simulation tool for performance evaluation of TCP traffic in OBS networks. *Annales Des Telecommunications*, 62:618–637, 2007.

31 NS-3 official website. https://www.nsnam.org/,.

32 V. Miletić, B. Mikac, and M. Dzanko. Modelling optical network components: A network simulator-based approach. In *2012 IX International Symposium on Telecommunications (BIHTEL)*, pages 1–6, 2012. doi: 10.1109/BIHTEL.2012.6412064.

33 T. Horvath, P. Munster, V. Oujezsky, M. Holik, and P. Cymorek. Time and memory complexity of next-generation passive optical networks in NS-3. In *2019 International Workshop on Fiber Optics in Access Networks (FOAN)*, pages 68–71, 2019. doi: 10.1109/FOAN.2019. 8933749.

34 Y. Nakayama and R. Yasunaga. ITU TWDM-PON module for NS-3. *Wireless Networks*, 2020. doi: 10.1007/s11276-019-02236-8.

35 Information Resources Management Association. *Networking and Telecommunications: Concepts, Methodologies, Tools, and Applications*. Idea Group Inc (IGI), USA, 2010. doi: 10.4018/978-1-60566-986-1.

36 sourceforge.net. omneteponmodule. http://omneteponmodule.sourceforge.net/dummy/PON/doc/ neddoc/index.html.

37 I.M. White, M.S. Rogge, K. Shrikhande, and L.G. Kazovsky. A summary of the hornet project: a next-generation metropolitan area network. *IEEE Journal on Selected Areas in Communications*, 21(9):1478–1494, 2003. doi: 10.1109/JSAC.2003.818838.

38 R.S. Tessinari, B. Puype, D. Colle, and A.S. Garcia. Elastico++: An elastic optical network simulation framework for OMNeT++. *Optical Switching and Networking*, 22:95–104, 2016. doi: 10.1016/j.osn.2016.07.001.

39 J. Oladipo, M.C. duPlessis, and T.B. Gibbon. Implementation and validation of an OMNeT++ optical burst switching simulator. In *2017 International Conference on Performance Evaluation and Modeling in Wired and Wireless Networks (PEMWN)*, pages 1–6, 2017. doi: 10.23919/ PEMWN.2017.8308024.

40 Inc. OPNET Technologies. Opnet network simulator. http://www.opnet.com/solutions/network_rd/modeler.html.

41 Z. Peng and P.j. Radcliffe. Modeling and simulation of ethernet passive optical network (epon) experiment platform based on opnet modeler. In *2011 IEEE 3rd International Conference on Communication Software and Networks*, pages 99–104, 2011. doi: 10.1109/ICCSN.2011.6013671.

42 T. Horvath, P. Munster, P. Cymorek, V. Oujezsky, and J. Vojtech. Implementation of ng-pon2 transmission convergence layer into opnet modeler. In *2017 International Workshop on Fiber Optics in Access Network (FOAN)*, pages 1–5, 2017. doi: 10.1109/FOAN.2017.8215254.

43 J. Zhang, H. Gu, and Y. Yang. A high performance optical network on chip based on Clos topology. In *2010 2nd International Conference on Future Computer and Communication*, volume 2, pages V2–63–V2–68, 2010. doi: 10.1109/ICFCC.2010.5497287.

44 Q. Cai, W. Hou, C. Yu, Pengchao Han, Lincong Zhang, and Lei Guo. Design and Opnet implementation of routing algorithm in 3d optical network on chip. In *2014 IEEE/CIC International Conference on Communications in China (ICCC)*, pages 112–115, 2014. doi: 10.1109/ICCChina.2014.7008253.

45 L. Zhu and H. Gu. A traffic-balanced and thermal-fault tolerant routing algorithm for optical network-on-chip. In *2019 18th International Conference on Optical Communications and Networks (ICOCN)*, pages 1–3, 2019. doi: 10.1109/ICOCN.2019.8934668.

46 A. Varga and R. Hornig. An overview of the OMNeT++ simulation environment. In *Proceedings of the 1st international conference on Simulation tools and techniques for communications, networks and systems and workshops*, 2008.

47 I.B. Martins, Y. Martins, F. Rudge, and E. Moschim. Importance of simulation tools for the planning of optical network. In *Education and Training in Optics and Photonics: ETOP 2015*, volume 9793, pages 248–255. International Society for Optics and Photonics, SPIE, 2015. doi: 10.1117/12.2223118.

48 A. Muhammad, G. Zervas, and R. Forchheimer. Resource allocation for space-division multiplexing: Optical white box versus optical black box networking. *Journal of Lightwave Technology*, 33(23):4928–4941, 2015. doi: 10.1109/JLT.2015.2493123.

49 M. Yaghubi-Namaad, A.G. Rahbar, and B. Alizadeh. Adaptive modulation and flexible resource allocation in space-division-multiplexed elastic optical networks. *IEEE/OSA Journal of Optical Communications and Networking*, 10(3):240–251, 2018. doi: 10.1364/JOCN.10.000240.

50 S. Zhang and K.L. Yeung. Dynamic service provisioning in space-division multiplexing elastic optical networks. *IEEE/OSA Journal of Optical Communications and Networking*, 12(11):335–343, 2020. doi: 10.1364/JOCN.396197.

51 Q. Yao, H. Yang, B. Bao, A. Yu, J. Zhang, and M. Cheriet. Core and spectrum allocation based on association rules mining in spectrally and spatially elastic optical networks. *IEEE Transactions on Communications*, 69(8):5299–5311, 2021. doi: 10.1109/TCOMM.2021.3082768.

52 L. Zhang, N. Ansari, and A. Khreishah. Anycast planning in space division multiplexing elastic optical networks with multi-core fibers. *IEEE Communications Letters*, 20(10):1983–1986, 2016. doi: 10.1109/LCOMM.2016.2593479.

53 C. Manso, R. Munoz, N. Yoshikane, R. Casellas, R. Vilalta, R. Martinez, T. Tsuritani, and I. Morita. Tapi-enabled SDN control for partially disaggregated multi-domain (OLS) and multi-layer (WDM over SDM) optical networks [invited]. *IEEE/OSA Journal of Optical Communications and Networking*, 13(1):A21–A33, 2021. doi: 10.1364/JOCN.402187.

54 F. Wang, B. Liu, X. Xue, L. Zhang, F. Yan, E. Magalhes, Q. Zhang, X. Xin, and N. Calabretta. Demonstration of SDN-enabled hybrid polling algorithm for packet contention resolution in optical data center network. *Journal of Lightwave Technology*, 38(12): 3296–3304, 2020. doi: 10.1109/JLT.2020.2976549.

55 Y. Xiong, Y. Li, B. Zhou, R. Wang, and G.N. Rouskas. SDN enabled restoration with triggered precomputation in elastic optical inter-datacenter networks. *IEEE/OSA Journal of Optical Communications and Networking*, 10(1):24–34, 2018. doi: 10.1364/JOCN.10.000024.

56 P. Selvaraj and V. Nagarajan. Match field based algorithm selection approach in hybrid SDN and PCE based optical networks. *KSII Transactions on Internet and Information Systems (TIIS)*, 12:5723–5743, 2018. doi: 10.3837/tiis.2018.12.007.

57 A.A. Daz-Montiel, B. Lantz, J. Yu, D. Kilper, D. Kilper, and M. Ruffini. Real-time QoT estimation through SDN control plane monitoring evaluated in mininet-optical. *IEEE Photonics Technology Letters*, 33(18):1050–1053, 2021. doi: 10.1109/LPT.2021.3075277.

58 S. Azodolmolky, M.N. Petersen, A.M. Fagertun, P. Wieder, S.R. Ruepp, and R. Yahyapour. Sonep: A software-defined optical network emulation platform. In *2014 International Conference on Optical Network Design and Modeling*, pages 216–221, 2014.

59 J. Mata, I. de Miguel, R.J. Durn, N. Merayo, S. Kumar Singh, A. Jukan, and M. Chamania. Artificial intelligence (AI) methods in optical networks: A comprehensive survey. *Optical Switching and Networking*, 28:43–57, 2018. doi: 10.1016/j.osn.2017.12.006.

60 R. Gu, Z. Yang, and Y. Ji. Machine learning for intelligent optical networks: A comprehensive survey. *Journal of Network and Computer Applications*, 157:102576, 2020. doi: 10.1016/j.jnca.2020.102576.

61 X. Liu, H. Lun, M. Fu, Y. Fan, L. Yi, W. Hu, and Q. Zhuge. AI-based modeling and monitoring techniques for future intelligent elastic optical networks. *Applied Sciences*, 10:363, 2020.

62 M. Aibin and K. Walkowiak. Monte carlo tree search with last-good-reply policy for cognitive optimization of cloud-ready optical networks. *Journal of Network and Systems Management*, 28:1722–1744, 2020. doi: 10.1007/s10922-020-09555-8.

Index

Optical Switching: Device Technology and Applications in Networks, First Edition. Edited by Dalia Nandi, Sandip Nandi, Angsuman Sarkar, and Chandan Kumar Sarkar.
© 2022 John Wiley & Sons, Inc. Published 2022 by John Wiley & Sons, Inc.

Printed and bound by CPI Group (UK) Ltd, Croydon, CR0 4YY

16/04/2025

14658426-0004